S0-BST-304

General College Mathematics

General College Mathematics

THIRD EDITION

WILLIAM L. AYRES
Professor of Mathematics,
Southern Methodist University

CLEOTA G. FRY
Associate Professor of Mathematics,
Purdue University

and

HAROLD F. S. JONAH
Late Professor of Mathematics,
Purdue University

McGRAW-HILL BOOK COMPANY

New York St. Louis San Francisco Düsseldorf London
Mexico Panama Sydney Toronto

GENERAL COLLEGE MATHEMATICS

Library of Congress Catalog Card Number 70-98482

02640

1 2 3 4 5 6 7 8 9 0 HDMM 4 3 2 1 0 6 9 8 7 0

This book was set in Baskerville by Black Dot, Inc., printed on permanent paper by Halliday Lithograph Corporation, and bound by The Maple Press Company. The designer was B. Kann; the drawings were done by Graphic and Industrial Design and AYXA, Inc. The editors were E. M. Millman and Sally Anderson. Stuart Levine supervised the production.

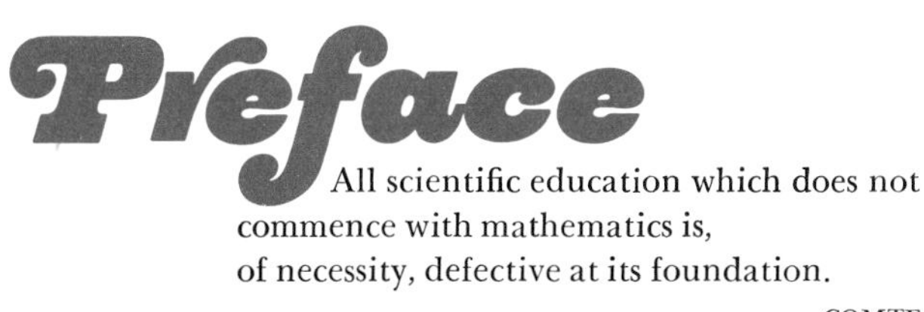

Preface

All scientific education which does not
commence with mathematics is,
of necessity, defective at its foundation.

COMTE

Most first-year courses are designed to provide a foundation for the calculus and other advanced mathematics courses. But since many students do not plan to go beyond the one-year course, taking this precalculus course is like going through preseason football training and then leaving the team just before the opening game. So this book was written for you if you plan to major in one of the nonmathematical fields such as the biological or social sciences or the humanities, or if you plan to terminate your education in a junior college.

In the course of the year, we will discuss some topics that will be slightly familiar to you, as well as some quite new ideas. We will dip back into arithmetic for another look at ratios and percentages, useful in many college courses, in business and finance, and in everyday life. We will look at linear and quadratic relations, since here are found most of the laws of the physical world around us. We will examine exponents and their first cousin, logarithms. Because a bit of trigonometry is needed in physics and other college courses, we will examine that subject in some detail.

Most of the topics mentioned so far are familiar to you, although another, deeper, look at them is desirable. (Historically these topics are very ancient. Arithmetic and algebra date back to the dawn of history and reached their modern form before the discovery of America. Plane geometry existed in almost its present form at the time of

Euclid, 300 years before Christ. There was a book on trigonometry as early as the thirteenth century, and some trigonometric facts were known to the early Babylonian astronomers. In the sixteenth century, trigonometric tables were calculated to ten decimal places.)

But we will also study some topics that may be new to you; in fact, some are rather recent in time. Linear programming, so essential in decision making in business and industry today, was developed during and after World War II. Topology was in its infancy about 1850 and most of it was discovered by persons still living today. Your library contains entire books on subjects discovered in mathematics since 1900. We will examine some of this "new mathematics" and discuss some problems that are not yet solved. Mathematics is very old, but new parts are being discovered today.

We shall spend considerable time on probability and statistics. Any study of social problems, economic situations, market possibilities, or scientific investigation employs the language of statistics.

We shall examine interest on money and its application to installment buying. We will look at the general laws of growth as applied to population studies, growth of bacteria, and decay of radioactive substances. And in the final chapter we will look at logic and the laws of precise reasoning.

Throughout the course emphasis will be on thinking things out rather than on memory, on understanding rather than drill and technique, and on problems from life rather than mathematical equations and formulas. We assume that you have had a year of algebra and know a little about plane geometry. Even if you have forgotten some of this, it will not be too serious, for you are expected to reason rather than to remember.

Finally, we will have failed unless you, the student, find something of interest, and fun, too, in these topics. Believe it or not, there is fun in mathematics! This book even contains a series of problems "Just for Fun!" Some are quite simple, and others will require careful reasoning. In most cases they require little or no formal mathematics but do require a thinking cap. It is hoped they will add some spice and excitement to the course. Try them all; you will find that they challenge and amuse you, and you will have some interesting puzzles to propose to your family or your roommate.

Answers to parts *(a)*, *(c)*, *(e)*, etc., of all the more formal exercises are given at the back of the book, along with most answers to odd-numbered exercises that do not contain parts. The exceptions are those answers that are simple graphs, proofs, and discussions.

In preparing this new edition, we have added more problem material and a few new topics and have rewritten some sections to make the explanations clearer. Major changes have been few, and we have tried to retain the informal flavor of the original edition.

We most regretfully report that Harold F. S. Jonah, who was so helpful during the original writing, died on July 22, 1960.

WILLIAM L. AYRES
CLEOTA G. FRY

Contents

General College Mathematics

CHAPTER ONE

Mathematics—What and Why

There is no problem in all mathematics that cannot be solved by direct counting. But with the present implements of mathematics many operations can be performed in a few minutes which without mathematical methods would take a lifetime. ERNST MACH

1 *Mathematics Is Born*

Somewhere in the early dawn of mankind, a cave man took a stick in his hand and made marks in the sand, one for each of his children or one for each of the animals he could see from where he was standing. These may have looked like // or ////. Or he may have touched the ends of his fingers on one hand with a finger of the other hand, one for each child or one for each animal in sight. This worked well as long as his sand pictures looked like // or //// or even /////, and as long as he didn't need more fingers than he had on his hand. But one day he saw some animals and his sand picture was ////////, which was very confusing. He tried it on his fingers without any better luck. His brother, a Stone Age Einstein, discovered that he could do this problem if he started over with his fingers and probably announced the answer as "a hand and middle finger." "But what do you do with my picture ////////?" This was a puzzler, but after some trial and error, our Stone Age thinker discovered he could make the marks in the sand ~~////~~ /// so that it agreed with his finger touching. At this moment mathematics was born. As long as the problems were simple, /// or //// would do. But //////////// was too difficult, and doing this by ~~////~~ ~~////~~ // was a real discovery and made the job easier.

From this crude start, down through the ages of history, man has added new ways of solving his problems whenever they became too difficult to solve by earlier methods. Each discovery of man was done to make his work easier. Every device of that complex structure of today which we call mathematics was invented at some time or other to make a difficult problem easier. If you are willing to duck hard problems and are still willing to work long hours, you need very little mathematics. But if you say, "There must be an easier way," and want shorter and easier ways to do your work, mathematics is your dish. Truly, mathematics is the paradise of the lazy. The more mathematics you know, the lazier you can be.

Recently a woman bought at a supermarket 48 water glasses that cost 17 cents each. The cashier figured the bill by punching 17 into the cash register time after time while she counted up to 48. Even with a cash register, this took several minutes. But it avoided multiplication, and you can avoid learning multiplication tables if you are willing to do such problems by such long additions. Actually we may consider nearly all mathematics as ways and means of doing difficult problems by easier ways. The more we know, the shorter and easier is our solution. Lazy man, thy name is mathematics.

How Much Is a Billion?

We often read that the national budget is so many billions. But can you realize how big a billion is? It is a thousand million, but this doesn't help much. Suppose you decided to count to a billion, one count every second, day after day. You would have to stop to eat and sleep, so we shall assume that a few of your friends help out to carry on while you are resting. If you start counting today and were eighteen years old, you would finish the count just before your fiftieth birthday.

Suppose a corporation was so badly managed that it lost a thousand dollars a day every day in the week. Suppose our corporation began losing this money at the birth of Christ and was to be bankrupt when it had lost a billion dollars. When would it go bankrupt? Well, it would still be in business today and would continue on for 700 years more!

Number Giants

There is a legend that an ancient Shah of Persia was so pleased with the game of chess that he called the inventor of the game to his throne room and offered to reward him. The inventor, being wise at matters mathematical as well as chess, asked for a chess board with one grain of wheat on the first square, two grains on the second, four on the third, and so on, doubling the number until all the squares were covered. The Shah agreed, although he thought it was a poor reward. Now the second square had 2 grains, the third 2^2, the fourth 2^3, so that the sixty-fourth square needed 2^{63} grains of wheat, which in round numbers is 9,180,000,000,000,000,000 grains. This is far,

256	512	1,024	2,048	4,096	8,192	16,384	32,768

FIG. 1

far more than the billion of the last section. In fact, if we estimate that a bushel of wheat contains about 540,000 grains of wheat, this would be 17,000,000,000,000 bushels. This is more than the wheat production of the entire world for centuries. The legend does not tell us what happened but it is likely that the Shah became so enraged at being tricked that he ordered the inventor's head cut off long before the sixty-fourth square was reached.

But even larger numbers are possible. It has been calculated that the number of molecules of water in a cubic centimeter is about 3 followed by 22 zeros. And the physicist Eddington has estimated that the number of electrons in the universe is 157 followed by 77 zeros.

However we can write a real giant of giants with just three nines as 9^{9^9}. This is 9 multiplied by itself more than 387 million times. Written as a number, it would stretch more than a thousand miles and is more than four million times as big as the number of electrons in the universe. This is a number so ridiculously large that it could not apply to anything and yet we need just three nines to write it.

A Perfect Hand at Bridge

You often read or hear of perfect hands at bridge. Assuming that one deal of the cards is just as likely to occur as another, let us see how often this perfect deal can happen.

If we take 2 cards, there are just 2 ways we may deal them, *AB* and *BA*. If we take 3 cards *A*, *B*, and *C*, we discover there are 6 ways *ABC, ACB, BAC, BCA, CAB*, and *CBA*. With 4 cards, the problem becomes more difficult, but we see that there are 24 ways as shown in Table 1. After somewhat more labor, we may see that 5 cards may be

TABLE 1

ABCD	*BACD*	*CABD*	*DABC*
ABDC	*BADC*	*CADB*	*DACB*
ACBD	*BCAD*	*CBAD*	*DBAC*
ACDB	*BCDA*	*CBDA*	*DBCA*
ADBC	*BDAC*	*CDAB*	*DCAB*
ADCB	*BDCA*	*CDBA*	*DCBA*

dealt in 120 different ways. Now 3 cards may be dealt in $6 = 2 \cdot 3$ ways, 4 cards in $24 = 2 \cdot 3 \cdot 4$ ways, 5 cards in $120 = 2 \cdot 3 \cdot 4 \cdot 5$ ways. Thus, it appears that the number of different ways a pack of 52 cards can be dealt is 2 times 3 times 4 times 5 and so on up to 52. This is really an enormous number for it turns out to be about 106 followed by 66 zeros. However, this is the number of ways the cards may be dealt and it does not matter in what order a player receives the cards, just so long as one player gets all the spades, another all the hearts, etc. The number of ways in which a player may be dealt the 13 spades is 2 times 3 times 4 and so on up to 13, which proves to be about 62 followed by 8 zeros. And we do not care which player gets the spades, which the hearts, etc.

Even dividing out these possibilities, we arrive at the chance of dealing a hand with one player getting all spades, another all hearts, etc., which is about one deal out of a number which has 57 zeros. If you play bridge every night in your lifetime, you are still not likely to see such a perfect hand.

▲ EXERCISES

1. If a pack contains 20 cigarettes and a man smokes a pack a day, how long will it take him to smoke a million cigarettes?
2. A mosquito is about $\frac{3}{8}$ inch long. How big would it be if it were enlarged a million times?
3. The heartbeat of a man is about 75 times a minute. How many times will it beat in a lifetime of 70 years?
4. If the round trip to the planet Mars covers 150 million miles and a man-carrying rocket travels 7 miles per second, how long will it take to make the round trip?
5. The diameter of the earth is about 8,000 miles. If it were diminished to

one-billionth of its size, how would it compare with the thickness of a hair? Take the thickness of a hair as 0.003 inch.

6. Dr. Edward Kasner of Columbia University invented a special name for the number 10^{100}. He called it a *googol.* The total number of electrons in the universe is how many googols?

The Greek philosopher Pythagoras caught one of his slaves stealing and punished him by telling him to walk back and forward past the seven columns of the temple of Diana, counting them, until he reached the thousandth column. He was to turn around at the seventh column, counting the sixth as the eighth, then again turning around at the first counting it as 13, count the second column as 14, and so on.

After he finished his punishment he was to tell Pythagoras which column he had counted as 1,000. Assuming that he neither cheated nor missed his count, which column did he report was 1,000?

April and November

"I was three times as old as my bride when I married 18 years ago. Now I am twice as old as she is."

You wonder what these ages were, and you might find out with a lot of trying and testing. However algebra gives us a lazy way to solve the problem.

Let x = age of wife when married
$3x$ = age of husband when married
$x+18$ = age of wife today
$3x+18$ = age of husband today

But the husband is twice as old as his wife today.

$$3x+18=2(x+18)$$
$$=2x+36$$
$$x=18$$

So the wife was 18 years old at the time of her marriage and her husband was 54 years old. Eighteen years later she is 36 and he is 72.

6 *Counting Sheep*

You have heard of the man who was riding past a flock of sheep and announced to his companion, "There are exactly 78 sheep in that field." His companion asked how he knew and he said, "I counted their feet and divided by four."

Now a few days later this feet counter said to his friend, "My neighbor has both chickens and cows. I counted and found they have 35 heads and 88 legs." How many chickens and how many cows does his neighbor have?

We could solve this problem by figuring that 35 chickens and 0 cows gives 70 legs, 34 chickens and 1 cow gives 72 legs, and so on until we arrive at the desired 88 legs. But this is long and tiring, and algebra again comes to our rescue.

Let $x=$ the number of cows
$35-x=$ the number of chickens
$4x=$ the number of legs on cows
$2(35-x)=70-2x=$ the number of chicken legs

$$4x+70-2x=88$$
$$2x=18$$
$$x=9$$

Thus the neighbor owns 9 cows and 26 chickens.

7 *Can You Think When the House Is on Fire?*

Mr. Fancy Dresser wakes in the middle of the night with the house on fire and the lights out. He has 5 pairs of blue socks and 8 pairs of brown socks in his drawer, but they are loose rather than tied in pairs. How many individual socks must he grab to be sure to be able to dress later with 2 socks of the same color?

There are 10 blue socks and 16 brown socks so we might say that he needed to pick up 12 socks to be sure of getting a pair. But let us reason this more carefully. If he picks up 2 socks, they might turn out to be *BlBl*, *BlBr*, or *BrBr*. In the first and third possibilities, he would have a pair but in the second he would not, so 2 is not the answer.

If he picks up 3 socks, what are the possibilities? They are *BlBlBl, BlBlBr, BlBrBl, BrBlBl, BlBrBr, BrBlBr, BrBrBl,* and *BrBrBr,* a total of eight possibilities. In the first four possibilities we have a pair of blue socks, and in the last four we have a pair of brown socks. So in every case we have a pair of the same color and our frightened but fashionable man need grab only 3 socks. Thus mathematics, or careful reasoning, has given us the answer.

8 *The Tortoise and the Hare*

A tortoise and a hare run a race. The hare can run twice as fast as the tortoise so he gives the tortoise a head start of 2 minutes. Does the hare pass the tortoise?

This problem resulted in many arguments among the Greek philosophers, some of whom reasoned as follows:

When the hare begins to run, the tortoise will be a certain distance ahead. By the time the hare reaches this point, the tortoise will have moved half this distance on ahead. And by the time the hare reaches this second point, the tortoise will have moved on ahead again half that distance. And when the hare reaches this third point, the tortoise has again moved half the distance on ahead. And so on. Hence the tortoise is always ahead, and the hare will never pass the tortoise. But our common sense tells us that the faster-moving hare must overtake and pass the lowly tortoise. Where is the flaw in the Greek philosophers' reasoning?

There are several ways of pointing out the trouble, but perhaps the simplest is to observe that running is a continuous process rather than the individual parts which concerned the Greek philosophers, since these parts come faster and faster. If the tortoise moves f feet per minute, then the tortoise has a head start of $2f$ feet and will be $(2f + nf)$ feet from the starting point after n minutes from the starting time of the hare. The hare will run $2nf$ feet in this same n minutes. We equate these two distances and find that the hare passes the tortoise after 2 minutes.

We may look at this same problem in another way. A man starts 1 mile from a river. He walks half the distance to the river, then half the remaining distance, then half of what is left, and so on. Will he ever reach the river?

That depends. If he walks halfway to the river in 30 minutes, then half of the remaining distance in 30 more minutes, then half of what is left in 30 minutes, and so on, he will never reach the river. He will die first. But if he walks at the same speed so that he gets halfway there in 30 minutes, half of the remaining distance in 15 minutes, half of what is left in $7\frac{1}{2}$ minutes, and so on, he will fall into the river exactly 1 hour after he starts his walk.

9 *Battle of Numbers*

Here is a game you can play with your friends. One of you chooses a number from 1 to 10. The other chooses any number 1 to 10 and adds it to the number chosen first. Then the first player takes a number 1 to 10 and adds it to the sum. The player who reaches the sum 100 wins the game.

For example, John chooses 6. Mary chooses 8, which gives 14. John chooses 5, which gives 19. Mary chooses 8 again, which gives 27. And so on.

Now let us see if John can arrange things so as to win. He must manage to reach a sum so that any number Mary chooses will not add to 100, but will leave the new sum close enough to 100 so that John can win. This number is 89. Regardless of which number Mary chooses, her sum will not be more than 99 and not less than 90. Then John can win.

We call 89 a winning position for John. Prior to this John wants a sum so that Mary cannot get to 89 but so that he can get to 89, no matter what number Mary chooses. This number is 78. From here Mary cannot go above 88 nor below 79. From each of these, or anything between, John can go to 89.

Thus there is a whole series of winning positions, 89, 78, 67, 56, 45, 34, 23, 12, and 1. If one player gains one of these positions, he is sure to win because he can move up from one to another despite anything the other player may do.

This is a very old game of strategy. Bachet de Méziriac wrote on the game in 1613 and suggested that you not grab a winning position too early in the game lest your "victim" catch on.

10 *Again Mathematics—What and Why?*

All the problems considered in this chapter, both useful and amusing, are examples of mathematics. But these are only a handful of the multitude of problems that involve mathematics. Mathematics involves numbers and easier methods to solve all kinds of problems with numbers. It tells us how to know how many eggs are in seven dozen without taking them out of their boxes and counting them. It tells us how to figure how many cups of flour to use in a recipe that calls for $2\frac{1}{2}$ cups of flour to feed eight persons if we wish to feed only six people.

But it is more than numbers. It tells the airplane designer the shape and size to make the wings for supersonic flight. It tells the surveyor how to measure distances and angles. It tells the bridge builder how large his girders should be to stand the expected load. It helps the astronomer predict the time of sunset and sunrise and the location of stars. Without mathematics or with only crude mathematics, all science and industry would grind to a halt and we should return to the life of prehistoric man, the cave man. Life would be simple, but neither pleasant nor profitable.

The cook, the carpenter, the salesman, the electronics engineer, the housewife—all find mathematics indispensable. Mathematics is a tool to be used in every occupation and activity in this highly industrial world of ours. Most persons look on it as a form of drudgery and accept it only because it is a useful tool.

But mathematics is more than a necessary tool, more than a system of tricks to solve problems. It is also a form of logic. It demands exact and rigorous thinking; any rigorous analysis of any situation may be called mathematics. This involves the careful use of words and some hard thinking, rather than the memorizing of tricks. In fact memory has little place in mathematics. Memory is used only to save time on facts that are used over and over. We could look up $7 \times 8 = 56$ every time we used it but this would be wasteful.

Mathematics differs from all other fields of knowledge in its emphasis on reasoning. *Facts* are established in other fields of science by observation and experiment. No mathematical *fact* is ever proved by observation of individual cases or experimentation alone. We are guided by observation and experimentation, but the final *proof* of a mathematical law must be pure reasoning. Such reasoning is the finest display of the mentality of man, the surest proof that he is *man* rather than *animal.*

▲ EXERCISES

1. What are the winning positions in the Battle of Numbers if the one who reaches 100 *loses* instead of winning?
2. What are the winning positions in the Battle of Numbers if you can choose any number from 1 to 19, and if the one who reaches 200 first wins?
3. A boy spent 50 cents on candy and bought 18 bars. Some were 2 for 5 cents and others were 3 for 10 cents. How many of each did he buy?
4. The two hands of a clock are together at 12 noon. At what time will they next be together?
5. "Where did you put that counterfeit silver dollar?" the cashier of the bank asked the teller. "Good gosh," said the teller, "I dropped it into

that pile of 8 good dollars. They look alike so you'll have a terrible time picking it out." "That's not difficult," said the cashier, and balancing a ruler across his finger, making a crude balance scale, he picked out the counterfeit from the pile of 9 dollars by just two weighings. Knowing that the counterfeit was lighter than the good dollars. how did he find the counterfeit?

6. A total of 82 players enter the tournament to determine the City Tennis Singles Championship. How many matches will it take to play the entire tournament? (If you require pencil and paper, you haven't found the simplest way to solve this.)
7. A customer in Smith's Shoe Store buys a pair of shoes for $14 and offers a $20 bill. Being short of change, Smith goes to Brown's Hardware Store next door and gets change for the $20 bill. Later in the day Brown comes back with the $20 bill and tells him it is counterfeit. Of course Smith has to redeem the bill by paying Brown $20. On top of this $20 lost, Smith has given the crook $6 in change. So Smith figures he is out a total of $26. Is he right or wrong?
8. A girl says, "I have the same number of brothers as sisters." Her brother says, "I have twice as many sisters as brothers." How many boys and girls are there in the family?

Problems just for fun

What part of the day has disappeared if the time left is twice two-thirds of the time passed away?

CHAPTER TWO

Ratio, Proportion, and Variation

Mathematical language is not only the simplest and most easily understood of any, but the shortest also. H. L. BROUGHAM

In our everyday conversation, we often make statements involving numbers and their relations to each other. For example, Mary is taller than John; this building is twice as tall as that building; my new car has 20 per cent more horsepower than last year's model; if I had 30 per cent more money in my purse, I could pay this bill; it is the same distance from my house to your house as it is from your house to your office.

We must know the meaning of certain statements about relative size, and we must be able to express these statements by mathematical equations or mathematical inequalities.

In mathematics we use the letters of the alphabet to represent numbers for which we may know the value or for which we hope to find the value by mathematical manipulation.

Letters $a, b, c, \ldots,$ near the beginning of the alphabet usually represent numbers for which we know the value and the letters w, x, y, z at the end of the alphabet represent the unknown values which we want to determine.

Order

When we speak of two numbers a and b, we instinctively compare them. That is, they are either equal or one is larger than the other. We need symbols to help us write these relations.

If a is equal to b, we write $a=b$.

If a is larger than b, we write $a > b$ and mean that the difference $a-b$ is a positive number.

For example:

$$6>4 \qquad \text{since} \qquad 6-4=2$$
$$-2>-7 \qquad \text{since} \qquad -2-(-7)=5$$

Let us find all positive integers which are less than 5. We want the solution to the inequality $x < 5$, where x is to be a positive integer.

The positive integers are 1, 2, 3, 4, 5, . . ., and we easily see that if $x = 1, 2, 3$, or 4, we have the solution to our problem.

One method that will help us visualize this order relation between real numbers is to think of the numbers as located on a number line. To construct a number line, we draw a straight line and locate a point 0 on it and then mark scales along this line:

Fig. 2

The numbers to the right of 0 are positive; the numbers to the left of 0 are negative. If we locate a number c on this line, we see that all the numbers to the right are greater than c and all numbers to the left are less than c.

It is now possible to compare any two real numbers. A little thought tells us that only one of the following three conditions can hold

$$a>b \qquad \text{or} \qquad a<b \qquad \text{or} \qquad a=b$$

Sometimes we have linear inequalities, that is, inequalities involving an unknown to be found. For example, for what value of x is the inequality $2x+8<0$ true?

This is equivalent to "for what value of x is $2x<-8$?" The solution is $x < -4$. We can easily check this result by referring to the number line. We then note that if x is less than -4, then $2x<-8$ and so $2x+8<0$.

Absolute Value

Many times we are not concerned with whether a number is positive or negative, but only with how large it is and how far it is from the point 0 on the number scale. We call this property the *magnitude or absolute value* of the number a and write $|a|$ for the magnitude.

Then we make the definition

$$|a| = a \quad \text{if} \quad a = 0 \quad \text{or if} \quad a > 0$$
$$|a| = -a \quad \text{if} \quad a < 0$$

We look at some examples:

$$|6| \ = +6$$
$$|-6| = -(-6) = +6$$

When we look at our number line we see that both 6 and −6 are 6 units from the zero point.

We might need to find values of x for which $|x| < 5$. Looking at the number line, we see that if $-5 < x < 5$, then $|x|$ is indeed less than 5. If we ask what values of x satisfy $|x| > 5$, then we see that $x > 5$ and $x < -5$ are correct solutions:

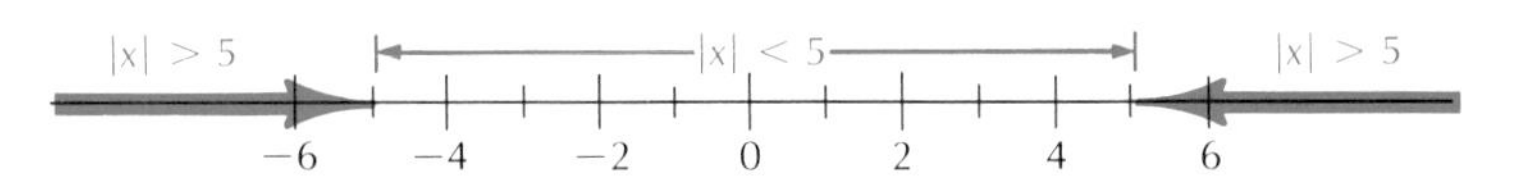

Fig. 3

It is always wise to consider the number line and "try out" our solution for inequalities to see if we have the correct solution. Always go back to the definitions for inequality between numbers and absolute value of numbers.

▲ EXERCISES

1. Write the absolute values of the following numbers

   a. 3 b. −3
   c. $\frac{1}{2}$ d. $-\frac{3}{2}$
   e. $-\frac{2}{3}$ f. −8

2. On the number line locate the numbers

   a. Whose absolute value < 3
   b. Whose absolute value > 3

3. If x represents a real number, locate all numbers such that

   a. $|x-2| < 3$ b. $|x-2| > 3$
   c. $3x+12 < 0$ d. $x-2 > 0$
   e. $3x+5 < x-3$ f. $2x+8 < 0$
   g. $|x| < 4$ h. $|x| > 4$
   i. $|x+1| > 3$ j. $|x-1| < 2$

4. Write an inequality that states

   a. x is within 3 units of 0
   b. x is within 3 units of 4
   c. x is within 4 units of -1

5. Write mathematical formulas that state that

   a. The distance from x to 2 is greater than 3
   b. The distance from x to -1 is greater than 3

6. Prove that if $a > 0$ and $b > 0$ then

   a. $a + b > 0$ b. $ab > 0$

Problems just for fun

A rectangle is inscribed in the quadrant of a circle as shown. What is the length of the diagonal AB?

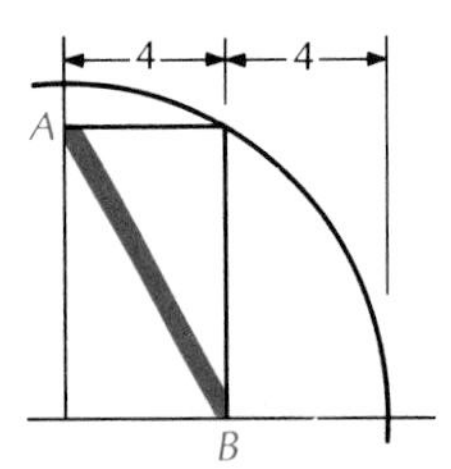

3 Ratio

If we have two books, one of which contains 486 pages and the other 162 pages, we can use the method of Sec. 1 to say that $486 > 162$. Also $486 - 162 = 324$ which tells us that the first book has 324 pages more than the second. However, there is still another way of comparing the two books by saying that the first book has three times as many pages as the second. This last method of comparison allows us to use smaller numbers and is usually the more useful of the methods.

It enables us to perform computations that cannot be performed by means of the method of differences and is used in making maps, blueprints, and house plans. It is also useful in the fields of science and engineering.

When the relation between two quantities is expressed by forming the quotient of the two quantities, the quotient is called the *ratio* of one quantity to the other quantity. Thus the ratio of A to B is written A/B. If, in the example above, we divide the number of pages in one book by the number of pages in the other book, we get

$$\frac{486}{162}=\frac{3}{1} \quad \text{or} \quad \frac{162}{486}=\frac{1}{3}$$

The quantities compared in this way must represent measures of the same kind and dimensions. We cannot compare weight and area, time and distance, or length expressed in feet and length expressed in yards. However, we can compare the two lengths if we convert both to feet or both to yards.

The ratio of two quantities is a pure number. For example, the areas of two circles whose radii are 4 inches and 9 inches are 16π square inches and 81π square inches. The ratio of the smaller area to the larger area is

$$\frac{16\pi \text{ square inches}}{81\pi \text{ square inches}}=\frac{16}{81}$$

As another example, the difference between the freezing point and the boiling point of water is $212° - 32° = 180°$ on the Fahrenheit scale and $100° - 0° = 100°$ on the centigrade scale. Then, between any two temperatures there are $\frac{180}{100} = \frac{9}{5}$ as many Fahrenheit degrees as centigrade degrees. That is, the ratio of the number of Fahrenheit degrees to centigrade degrees is $\frac{9}{5}$.

For computational work, ratios should be reduced to the simplest form.

Proportion

A proportional equation is formed by joining two equal ratios by an equality sign. For example,

$$\frac{2}{3}=\frac{12}{18} \quad \text{and} \quad \frac{3}{5}=\frac{24}{40}$$

When two ratios a/b and c/d are set equal to each other, a proportion is formed which states that a divided by b is equal to c divided by d.

The proportions determined by the ratios found in the problems in the preceding section are

$$\frac{\text{Number of pages in large book}}{\text{Number of pages in small book}} = \frac{3}{1}$$

$$\frac{\text{Area of the small circle}}{\text{Area of the large circle}} = \frac{16}{81}$$

$$\frac{\text{Number of Fahrenheit degrees}}{\text{Number of centigrade degrees}} = \frac{9}{5}$$

There are many ways in which proportions may be arranged. It is sufficient to notice that the proportion is simply an equation, and as such, we can deal with it as we would any other equation in algebra.

■ **Example.** Find the area of a room if the length is 15 feet and the ratio of the length to the width is $\frac{4}{3}$.

□ ***Solution.*** $$\frac{\text{Length of room}}{\text{Width of room}} = \frac{4}{3} = \frac{15}{\text{width of room}}$$

Solving, we find that the width of the room $= \frac{45}{4} = 11\frac{1}{4}$ feet, and the area of the room is $15 \times 11\frac{1}{4}$ square feet.

There are many types of problems that lead to proportions. For instance, the cost of a number of objects increases in proportion to the number of objects, so that if the number is doubled, the cost is doubled, and so on. Sometimes a quantity depends simultaneously on two different things. For example, the quantity of work done is in proportion to the number of workmen and the number of hours each man works per day.

5 *A Word of Warning*

Particular attention should be called to the fact that the theory of proportion applies only to constants and to things that increase or decrease in *constant* ratio. Careful consideration must be given to this fact. A few examples will illustrate this. If one man can do a certain amount of work in one day, will two men working together do twice the amount in one day? It depends on the job. In some jobs the two men would get in each other's way. If a tank contains 10 gallons of water and it all runs out through a hole in the bottom in 1 hour, would it require double that time to empty the tank if the tank contained 20 gallons? The answer is that it would require less than double the time. If 1 horse can pull a load of 1 ton, could 10 horses pull 10 tons? Only if all horses pull alike in amount and manner. We turn our attention now to problems which involve constant ratios.

6 *Problems in Proportion*

In order to write the correct proportion for a problem, we must fully understand the problem. Read the problem slowly and be certain you understand it.

■ **Example.** If 36 grams of water yield 32 grams of oxygen, how many grams of water will be required to yield 10 grams of oxygen?

□ ***Solution.*** We assume that the amounts of oxygen produced are proportional to the number of grams of water. Hence twice as much water should yield twice as much oxygen. That is, 72 grams of water should yield 64 grams of oxygen. Half of 36, or 18, grams of water should yield 16 grams of oxygen. Notice that

$$\frac{72}{64}=\frac{18}{16}=\frac{36}{32}$$

i.e., all the ratios are equal to the ratio $\frac{36}{32}$. In fact, to say that the amount of oxygen produced is proportional to the amount of water is equivalent to saying that the ratio

$$\frac{\text{Amount of water}}{\text{Amount of oxygen produced}}=\frac{36}{32}$$

Now in our problem, let x be the number of grams of water which will yield 10 grams of oxygen. The previous proportion then gives at once

$$\frac{x}{10}=\frac{36}{32}$$

so that

$$x=10\cdot\frac{36}{32}=\frac{45}{4}=11\frac{1}{4}\text{ grams of water}$$

We can solve this problem in another way. Since 36 grams of water yield 32 grams of oxygen, then it takes $\frac{36}{32}$ grams of water to yield 1 gram of oxygen, and so it will take 10 times this much to yield 10 grams of oxygen, i.e.,

$$10\cdot\frac{36}{32}=11\frac{1}{4}\text{ grams of water}$$

▲ **EXERCISES**

1. At the time a vertical pole 6 feet tall casts a shadow 7.5 feet long, a tree casts a shadow 30 feet long. How tall is the tree?
2. If 180,000 calories of heat is obtained from 20 pounds of a certain grade

of coal, how many calories of heat will be released from 1 ton of this coal?

3. The ratio of the weight of zinc to copper in brass is $\frac{2}{3}$. What is the weight of copper in a piece of brass which weighs 7.82 pounds?
4. When decomposed by an electric current, 18 grams of water yield 2 grams of hydrogen. How much hydrogen could be obtained from 30 grams of water?
5. A piece of alloy contains 3.72 pounds of copper and 2.52 pounds of zinc. How much zinc and copper make up a piece of alloy ten times as heavy?
6. Neglecting friction in pushing a weight up an inclined plane, the ratio of the weight to the required force is equal to the ratio of the distance to the height the weight is moved and lifted. Find the force required to move a weight of 2,000 pounds 60 feet up an incline to a height of 12 feet.
7. Divide the number 960 into two parts which will be in the ratio of $\frac{7}{5}$.
8. If soft solder is composed of two parts of tin and one part of lead, how much lead and how much tin are needed to make 40 pounds of solder?
9. On a farm, the area of land in cultivation exceeds the area of the grazing land by 200 acres. If the ratio of cultivated land to grazing land is $\frac{7}{2}$, how many acres make up this farm?
10. The scale of a map is 1 to 100, i.e., 1 inch represents 100 miles. What is the actual area in square miles of a section shown on a map as a rectangle 1.5 by 3 inches?
11. A university has an enrollment of 21,300 students. The ratio of men students to women students is $\frac{8}{7}$. How many "Eds" and "Coeds" are enrolled?
12. A recipe for pickling syrup for fruits calls for 3 quarts of sugar and 2 quarts of vinegar. If we assume that this produces 5 quarts of syrup, thus ignoring the spaces between the grains of sugar, how much sugar and vinegar are needed for 2 gallons of syrup?

7 *Percentage*

Ratios and proportions are frequently expressed by *percentages.* The word "percentage" or "per cent" is derived from the Latin words *per centum,* meaning "out of a hundred," or "per hundred." If a club had 100 members and 90 attended a meeting, the attendance would be 90 out of 100, or 90 per cent, for that meeting. If a student answers 8 questions correctly out of 10 questions, his score is 80 per cent correct. The symbol % is used to denote per cent. One per cent is written as 1% and twenty per cent is written as 20%.

Per cent is a ratio of something. We must always be clear on the question "per cent *of what?*" even when it is not stated. Otherwise we fall into errors. A dealer says he makes 25% profit on an article. Twenty five per cent of what? He means his profit is 25% of his wholesale cost.

For purposes of calculation, per cent must be changed into either decimal or fraction ratios. If there were 30 questions on a test and a student answered 70% of them correctly, he would have answered

$$\frac{70}{100}(30)=21 \text{ questions}$$

We could have obtained this answer by multiplying 30 by 0.7, since 70% = 0.7.

Common fraction ratios may be changed into per cents. For example,

$$\frac{3}{4}=0.75=\frac{75}{100}=75\%$$
$$\frac{1}{8}=0.125=\frac{12.5}{100}=12.5\%$$

The eight digits are arranged in two columns as shown.

3	6
2	4
7	5
9	8

Can you rearrange the numbers, moving as few as possible, so that the two columns will add to the same number?

▲ EXERCISES

1. Change the following to fractions and decimals:

a. $62\frac{1}{2}\%$ b. 33.3%
c. 125% d. 37.5%
e. 12.5% f. 200%

2. Change the following fractions into per cent:

   a. $\frac{3}{8}$ b. $\frac{5}{6}$
   c. $\frac{2}{3}$ d. $\frac{2}{5}$
   e. $\frac{2}{6}$ f. $\frac{1}{4}$

3. It rained on 20% of the days of April. On how many days in the 30-day month did it rain?
4. A dealer sells a house for $36,000 and makes a 6% commission. How much does the dealer get?
5. Seventeen per cent of the students in a college are girls. How many boys are there in this college if the total enrollment is 12,000?
6. The rainfall this year is 15% more than the rainfall of 32 inches last year. What is the rainfall this year?
7. When the owner of a shop returned from lunch, the clerk said, "I sold those last two vases from the shipment from the Eastern Art Company. The customer objected to the price on the red vase and offered $1.20. I know this was below cost, but he agreed to take the yellow vase at the same price. We took a 20% loss on the red vase but made 20% on the yellow vase, so we broke even." Why was the boss unhappy?
8. A union negotiates a new contract calling for a 6% increase in pay. If a worker makes $2.70 an hour under the old scale, what will his new hourly rate be? (Round off to the nearest cent.)
9. Thirty-two of a class of forty students are freshmen. What per cent of the class are upperclassmen?
10. Fifteen per cent of the students in a class failed a test. If 68 of the students passed the test, how many were in the class?
11. A bracelet which cost $15 wholesale has been marked to sell at $24. For a clearance sale, what per cent may the jeweler mark down the price and still not lose money on the bracelet?
12. A seller offers 5% discount for cash. A customer buys a TV set and writes a check for $339.15. What was the list price of the TV set?

Find the number which when divided by 4 more than itself has a quotient of $\frac{3}{4}$.

8 *Variation*

In studying the statements expressing the dependence or relation between two quantities, we discover that these statements often contain the words "is proportional to," "varies directly," "varies jointly," etc. For example, the distance that a person can see on a desert is found to vary with the height of the observer above the ground; the stretch of a spring varies with the weight placed on the spring; the circumference of a circle is proportional to the diameter; the volume of an enclosed gas varies inversely with the pressure if the temperature of the gas is kept constant; and the height to which a liquid rises in a capillary tube is inversely proportional to the radius of the tube and directly proportional to the surface tension of the liquid.

The above remarks, with respect to the variation between elements, suggest that we discuss the topic of variation, which has a language of its own. We need to understand and translate this language into mathematical equations. Variation is usually classed as *direct, inverse,* or *multiple. Each of the three types of variation is described by standardized phrases, and each type has a standard equation.*

9 *Direct Variation*

The phrases that characterize direct variation are "varies as," "varies directly as," "is proportional to," and "is directly proportional to."

The statement that one quantity A is proportional to or varies as another quantity B means that the quantity A divided by the quantity B is a constant k. That is,

$$\frac{A}{B}=k \qquad \text{or} \qquad A=kB$$

The equation $A = kB$ is the *standard equation* for direct variation. The constant k is called the constant of *proportionality*, or *variation.*

The constant of proportionality k is very important. Even though the value of k is constant, it has the units of the quantity A divided by the quantity B. For example, if the distance in feet that a car travels is proportional to the time in seconds, we have

$$\frac{\text{Distance}}{\text{Time}}=k$$

and the constant k has units of feet per second. On the other hand, if the time is measured in minutes, the constant k will be in units of feet per minute and will be 60 times larger than the previous case.

We leave out the constant of proportionality only when the

variable A is equal to the variable B, or $A = B$. This is rarely possible, as the following examples will illustrate.

Example 1. The cost of a beef roast varies as the weight of the roast. The cost of the roast does not equal the weight, but

$$\frac{\text{Cost}}{\text{Weight}} = k \qquad \text{or} \qquad \text{cost} = k \text{ times weight}$$

In this case, the constant of proportionality k is the price per unit of weight of beef roast.

Example 2. The weight of gold nuggets is proportional to the volume. The weight of the nuggets does not equal the volume, but

$$\frac{\text{Weight}}{\text{Volume}} = k \qquad \text{or} \qquad \text{weight} = k \text{ times volume}$$

Here k, the constant of proportionality, is the density or weight per unit volume of gold.

Example 3. The distance d in feet that a body falls from rest varies directly with the square of the time of fall t in seconds. The exact relation between d and t is given by

$$\frac{d}{t^2} = k \qquad \text{or} \qquad d = kt^2$$

The constant k has units of feet per second per second, or feet per $(\text{second})^2$.

Never leave out the constant of proportionality. On a few rare occasions you may find it is numerically equal to 1.

10 Inverse Variation

The phrases that characterize inverse variation are "varies inversely as" and "is inversely proportional to."

The statement that one quantity A varies inversely or is inversely proportional to another quantity B means that A varies with the reciprocal of B, which is $1/B$. Then we may write

$$\frac{A}{1/B} = k \qquad \text{or} \qquad AB = k$$

and k is again the constant of proportionality. The equation $AB = k$ is the standard equation for inverse variation.

The equation $AB = k$ states that if A gets larger, B must become smaller in the same ratio in order that the product AB remain a constant. In this case, if we triple A, we must reduce B to one-third its original value. We can always find the value of k, the constant of proportionality, if we know the value of one quantity A corresponding to any value of the other quantity B.

■ **Example.** The illumination I is inversely proportional to the square of the distance d from the source of light. The exact relation between I and d is given by

$$Id^2 = k \qquad \text{or} \qquad I = \frac{k}{d^2}$$

Problems Involving Direct and Inverse Variation

■ **Example 1.** Under constant temperature, the density of a gas D in a container varies as its pressure P. By actual measurement it was found that when the density of the gas was 0.075 pound per cubic foot, the pressure was 2,000 pounds per square foot. Find the density when the pressure is 1,500 pounds per square foot.

□ ***Solution.*** Since the pressure P varies as the density D, we can write

$$\frac{P}{D} = k \qquad \text{or} \qquad P = kD$$

But $P = 2{,}000$ pounds per square foot when $D = 0.075$ pound per cubic foot, therefore

$$k = \frac{2{,}000}{0.075}$$

Then the relation between pressure P and density D becomes

$$P = \frac{2{,}000}{0.075}D$$

To find D when $P = 1{,}500$, substitute this value into the equation above.

$$1{,}500 = \frac{2{,}000}{0.075}D$$

Solving for D we get

$$D = 1{,}500 \cdot \frac{0.075}{2{,}000} = 0.05625 \text{ pound per cubic foot}$$

Therefore the density is 0.05625 pound per cubic foot when the pressure is 1,500 pounds per square foot.

There is another way to solve this problem without finding the exact value of the constant of proportionality.

We know from the statement of the problem that

$$P=kD$$

But

$$P=2{,}000 \text{ pounds per square foot}$$

when

$$D=0.075 \text{ pound per cubic foot}$$

Therefore

$$2{,}000=k(0.075)$$

We want to find the value of D when $P=1{,}500$ pounds per square foot, or

$$1{,}500=kD$$

If we divide the last two equations, we get

$$\frac{1{,}500}{2{,}000}=\frac{kD}{k(0.075)}=\frac{D}{0.075}$$

We can solve at once for D, getting

$$D=\frac{1{,}500}{2{,}000}(0.075)=0.05625 \text{ pound per cubic foot}$$

Note. The student should pay particular attention to the units to be attached to his final answer when he has solved practical problems such as the example discussed above.

■ **Example 2.** Since the distance d in feet that a body falls from rest is directly proportional to the square of the time of fall t in seconds, we write

$$d=kt^2$$

Find the distance fallen in 4 seconds if a body falls 16 feet in 1 second.

□ ***Solution.*** Since the body falls 16 feet in 1 second, we can write

$$16=k(1)^2$$

The distance the body falls in 4 seconds is given by

$$d = k(4)^2$$

If we divide these two equations, we get

$$\frac{d}{16} = \frac{k(4)^2}{k(1)^2} = 16$$

Then

$$d = 16(16) = 256 \text{ feet}$$

In a problem dealing with variation between two quantities A and B, in which we know the value of A corresponding to a particular value of B and desire to find the value of A for another value of B, it is not necessary to find the constant of variation. It is quicker to set up the equations for each case and solve for the unknown by dividing the two equations.

▲ EXERCISES

In the following examples, set up the equations for the relations between the variables and find the constant of proportionality when enough information is given.

1. The area of a circle is proportional to the square of the diameter.
2. The circumference of a circle is proportional to the diameter.
3. The intensity of sound is inversely proportional to the square of the distance from the source of sound.
4. The daily wage is proportional to the number of hours worked.
5. The volume of an enclosed gas is inversely proportional to the pressure if the temperature of the gas is kept constant. If a tank contains 8,000 cubic feet of gas at a pressure of 18 pounds per square inch, what is the volume if the pressure is increased to 22 pounds per square inch?
6. The pressure on a submarine is proportional to its depth below the surface of the ocean. Find the pressure at a depth of $\frac{1}{2}$ mile if the pressure at 100 feet is 64 pounds per square foot.
7. A body which is dropped from a height strikes the ground with a speed which is proportional to the square root of the height. An object falling from 16 feet strikes the ground with a speed of 32 feet per second. How far must an object fall to hit the ground with a speed of 60 feet per second?
8. The pressure P in pounds per square foot exerted by a wind against a wall varies as the square of the wind velocity V in miles per hour. If $P = 30$ pounds per square foot when $V = 90$ miles per hour, find the pressure when the wind velocity is 50 miles per hour.
9. The amount A in grams of a substance digested in Q grams of pepsin in 1 hour varies as the square root of Q. If $A = 7$ grams when $Q = 100$

Problems just for fun

'Tis the time for the annual Freshman Hop! The guys are backward and the gals are bashful. The person delegated to see that everyone dances finds three girls and three boys who just can't get together. Trying to remedy this situation, the well meaning get-gals-and-guys-together person arranges the six people as shown, with a space between the boys and girls. "Let's pair off!" he calls. The plaid couple, the polka-dot couple, and the plain couple are to get together. *But it must be done according to the rules.* At each call, one person moves into an empty space by stepping into it or leapfrogging over one person. Boys move to the left and girls move to the right. Can do in six moves. You call 'em!

grams, find the amount digested in 150 grams of pepsin in 1 hour.

10. A current flowing in an electric circuit varies inversely as the resistance of the circuit. If the current I is 10 amperes when the resistance R is 11 ohms, write an equation for the variation. Find the current when the resistance is 5 ohms.
11. The distance required to stop a certain car varies as the square of the velocity. If the car with a speed of 35 miles per hour can stop in 45 feet, how many feet are necessary to stop a car traveling at 50 miles per hour? At 60 miles per hour?
12. The horsepower required to drive an airplane varies with the fourth power of the plane's velocity. If two 150-horsepower motors will drive a plane at a speed of 75 miles per hour, how much horsepower is necessary to drive this plane at 100 miles per hour? At 120 miles per hour?

Multiple Variation

We have just learned to express as an equation the relation between two quantities or variables when the relationship was given in certain terms. Now we know that one quantity may depend upon more than one other quantity. For example, the size of a man's pay check depends both on the number of hours worked and the hourly rate of pay; the brightness in a corner of a room depends upon the number of lights, the intensity of each light, and the distances of the corner from the various lights.

The phrases that characterize multiple variation are "varies jointly as," "is directly proportional to," and "varies directly as ______ and inversely as ______."

The statement that one quantity A is directly proportional to B and C means that A divided by the product of B and C is a constant k. That is,

$$\frac{A}{BC}=k \qquad \text{or} \qquad A=kBC$$

For example, the area A of a triangle varies with the height h and the base b. The relation between the area, height, and base of the triangle is given by the equation

$$A=kbh$$

The volume V of a cylinder varies as the height h and the square of the radius r of the cylinder. The exact relation between V, h, and r is given by

$$V=khr^2$$

If a quantity A varies directly as quantity B and inversely as C, the equation for this variation is

$$A = \frac{kB}{C}$$

The amount of illumination I on a table varies directly as the brightness c of the light source and inversely as the square of the distance d from the light. The equation relating I, c, and d is

$$I = \frac{kc}{d^2}$$

From the equation, we notice that the stronger the light source the greater the amount of illumination, but the farther away the light, the less the illumination.

▲ EXERCISES

In the following exercises, set up the equations for the relations involved and evaluate the constant of variation when enough information is given.

1. The area of a rectangle varies directly as the length and width of the rectangle.
2. The intensity of sound varies directly as the strength of the source and inversely as the square of the distance from the source of the sound.
3. The force of wind on a sail is directly proportional to the area of the sail in square feet and the square of the wind's velocity in miles per hour. If the force on 1 square foot of sail is 1 pound when the wind velocity is 15 miles per hour, find the force on 1 square foot of sail when the wind velocity is 10 miles per hour.
4. Using the data of Exercise 3, what must be the value of the wind velocity to exert a force of 20 pounds on 1 square foot of sail?
5. The mass of metal deposited by passing a current of I amperes through an electrolytic solution varies directly as the atomic weight of the metal deposited, the current, and the time. If a current of 10 amperes deposits 1.08 grams of silver from a silver chloride solution in 96.5 seconds, how long will it take a current of 2 amperes to deposit 0.56 gram of silver?
6. The heat lost per hour through a glass window of a house on a cold day varies directly as the difference between the inside and outside temperature and the area of the window, and also varies inversely as the thickness of the window. If 6000 calories of heat is conducted through a window 100 by 150 centimeters, $\frac{1}{4}$ centimeter thick, in 1 hour when the temperature difference is 40°C, how many calories is conducted in 1 hour through a glass window $\frac{1}{3}$ centimeter thick having the same area, when the temperature difference is 20°C?

Problems just for fun

Something is wrong!

$$1 \quad 2 \quad 3 \quad 4 \quad 5 \quad 6 \quad 7 \quad 8 \quad 9 = 100$$

The typist left out two plus signs and two minus signs because they didn't look important! Can you insert them in the equation in order to make it correct?

Review Exercises

1. Write an inequality which states that

    a. x is within 3 units of 5
    b. x is within 2 units of -5
    c. $2x$ is within 3 units of 5
    d. $\frac{x}{2}$ is within 1 unit of 4
    e. Distance from $4x$ to -5 is greater than 2
    f. Distance from x to 3 is greater than 4

2. Write a set of inequalities expressing the fact that x lies inside the line segment bc where $b < c$.
3. Write a set of inequalities expressing the fact that x lies outside the line segment bc if $b < c$.
4. Solve the following inequalities

    a. $3x + 5 < x - 7$
    b. $x + 8 \geq 5x - 12$
    c. $|2x - 5| < 1$
    d. $|x - 3| < 2$
    e. $|x| > 6$

5. A group of 108 people gathered for a meeting. The ratio of the number of men to women was $\frac{7}{2}$. How many men and how many women were there in the group?
6. Roughly speaking, the demand for eggs varies inversely as the

price. If a grocer sells 120 dozen eggs a week when they are 50 cents a dozen, how many should he expect to sell in a week when the price is 60 cents a dozen?

7. Divide the number 850 into two parts which will be in the ratio of $\frac{2}{3}$.
8. Represent the following statements in equation form:

   a. The circumference of a circle is proportional to the radius.
   b. The velocity with which a ball strikes the ground varies directly as the square root of the height.
   c. The gravitational force of attraction between two bodies is directly proportional to the two masses m_1 and m_2 and is inversely proportional to the square of the distance between the centers of the masses.

9. A ball in part *b* of Exercise 8 falls from a height of 100 feet and strikes the ground with a speed of 80 feet per second. With what speed will a ball strike the ground if it is dropped from a height of 49 feet?
10. Choose suitable letters, then express the following statement in equation form: The resistance in a wire conductor varies inversely as the square of the diameter of the wire.
11. If V is the volume of a cylindrical container, r is the radius of its base, and h is its height, then $V = kr^2h$. Give a statement in the language of variation which is equivalent to this equation.
12. The period of a pendulum is directly proportional to the square root of its length. A pendulum 9 feet long has a period of 4 seconds. What will be the period if the pendulum is shortened to 4 feet?

CHAPTER THREE

The Concept of Function

...that flower of modern mathematical thought—the notion of a function.

THOMAS J. MCCORMACK

1 *Functions*

In the last chapter, we found that we were able to express the relationship between two quantities in words and were able to write the equation that expressed the relationship or variation between the quantities.

When two quantities or two variables x and y are so related that for each value of the first variable a corresponding value of the second is determined, then we say that the second variable is a function of the first variable. Mathematicians say that there is a functional relation between the variables and they have in mind some mechanism that assigns values to the second variable corresponding to given values of the first variable.

To make the meaning clearer let us consider the following examples:

1. The cost of shipping freight between two points is a function of the weight.
2. The area of a circle is a function of the radius.
3. The height of the mercury in a thermometer is a function of the temperature.
4. If $y=x^2$, the value of y is a function of the value of x.

We also know that one quantity may depend upon more than one variable. For example, the cost of manufacturing an article is a function of the cost of the material used, the cost of the labor required, and the overhead. The cost of any object depends upon the cost of production, transportation, handling, etc. On the other hand, the price of an object depends upon profit, supply, and demand, as well as cost. The rate at which the population of a country changes is a function of the birth and death rate, immigration, and emigration.

Symbols for Functions

In mathematics, it is found convenient to invent a symbolic way of indicating a functional relationship between two variables. Consider the first example above: it states that the cost C of shipping freight is a function of the weight W. That is, C is a function of W, and we abbreviate the statement still further by writing

$$C = f(W)$$

The symbol $f(W)$ stands for the phrase "function of W." If we wish to speak of a second function, we use the symbol $F(W)$, $g(W)$, or $\phi(W)$, etc. If, in the second example, we let A be the area and r the radius, then the statement "The area of a circle is a function of the radius" can be replaced by

$$A = F(r)$$

The number of miles m that an automobile can travel without stopping is a function of the number of gallons g of gasoline in the tank at the start of the trip. That is, $m = f(g)$. The height h to which a ball will rise when thrown upward is a function of the speed v with which it is thrown, or $h = g(v)$.

Definition of a Function

We are led in this way to formulate the following definition: *One variable is said to be a function of a second variable if to each possible value of the second variable there corresponds a value of the first variable.* The second variable is called the *independent* variable, and the first variable which depends on the second is called the *dependent* variable. Thus, in the last example above, v is the independent variable and h is the dependent variable.

The serious student may object to this definition because it does not say that one variable *causes* the other variable to change. However there are many functional relations in which a change in the independent variable may leave the other variable unchanged. For example, the postage on each ounce or fraction of an ounce of first-class mail is 6 cents. The dependent variable is the amount of postage, and the independent variable is the weight of the letter. If we change the independent variable (weight of the letter), we do not necessarily produce a change in the dependent variable (cost of postage).

It is sometimes a matter of choice which variable is called the independent variable and which the dependent variable. For example, the area A of a circle is given by the expression

$$A = \pi r^2 \qquad \text{where } A = f(r)$$

but the radius r may be a function of the area, that is,

$$r = g(A) \qquad \text{or} \qquad r = \sqrt{\frac{A}{\pi}}$$

If we know the area of the circle, we can calculate the radius, and conversely. However, there are many cases in which one variable is expressed in terms of another variable, but we cannot reverse this situation. For example, if we know the weight of first-class mail, we can calculate the postage. But the postage does not determine the weight. Why?

▲ EXERCISES

1. Think up three more examples of functional relations between two quantities. Identify your independent and dependent variable in each case.
2. Express in functional notation the notations involved in Exercises 1 to 12 at the end of Sec. 11 of Chap. 2. In each case identify the dependent and independent variables.
3. Each of the following statements gives a functional relationship. State them by means of a symbolic equation and name the independent and dependent variable in each case.

   a. A man's weekly pay check for a 40-hour week depends upon his hourly wage.
   b. The energy released in an atomic reaction increases with the mass of the transformed material.
   c. The pressure on a submarine increases with the depth.
   d. The cost of coal increases with the freight charges.

4. The cost C of finishing the exterior of a house depends upon the number of bricks B used, and the number of bricks used depends upon the external surface area S to be faced with bricks.

   a. Discuss the relation between C and B.

b. Discuss the relation between B and S.
c. Discuss the relation between C and S.
d. Indicate which are the independent and dependent variables.
e. Can the dependent and independent variables be interchanged?

5. The time T that it takes to silver plate an object depends upon the mass M of silver deposited, and the mass of silver deposited depends upon the surface area A of the object to be covered.

   a. Discuss the relation between T and M.
   b. Discuss the relation between M and A.
   c. Discuss the relation between T and A.
   d. Indicate which are the independent and dependent variables.
   e. Can the dependent and independent variables be interchanged? Why?

Problems just for fun

From a deck of cards select two aces, two deuces, three threes and four fours. Arrange these eleven cards in a row in such a way that there is one card between the two aces, two cards between the two twos, three cards between the three threes and four cards between the four fours.

Domain and Range for a Function

The set of possible values that the independent variable can assume is called the *domain of definition* of the function and the set of resulting values that the dependent variable takes on is called the *range of values* for this function.

For $m = f(g)$ above then, g, which stands for the number of gallons of gasoline in the car's tank, is the independent variable and can take on any value between 0 and 23 if the tank holds a maximum of 23 gallons. The domain of definition for the function $m = f(g)$ is then $0 \leqq g \leqq 23$. For another car the domain of definition may be different. The range, that is the number of miles that the car can travel, depends on the relation $f(g)$ and will be $0 \leqq f(0) \leqq f(23)$ where $f(0)$ is the value of the function for $g = 0$ gallons and $f(23)$ is the value of the function for $g = 23$ gallons.

A common mistake made by students is not noting the restrictions on the independent variable, that is, the domain of definition of the function, and then forgetting to check the range of values that the dependent variable can assume.

5 *The Important Mathematical Problem*

The important problem for us will be to discover when one variable is a function of one or more other variables and then to express this relation in such a way as to make its character apparent. It is usually a simple matter to observe the existence of a functional relation between two variables, but it may be a much harder problem to discover precisely what the character of the relation between the variables is. For example, it is easy to see that the area of a circle is a function of the radius. But it was no easy job to discover that the area of a circle is precisely $A = \pi r^2$. Indeed, it is sometimes not possible to find any satisfactory relation between the variables. For example, we know that the temperature of a room is a function of time, but we cannot state this relationship precisely. There are many ways of expressing precisely the functional relation between two variables. The most important ways are by:

A table of values
A graph
An equation, theoretically or empirically obtained
A precise verbal statement

In each of these cases we should be sure which variable is the independent variable and which variable is the dependent variable. Always check the domain of definition for the independent variable and the corresponding range of values for the dependent variable.

Expression of a Functional Relation by Tables and Graphs

By actual measurement, the distance D in miles that a man can see on a desert is found to vary with the height h in feet of the observer's eye above the ground, according to Table 2. It is hard to draw any definite conclusions about the functional relation between h and D simply by studying Table 2. We can conclude that D increases as h increases, but from the table we cannot answer such questions as,

TABLE 2

h, feet	0	10	50	100	150	200	300	400
D, miles	0	3.9	8.7	12.3	15.1	17.4	21.3	24.6

"How far can we see if $h = 250$ feet?" To answer such questions and to exhibit better the functional relation, we resort to graphing. The domain of definition is $0 \leqq h \leqq 400$ feet. The range of values is $0 \leqq D \leqq 24.6$ miles.

The graph in Fig. 4 is constructed from the numbers in Table 2. The independent variable h is plotted along the horizontal base line, and the dependent variable D is plotted as the vertical distance above each of these points, thus locating the eight points shown in Fig. 4.

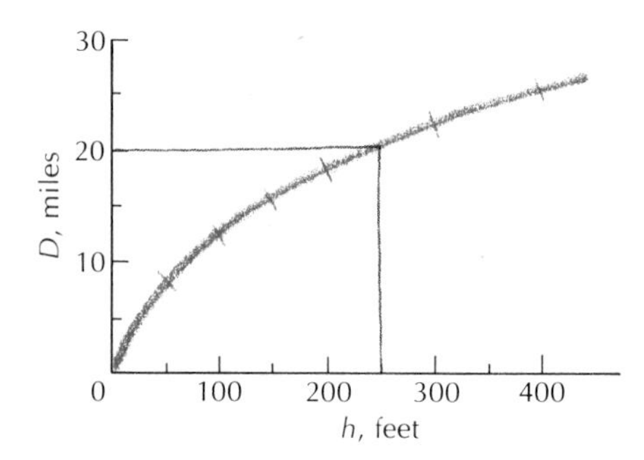

FIG. 4

Since we believe that distance D varies gradually as the height h changes, we draw a smooth curve through the eight points. However we should not draw a smooth curve for the graph of the postage on a letter as a function of the weight. Why? In place of a smoooth curve, how would the proper graph of the postage function look?

Although we do not know the exact relationship between h and D (later we will determine this relationship), this graph can be used to answer many questions regarding the behavior of D relative to h. For example, when $h = 250$ feet, D is equal to approximately 20 miles. How much farther can we see at a height of 270 feet than at 70 feet?

We cannot use the graph to answer questions about the distance D that we can see on a desert if our eyes are more than 400 feet above the ground. Why?

In drawing a graph it is important to explain carefully the meaning of the symbols used and to explain the scale. In the following problems, take particular care to label the figure in such a way that a perfect stranger could pick up your paper and make sense out of it.

▲ EXERCISES

1. Which quantities would you plot vertically and which horizontally if you wished to show the relation between the death rate of bacteria and the amount of sunshine to which they are exposed? Between the pressure on a submarine and its depth below the surface of the water?
2. The area A of a wound in square centimeters decreased with the time t in days as shown in Table 3. Plot the curve of healing. From the graph find the area A after 10 days.

TABLE 3

t	0	4	8	12	16	20	24
A	16.3	10.8	6.6	3.9	2.2	1.1	0.4

3. Table 4 gives the boiling point of water in T degrees centigrade at various pressures in P millimeters of mercury. Plot the pressure-temperature curve. Find T when $P = 825$, also P when $T = 100.8$.

TABLE 4

P	760	787.7	816	845
T	100	101	102	103

4. The amount of water, or weight of water vapor, that a cubic meter of air can hold depends upon the temperature of the air. Table 5 shows the amount of water vapor W that a cubic meter of air can hold at the various temperatures T. Plot the data in Table 5. What is the amount of water vapor in a cubic meter of air when the temperature is 28°? When it is 12°? What is the temperature of the air if a cubic meter of air holds 16.2 grams of water vapor?

TABLE 5

T, °C	*W*, grams	*T*, °C	*W*, grams
−20	1.0	15	12.8
−15	1.5	20	17.2
−10	2.3	25	22.9
−5	3.4	30	30.1
0	4.9	35	39.3
5	6.8	40	50.9
10	9.3	45	64.3

5. Table 6 gives the velocity v in feet per second with which a ball strikes the ground t seconds after being dropped, if air resistance is neglected. What is the velocity after 1.5 seconds, 7 seconds of fall? How long does it take for the ball to obtain a velocity of 200 feet per second?

TABLE 6

t	1	3	5	8	12
v	32	96	160	256	384

6. Table 7 gives the velocity v in miles per hour of a car which is slowing down t seconds after the brakes are applied. How fast is the car traveling after 3 seconds? How long does it take before the car is moving 8 miles per hour?

TABLE 7

t	1	2	4	5
v	25	20	10	5

7. Table 8 gives the distance s in feet that a body falls from rest after t seconds of fall. How far has the body fallen after 2.5 seconds? How long does it take to fall 80 feet?

TABLE 8

t	0.5	1	2	3	4
s	4	16	64	144	256

Problems just for fun

Mr. Smith had to pay 75 cents a square foot for glass to replace a broken window which was square and measured 4 feet from top to bottom. Yet Mr. Smith paid $6 for the single sheet of glass required to fix the window. Account for this total.

CHAPTER FOUR

Linear Equations and Relations

The human mind has never invented a
labor-saving machine equal to algebra.

THE NATION, vol. 33, p. 237

1 *The Cartesian Rectangular Coordinate System*

Before proceeding to a more general discussion of graphing, let us pause for a brief look at the framework involved. Let us draw a horizontal line and a vertical line which intersect at right angles at the point O. These two lines form the coordinate axes of the reference system. The horizontal line is called the x axis and the vertical line is called the y axis. The two axes divide the plane into four pieces which are called *quadrants*. For easy reference these are named by numbers, first quadrant, second quadrant, etc., as shown in Fig. 5. The point of intersection of the two axes is called the *origin*. Scales are marked along the x and y axes. By convention, positive numbers are attached to the points to the right of the origin on the x axis and upward from the origin on the y axis. Likewise, negative numbers are assigned to the points to the left of the origin on the x axis and downward from the origin on the y axis. This reference system is known as the *cartesian rectangular coordinate system.*

2 *Locating a Point in a Plane*

Now, the position of any point P in the xy plane can be specified conveniently by referring it to the cartesian coordinate system. Through a point P draw a line PN parallel to the x axis. The distance ON is called the y *coordinate*, or *ordinate*, of the point P and is designated by y. Likewise, the line PM drawn parallel to the y axis determines the distance OM, which is called the x *coordinate*, or *abscissa*, of P. It is convenient to indicate the coordinates of P on the graph near the point by the symbol $P(x,y)$ or just (x,y). The point P shown in Fig. 5 has coordinates $OM = x = 3.6$ and $ON = y = 2.2$ or $(3.6,2.2)$. Similarly we may find the coordinates of any other point in the plane or locate the point when its coordinates are given.

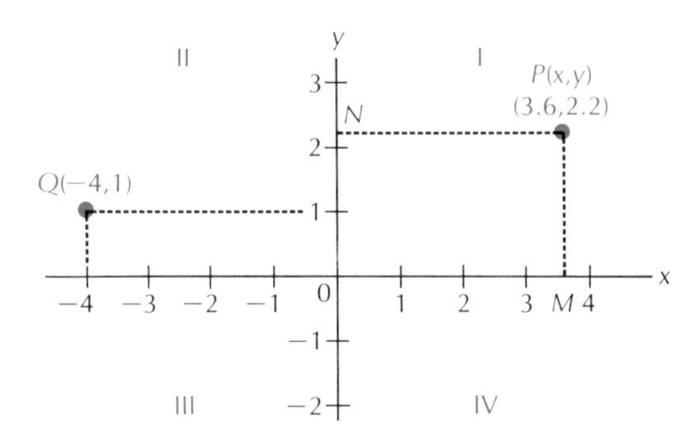

FIG. 5

Let us practice a bit. Let us locate the point Q that has coordinates $(-4,1)$. The x coordinate of Q is -4, so we must go 4 units along the x axis to the left of the origin. Since the y coordinate of Q is 1, we move 1 unit upward from the x axis. *Remember that the x coordinate is always written first.*

The scale on the y axis need not be the same as that on the x axis. We choose suitable scales to make the graph as large as desired. This point will be discussed again in the following sections. It is obvious that graph paper with vertical and horizontal lines was made especially for work with the rectangular coordinate system.

3 *Historical Notes*

The coordinate system described in Sec. 2 above was introduced by the French mathematician and philosopher René Descartes (1596–1650). This coordinate system joined the fields of algebra and geometry together into the field of analytical geometry. This new field of analytical geometry gave mathematicians new techniques for solving many problems that had defied solution.

Before the time of Descartes, the existence and use of negative numbers was a highly controversial subject among mathematicians. With the advent of the theory of analytical geometry, which united

Problems just for fun

Miss Swank, Miss Social, and Miss Highhat descended upon the hatter and demanded the latest creations in chapeaux. The hatter asked the three ladies to take three chairs. He had, however, arranged the chairs in such manner that Miss Highhat could see the hats that Miss Swank and Miss Social tried on, and Miss Swank could see Miss Social's hat, but Miss Social could not observe the other two ladies' hats. The hatter showed the ladies five hats, three of which were green and two of which were pink. He mixed the hats and from the rear placed a hat upon each lady's head. He asked Miss Highhat the color of her hat. She replied that she did not know. He asked Miss Swank the color of her hat. She replied that she did not know. When Miss Social was asked the color of her hat, she could tell the hatter the correct color. What was the color of Miss Social's hat?

the concepts of direction and of numerical distances, it was found that the negative numbers had meaning and a definite place in the number system. The negative numbers were accepted by mathematicians without further argument.

▲ EXERCISES

1. Use graph paper and take the vertical scale equal to the horizontal scale. Plot the following points:

a.	(1,2)	b.	(6,2)
c.	(−1,3)	d.	(4,−3)
e.	(−2,−5)	f.	(−3,7)
g.	(2,−4)	h.	(3,7)
i.	(−3,−3)	j.	(−1,−6)

2. Take the vertical scale twice the horizontal scale and plot the points in Exercise 1.
3. Draw a cartesian coordinate system and shade the region or regions in which the point (x,y) can be found if the following is known:

a.	$x > 0$	and	$y > 0$	b.	$x > 0$	and	$y < 0$
c.	$x < 0$	and	$y > 0$	d.	$x < 0$	and	$y < 0$
e.	$0 < x < 3$	and	$2 < y < 3$	f.	$-1 \leqq x < 0$	and	$y \geqq 0$
g.	$\|x\| \geqq 2$	and	$\|y\| \geqq 1$	h.	$\|x\| \leqq 1$	and	$\|y\| \leqq 2$

4 *Distance between Two Points*

We have just described the cartesian coordinate system that establishes a correspondence between the geometric system (the points in the plane) and an algebraic system [the pairs of numbers (x,y)]. This correspondence enables us to solve geometric problems algebraically and also certain algebraic problems geometrically.

If we know that two points P_1 and P_2 have coordinates (2,3) and (−2,−2) respectively, we can easily find the distance d between them. We first draw a cartesian coordinate system of axes and locate the two points (see Fig. 6).

Next from the point (2,3) we draw a line parallel to the y axis, and from the point (−2,−2) we draw a line parallel to the x axis. These two lines intersect at the point P_3 whose coordinates are (2,−2). The distance from P_1 to P_3 is 5 and the distance from P_2 to P_3 is 4. We notice that the triangle $P_1P_2P_3$ is a right triangle and the distance d is given by $d = \sqrt{5^2 + 4^2} = \sqrt{41} = 6.40$.

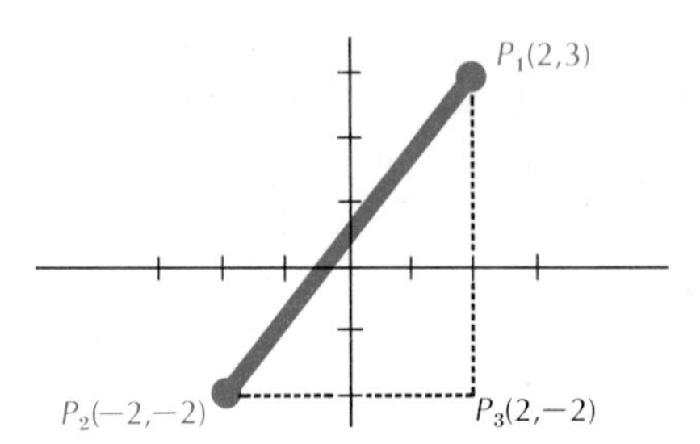

FIG. 6

If P_1 and P_2 have coordinates (x_1,y_1) and (x_2,y_2) respectively, the distance P_1P_2 between them is given by

$$\overline{P_1P_2} = \sqrt{(x_2 - x_1)^2 + (y_2 - y_1)^2}$$

▲ EXERCISES

1. Draw a coordinate system and prove that the distance equation above is true.
2. Find the distance between the following pairs of points:

   a. (−3,2), (2,−3)
   b. (1,1), (−3,−2)
   c. (0,0), (3,4)
   d. (0,0), (−3,−4)
   e. (1,−1), (1,−6)
   f. (1,−1), (7,−1)
   g. (−4,2), (−3,−2).

3. Find the area of the triangle whose vertices are the points (−1,2), (3,2), and (−4,6).
4. Use the distance formula to determine if the triangle whose vertices are (1,1), (3,2), and (2,12) is a right triangle.

Problems just for fun

It takes 50 minutes to clean 6 rectangular windows which are 9 by 4 feet. How long will it take to clean 6 mirrors which are 9 by 4 rectangles?

The Graph of the Equation $y = kx$

In Sec. 6 of Chap. 3, we learned that a table giving the functional relation between two quantities does not picture this relation as well as a graph does. In Chap. 2 we learned to write an equation for the relation between two quantities when they varied together in certain ways. We now ask whether a graph shows better than an equation or statement the dependence of one quantity upon another quantity. The answer to this question is that, although a graph may not be better than the equation, it will give a picture of the relation under consideration, and the student will be better able to visualize the variation under discussion.

What type of curve or graph represents the statement that the quantity y is proportional to the quantity x? Since we can write the equation $y = kx$ for this relation, we must find the graph of the equation $y = kx$. Of course, the curve will depend upon the value of k, the constant of proportionality. In order to get some knowledge of the shape of the curve represented by the equation $y = kx$ and the effect that k has on the curve, let us plot the curve for $y = kx$ when $k=\frac{1}{2}$, 1, 2, -1, and -2. That is, we shall plot the curves for the equations $y=x/2$, $y=x$, $y=2x$, $y=-x$, $y=-2x$.

In order to plot the equation $y=x$, we observe that y is a function of x, and if we change the value of x, we change the value of y. Let x take on different values (both positive and negative), and find the corresponding value for y. For example, when $x = 2$, $y = 2$. We tabulate our results in Table 9.

TABLE 9

x	0	1	2	-1	-2
y	0	1	2	-1	-2

In order to plot the values in Table 9, we treat the pairs of values, (0,0), (1,1), (2,2), . . ., as the coordinates (x,y) of points in the plane. We locate these points on a cartesian graph and join the points by a smooth curve. The graph of the equation $y=x$ is a straight line passing through the origin.

We plot on the same coordinate system (Fig. 7) the curves for the equations $y = x/2$, $y = 2x$, $y = -x$, and $y = -2x$ in the same way. We see that the graph of each equation is a straight line which passes through the point (0,0). Changing the value of the constant k in the equation $y = kx$ changes the steepness or the slope of the lines.

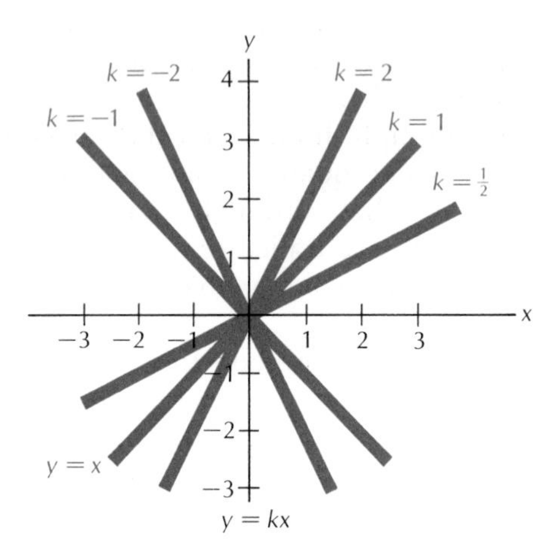

FIG. 7

We can conclude that the equation $y = kx$ represents a straight line that passes through the point (0,0). In the future, when we find a graph which is a straight line passing through the origin, we will know that the equation of the functional relation between the variables is of the form $y = kx$. We can determine the value of the constant k and have the exact expression for the functional relation given by the graph.

For example, when weights are hung on a spring, it is found that the spring is stretched by different amounts. Table 10 gives the observed stretch of the spring for different weights.

TABLE 10

W (weight in pounds)	0	1	2	3	4	5	6
S (stretch in inches)	0.0	0.5	1.0	1.5	2.0	2.5	3.0

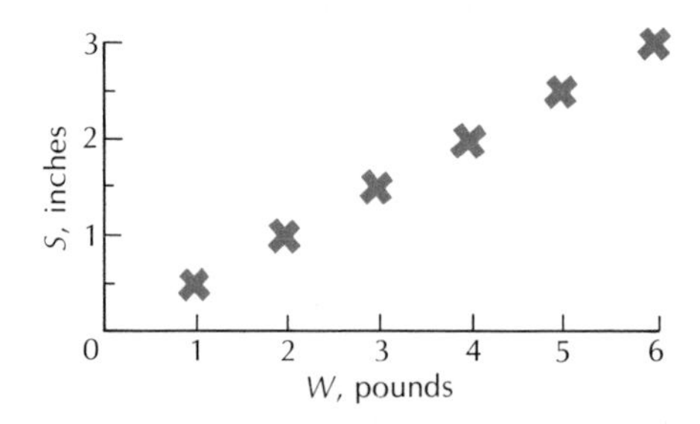

FIG. 8

Since a given weight W produces a given stretch S, we have $S = f(W)$ and we plot the values as shown in Fig. 8. If we draw a smooth curve through the points, we observe that it is a straight line through the point (0,0). So we know that the relationship between S and W is of the form $S = kW$. To determine k, we notice that $S = 2$ inches when $W = 4$ pounds, so that $2 = k4$ or $k = \frac{2}{4} = \frac{1}{2}$. The equation becomes $S = W/2$ and we observe that any of the values given in Table 10 fit this equation. But we cannot conclude that we can extend these values. Why?

6 *The Graph of the Equation* $y = kx + b$

If the functional relation between two variables y and x is given by the statement that y minus a constant b is proportional to x, we write

$$\frac{y-b}{x} = k \qquad \text{or} \qquad y - b = kx \qquad \text{or} \qquad y = kx + b$$

where k is the constant of proportionality. In this case, the quantity $y - b$ *is proportional to* x. If $b = 0$, then y is proportional to x, and we have $y = kx$, which is the case just discussed in Sec. 5.

What change in the curves for $y = kx$ is made by adding the constant b to the right-hand side of the equation? Do you expect the graph of the equation $y = kx + b$ to be a straight line? Of course, we observe that the graph of this equation will depend upon the value of the two constants k and b.

In order to answer these questions, let us plot the graphs for $y = 2x$, $y = 2x + 1$, $y = 2x + 2$, and $y = 2x - 2$. We make a table similar

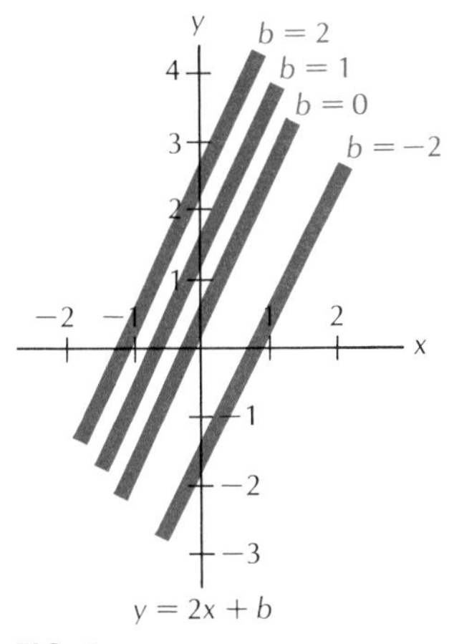

FIG. 9a

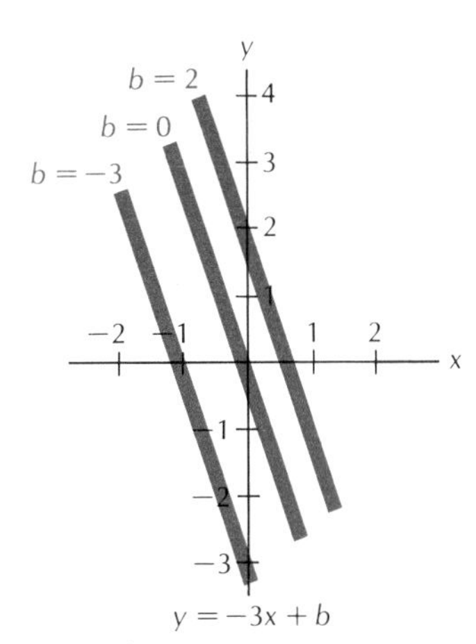

FIG. 9b

to Table 9 for each equation and plot the pairs of values (Fig. 9*a*). We see that all the curves are straight lines but that only one of them passes through the point (0,0), namely, the one in which $b = 0$. Furthermore, the lines are all parallel; that is, they have the same slope.

If we plot graphs for the equations $y = -3x - 3$, $y = -3x$, and $y = -3x + 2$, we find that only one passes through the point (0,0) and that all the lines are parallel (Fig. 9*b*).

▲ EXERCISES

Plot the curves for the following equations:

1. $y = 4x$
2. $y = -3x$
3. $y = \frac{x}{2} - 3$
4. $y = x + 1$
5. $y = 6x$
6. $y = 6x + 2$
7. $y = 6x - 2$
8. $y = 2x$
9. $y = -2x - 7$
10. $y = -2x + 1$

Find a number that will be doubled if 24 is added to $\frac{2}{3}$ of the number.

The Graphs of the Equations $y = b$ and $x = c$

In the relation $y = kx + b$, let $k = 0$, and the equation becomes $y = b$. This is the equation of a line every point of which has a y value equal

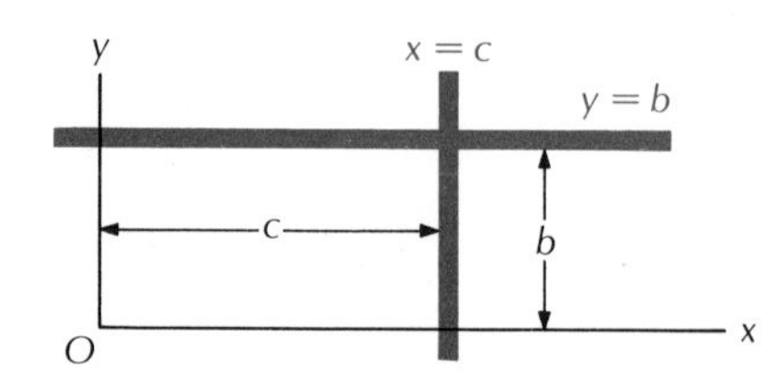

FIG. 10

to b; that is, $y = b$ is a line parallel to the x axis and b units from it. Likewise, the equation $x = c$ is a line parallel to the y axis and c units from it (Fig. 10).

▲ EXERCISES

1. Plot the lines $y=3$; $y=-2$; $y=7$; $x=5$; $x=-3$; $x=0$; $y=0$.
2. Given the equation $y=ax+b$.

   a. Tell what happens to the graph of the equation if we keep a fixed and vary b.
   b. Tell what happens to the graph of the equation if we keep b fixed and vary a.
   c. Tell what happens to the graph of the equation if $a=0$ and we vary b.

   Sketches may be useful to illustrate your discussion.

3. a. Plot the equation $y=-2x+3$.
   b. At what point does the curve cross the x axis?
   c. At what point does the curve cross the y axis?

Can you support a dime on three matches, if the dime must touch all three matches but not touch the table nor a match head, and no match head is to touch the table?

8 *Inequalities Involving Two Unknowns*

We should not expect to be lucky always and find our relations expressed in equation form. We may encounter a relation between two variables or quantities that is given by an inequality. For example, for what values of the variables x and y is $x+2y>1$? This is the same as $x>1-2y$ or $2y>1-x$.

When we considered inequalities involving one variable in Chap. 2, we sometimes found an infinite number of solutions, that is, values of the unknown that satisfied the given inequality. Thus we should not be surprised if we again find many pairs of numbers x and y for which $x+2y>1$ is true.

In Chap. 2 we found that an easy way to solve relations involving inequalities was to look at them from a graphical point of view.

Let us use this same approach here.

If we had $x+2y=1$, we could plot the graph of the equation (Fig. 11*a*).

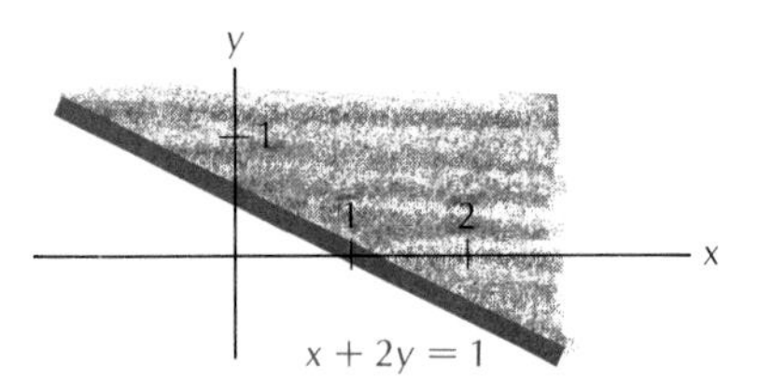

FIG. 11a

Now we can answer the question: Which values of x and y will make $x+2y>1$?

If any point (x,y) on the line will satisfy $x+2y=1$, then any point (x,y) lying in the shaded area will satisfy $x+2y>1$.

Had the inequality been $x+2y<1$, the solution would be any point in the unshaded region of the plane.

If we use the graphical method and are careful in choosing the proper regions, this type of problem is easy and fun to solve.

■ **Example.** We want the solution or solutions to the two inequalities

$$x+2y>1 \qquad \text{and} \qquad x-y<1$$

That is, we are asking for all pairs of values (x,y) or points, since we chose the graphical method, that will satisfy both inequalities at the same time. We find the solution to $x-y<1$. The shaded area of Fig. 11*b* gives the values of (x,y) such that $x-y<1$.

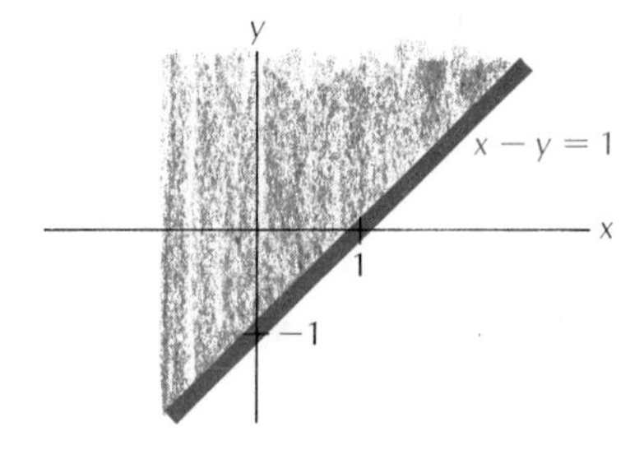

FIG. 11b

Now we combine the two figures above and get Fig. 11*c*. The points lying in the overlap of the two areas will then satisfy the two inequalities

$$x + 2y > 1 \qquad \text{and} \qquad x - y < 1$$

If we had $x + 2y \geqq 1$ and $x - y \leqq 1$, then we would include the points on the boundaries also.

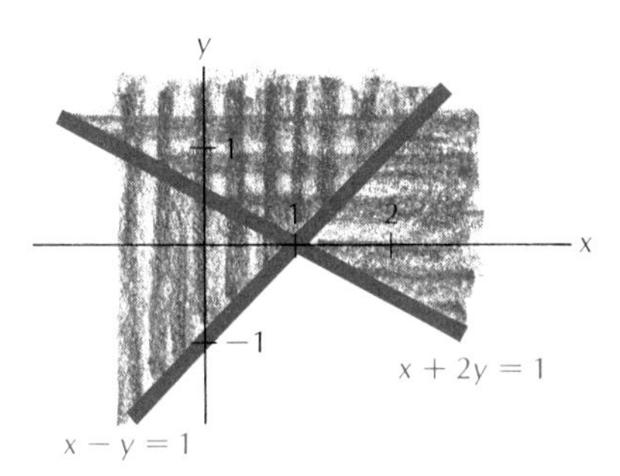

FIG. 11c

We may have three or more relations involving inequalities between two variables and ask for a solution or a set of solutions. For example, what pairs of values (x,y) or points will satisfy the three relations

$$x + 2y > 1 \qquad x - y < 1 \qquad y \leqq \tfrac{1}{2}$$

▲ EXERCISES

1. Find the solutions to

   a. $x \geqq \frac{1}{2}$
   b. $x - 1 < 2$
   c. $y - x \leqq 1$
   d. $y + x \geqq 3$
   e. $2y - 3x \leqq 6$
   f. $1/x \geqq 1/y$
   g. $y \leqq 4x + 2$

2. Show that the solution to the three inequalities at the end of the section above is the area bounded by the lines $y = \frac{1}{2}$, $y = x - 1$, $2y = 1 - x$. Which of the lines may be included?
3. Find the solution to the pairs of inequalities

   a. $x \geqq y$, $x < 0$
   b. $x + y \geqq 1$, $x + 2y < -2$
   c. $x + y \leqq 1$, $x < y + 1$

4. Find the solutions to the three inequalities

a. $x \geqq y, x < 0, x + 2y \leqq 1$
b. $x + y \geqq 1, x + 2y < -2, x < y$
c. $x + y \geqq 1, x + 2y < -2, x \leqq 1$
d. $x + y \geqq 1, x < y + 1, x \geqq 0$

9 *The Equation for a Functional Relation Exhibited by a Table*

Just as before, we can conclude that when the graph is a straight line, the functional relation between the two variables y and x is $y = kx + b$, where k and b are arbitrary constants which can be determined.

■ **Example.** Table 11 shows several temperatures, °C, on the centigrade scale with their equivalent temperatures, °F, on the Fahrenheit scale.

TABLE 11

°C	0	10	60	100	160	200
°F	32	50	140	212	320	392

□ ***Solution.*** In order to visualize the relation between the two temperature scales, we plot the data in Table 11 (Fig. 12). Since the graph is a straight line, we conclude that the functional

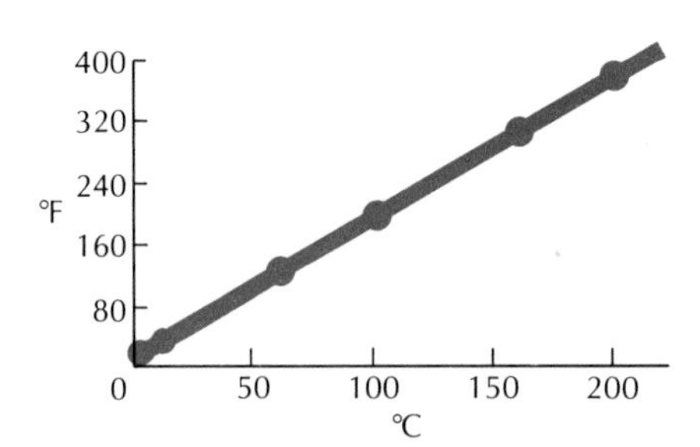

FIG. 12

relation between C and F is given by $F = kC + b$. In order to determine the two constants k and b, pick two corresponding pairs of values, for example, $C = 10°$ when $F = 50°$ and $C = 100°$ when $F = 212°$, and substitute them into the equation

$$F = kC + b \tag{1}$$

obtaining the two equations

$$50 = 10k + b \tag{2}$$
$$212 = 100k + b \tag{3}$$

We need to determine a pair of values for k and b which when substituted into equations (2) and (3) will make the right-hand sides of the equations equal to the left-hand sides.

Clearly, $k = 5$ and $b = 0$ is a solution of equation (2). However, this is not a solution of equation (3), for $100 \times 5 + 0 = 500$ instead of 212. Likewise, $k = 0$ and $b = 212$ is a solution of equation (3). But this does not satisfy equation (2), for $10 \times 0 + 212 = 212$ and not 50. We can find many values of k and b which satisfy one of the two equations. However, our problem is to find the values of k and b which will make both equations true. In order to find this common solution, we have to solve the two equations simultaneously.

We subtract equation (2) from equation (3)

$$\begin{array}{l} 212 = 100k + b \\ \underline{50 = 10k + b} \\ 162 = 90k \end{array}$$

and get

$$k = \frac{162}{90} = \frac{9}{5}$$

We multiply equation (2) by 10 to get

$$500 = 100k + 10b$$

from which we subtract equation (3)

$$\begin{array}{l} 500 = 100k + 10b \\ \underline{212 = 100k + b} \\ 288 = 9b \end{array}$$

and get

$$b = \frac{288}{9} = 32$$

Substituting these values for k and b into equation (1), we find that the exact functional relation between the temperatures read on the centigrade and Fahrenheit thermometers is

$$F = \tfrac{9}{5}C + 32 \tag{4}$$

Problems just for fun

Alice skipped down to the spring, gaily swinging two pails. Exactly four quarts of water to bring. "Alas! Alack!" she wails, "one pail holds three quarts and the other five quarts. How, oh how, to measure four quarts?"

▲ EXERCISES

Plot the data given in the tables for the following problems. If the graph is a straight line, find the exact functional relation between the variables. Do your equations have meaning for all values of the dependent variable? Discuss.

1. Table 12 shows the amount of potassium iodide in grams W which will dissolve in 100 grams of water at various temperatures T.

TABLE 12

T	10	20	30	40	50
W	136	144	152	160	168

2. Table 13 gives the volume V in cubic centimeters of a quantity of gas at T degrees centigrade. Find V when $T=50°$.

TABLE 13

T	−33	12	27	42
V	160	190	200	210

3. Table 14 gives the pressure P in pounds per square foot due to water at various depths h in feet below the surface of the ocean.

TABLE 14

h	1.56	2.34	3.90	10.92	15.60
P	100	150	250	700	1,000

4. The cost C in dollars of publishing a small pamphlet will vary with the number N to be printed, as shown in Table 15.

TABLE 15

N	1,000	2,000	3,000	4,000
C	150	170	190	210

5. The weight W in pounds of a barrel containing water is a function of the number of gallons g of water. Table 16 gives the relation between g and W. From the equation obtained, find the weight W of $4\frac{1}{2}$ gallons of water.

TABLE 16

g	1	2	3	3.5	4
W	18	26	34	38	42

6. The profit that a filling station makes on the sale of gasoline is shown in Table 17. How many gallons of gasoline would have to be sold to give a profit of 35 cents?

TABLE 17

No. of gallons sold	2	5	8	12	16
Profit in cents	−5	10	25	45	65

7. The velocity in miles per hour of a car which is slowing down is given in Table 18, where t represents the number of seconds of braking. How long did it take the car to come to rest?

TABLE 18

t	1	2	4	5
v	25	20	10	5

8. The length of nails of various sizes (in pennies) is given in Table 19. What is the length of a 10-penny nail?

TABLE 19

Size of nails in pennies	2	4	6	8
Length of nails in inches	1	1.5	2	2.5

9. Construct a table to display values of y corresponding to x for $-2 \leqq x \leqq 4$ for the equation $y=3x-14$.

10. y is a function of x and the relation is given in Table 20. Find the relation between y and x. What value of x gives $y=4$?

TABLE 20

x	−2	−1	0	1	2
y	1	3	5	7	9

10 *The Equation of a Straight Line through Two Points*

We have seen that we can determine the exact functional relation between two variables if the graph of the variation is a straight line. We did this by assuming that the relation was given by an equation of the form $y=kx+b$, where k and b were constants to be determined. We picked two pairs of corresponding values for the variables x and y and substituted them into the equation $y=kx+b$. We then solved these two equations in order to get the values of k and b. We were able to determine the values of k and b by knowing two points on the graph. We need just two points to determine a line, for if we have two points, we can draw a line through them.

If we draw a line through two points (5,4) and (1,−6), we know that the equation of the line will be given by $y=kx+b$. Let us determine the constants k and b. We substitute the two points into the equation and get

$$4=5k+b$$
$$-6=k+b$$

If we subtract the second equation from the first we get

$$10=4k$$

and

$$k=\frac{10}{4}=2.5$$

We substitute 2.5 for k in the second equation to get

$$-6=2.5+b$$

Solving for b, we get $b=-8.5$. The equation of the line through the two points (5,4) and (1,−6) is

$$y=2.5x-8.5$$

We have concluded that the equation $y = kx + b$ yields a straight-line graph when plotted. Any equation of this type which involves only the first power of the variables is called a *linear equation.*

▲ EXERCISES

Find the equation for the lines through the following pairs of points. Use the equations $y = kx + b$ or $x = c$, and determine the value for k, and b or c in each case.

1. $(0,0)$, $(-1,3)$
2. $(2,-1)$, $(-3,4)$
3. $(-3,-1)$, $(4,5)$
4. $(-6,2)$, $(8,7)$
5. $(5,0)$, $(5,9)$
6. $(-2,4)$, $(8,4)$
7. (x_1,y_1), (x_2,y_2)
8. $(-3,-1)$, $(-3,3)$

Mary selected three bananas and two oranges from a fruit peddler's cart. After she paid the peddler 28 cents for the fruit, she changed her mind and exchanged an orange for a banana. "That'll cost you another penny," said the peddler. Can you help Mary figure out the price of oranges and bananas?

CHAPTER FIVE

Translation into Equations

We can always depend upon it that algebra which cannot be translated into good English and sound common sense is bad algebra.

W. K. CLIFFORD

Only in the classroom will we encounter equations tailor-made for us to solve. If our knowledge of mathematics is confined to the solution of such equations, it will be worthless. In life we meet problems stated in words and not in x's and y's. We must translate these words into symbols, which are the mathematical language, before our mathematics becomes useful. In this translation of words into equations and subsequent solution, we see the truly practical nature of algebra.

To solve problems in algebra, we represent by a letter the unknown number which we are trying to find. The statement of the problem leads to the formulation of an equation involving this letter. We try to solve this equation for the unknown quantity. The power of this method lies in the fact that *we may treat this letter just as though we knew its value.*

1 *A Game*

There is a common tendency for students to regard algebra as a set of rules of operation which they use to build certain combinations of letters and numbers. By taking this narrow view of the subject, they lose sight of the real meaning of algebra. The development of algebraic expressions has a practical use when it is applied to solving problems.

Consider the following game. I tell you to think of a positive number, and you think of the number 6; but, being algebraically inclined, I think of *any* number x. Then I tell you to add 3, and I add 3 to x

$$6+3=9 \qquad x+3$$

Now square the result:

$$9^2=81 \qquad (x+3)^2=x^2+6x+9$$

Subtract 6 times the original number:

$$81-(6)(6)=81-36=45 \qquad x^2+6x+9-6x=x^2+9$$

Subtract 9:

$$45-9=36 \qquad x^2+9-9=x^2$$

Now, I ask you to give me your answer, 36. Comparing this with my general result x^2, I conclude that $x^2=36$, or $x=6$; hence your number must have been 6.

I have forced you in this simple game to perform certain arithmetical steps, while I have been performing algebraic operations. The only difference between your calculations and mine is that you operate with a specific number 6, while I operate with *any* number x. I have gone through the same operations of addition, subtraction, multiplication, etc., as you have. In a certain sense, I have generalized your calculation by using a general (any) number x instead of the specific number 6. In this respect, we might call algebra *generalized arithmetic*. But what have I gained? To some, I grant that I have made things harder; but in reality, I am saving myself a lot of hard work. Suppose I tell you to think of another number, and ask you to perform the same calculations as before. At the end you give me the number 81. Then, without performing any calculations at all, I merely set $x^2=81$ and conclude that your number was 9. My algebraic calculation and final result x^2 holds for whatever positive number you choose, for I let x be any number that you could choose.

■ **Example.** How to tell a person's age and month of birth. (Have the person whose age is to be discovered do the figuring.)

□ ***Solution.*** For example, suppose a coed is 19 and was born in December. Let her put down the number of the month in which she was born and proceed as follows:

Number of month of birth	12
Multiply by 2	24

Add 5	29
Multiply by 50	1,450
Then add her age, 19	1,469
Subtract 365, leaving	1,104
Now add 115	1,219

She then announces the result 1,219, whereupon she may be informed that her age is 19, and December, or the twelfth month, is the month of her birth. The two figures on the right in the result will always indicate the age, and the remaining figure or figures the month in which her birthday comes.

The Method of Procedure

Experience shows that the translation of the problem into the language of algebra is usually the most difficult part of the work. For that reason, we suggest a series of steps which will help the student.

1. *Read the problem carefully.* Unless the statement of a problem is clearly understood, it is useless to try to solve the problem. Careful thought is an essential part of this reading.
2. *Write an accurate statement of what the unknown variable is to represent.*
3. *Express the given data in terms of the unknown variable.*
4. *Write the relations of the problem in terms of the unknown.* This will form an algebraic equation.
5. *Solve the resulting equation.*
6. *Formulate the answer in the terms first used.*
7. *Check your results in the original statement of the problem.* This will catch errors. Also, sometimes the algebra will give numbers which are physically impossible as answers, for example, a negative length.

Success in this work *requires clear* and *orderly reasoning.* An *orderly arrangement* of the written work not only aids in the explanation of the reasoning but is valuable help in securing clear and accurate thought.

Problems Involving One Unknown

We are now ready to translate some simple problems into the language of algebra and solve the resulting equations.

■ **Example 1.** A student club has a cover charge of 35 cents per person. This charge is to help defray the expense of operation and entertainment. If one wishes to drink Cokes, he may order them at 10 cents per Coke. If the bill for a party of two came to \$1.10, how many Cokes did this couple drink?

□ ***Solution***

Let x=the number of Cokes consumed

Then

$10x$=cost of these Cokes in cents

Also

$35+35=70$ cents=the cover charge for this couple

Since the total cost = \$1.10 = cover charge + cost of Cokes, we have

$$110=70+10x$$

which is the algebraic equation which expresses the relations stated in the problem. To solve this equation for x, subtract 70 from each member and then divide by 10:

$$110-70=10x$$
$$40=10x$$

and

$$x=4$$

Thus we find that the couple ordered 4 Cokes.

Let us check this answer: 4 Cokes at 10 cents per Coke cost 40 cents; the cover charge for two people is 70 cents, and the bill is equal to $40+70=110$ cents.

▲ EXERCISES

Work the following problems using the method outlined above. *Always* check your answer.

1. How many Cokes at 10 cents each were ordered by a party of three if the bill was \$1.75, and the cover charge was 35 cents per person?
2. A party service delivers Cokes. The delivery fee is 25 cents. Each Coke costs 10 cents, and the bottle deposit is 2 cents. Find an algebraic expression for the cost of having n bottles of Coke delivered.
3. If there were 24 Cokes delivered in Exercise 2, find the total charge.
4. What would be the bill in Exercise 3 if there were 10 empty bottles to be exchanged?
5. A well-known pitcher signs up with a ball club. His contract calls for \$50,000 plus a \$1,000 bonus for each game he wins. Write an algebraic

expression for this player's salary if he wins n games in one season.

6. What will be the pitcher's salary in Exercise 5 if he wins eight games during the season?
7. A university contributes an amount equal to 10% of an employee's salary into his pension retirement fund, and the employee contributes 5% of his salary. In addition, the university takes out \$6.65 for hospital insurance and \$7.20 for life insurance each month. What is the employee's monthly salary if \$61.35 is the total amount withheld each month?
8. How much money is placed in the retirement fund each month for the employee in Exercise 7?
9. It costs a farmer \$8 for each baby pig weighing 15 pounds, and it costs 10 cents a pound to fatten the pigs. If a farmer gets 18 cents per pound for his hogs, how much must each hog weigh if the profit per hog is to be \$10?
10. A grocer needs 13% of his sales to cover profit and overhead. There is a fixed overhead of \$700 a month for rent, heat, electricity, etc., and a variable overhead of 3 cents of every sales dollar for checking, cashiers, stocking shelves, etc.

    a. Find the sales volume needed to break even.
    b. Find the sales volume needed to make a monthly profit of \$600.

11. The combined age of two persons is twenty-five years. One is three years older than the other. How old is each?
12. A corporation has an order for 3,600 TV picture tubes. If 10% of the tubes manufactured are rejected because of defects, how many must be manufactured to meet the order?

■ **Example 2.** A salesman makes a commission of \$75 on one type of car and \$100 on another type. One year this salesman sold 39 cars and received commissions totaling \$3,300. How many sales of each type did he make?

☐ ***Solution***

$$\begin{aligned} \text{Let } x &= \text{number of cars sold at \$100 commission} \\ 39 - x &= \text{number of cars sold at \$75 commission} \\ 100x &= \text{number of dollars commission at \$100 per car} \\ 75(39 - x) &= \text{number of dollars commission at \$75 per car} \end{aligned}$$

The total number of dollars commission is

$$100x + 75(39 - x) = 3{,}300$$

or

$$\begin{aligned} 100x + 2{,}925 - 75x &= 3{,}300 \\ 25x &= 3{,}300 - 2{,}925 = 375 \end{aligned}$$

Problems just for fun

On his last birthday, Grandfather claimed that he had lived one-fourth of his life as a boy, one-sixth as a youth, one-third as a man, and had spent 13 years in his dotage. How old was Grandfather?

Then

$$x = \frac{375}{25} = 15$$

and

$$39 - x = 24$$

This salesman sold 15 cars for which he got \$100 per car commission, $15 \times 100 =$ \$1,500, and 24 cars for which he got \$75 per car commission, $24 \times 75 =$ \$1,800; \$1,500 + \$1,800 = \$3,300, which was his total commission.

▲ EXERCISES

Work the following problems. Be sure to check your answers.

1. In a 1-year period an automobile salesman sold 39 cars. He sold twice as many cars on which he got a commission of \$75 per car as cars on which he got a commission of \$100. Find the total commission made by this salesman.
2. A salesman sold a total of 60 cars of two types during a 1-year period. He received a commission of \$80 per car for three-fourths of the cars sold. How much commission did he receive on each of the other cars if his total commission amount to \$4,500?
3. At a sale a housewife bought 60 cans of food. She paid 20 cents per can for some and 12 cents per can for the remainder. How many cans at each price did the housewife buy if she paid \$7.84?
4. Find three consecutive integers whose sum is 54.
   Hint: $N =$ one number, $N + 1 =$ second number, and $N + 2 =$ third number.
5. Find three consecutive even integers whose sum is 54.
   Hint: $2N =$ one number, $2N + 2 =$ second number, and $2N + 4 =$ third number, where N is an integer.
6. A farmer receives \$16.90 for his eggs. He gets 46 cents per dozen for some and 32 cents per dozen for the remainder. How many dozen of each grade did he sell if he sells 48 dozen eggs?
7. An art dealer bought 10 paintings at an auction. Some were classical and some modern. The total cost was \$1,387.90. The average cost of the classical paintings was \$212.50, and the average cost of the modern paintings was \$107.20. How many of each type did the dealer buy?
8. The weight in pounds that can be lifted by a balloon filled with helium is equal to 0.069 times the volume of the gas in cubic feet. If the bag, gear, ropes, and basket weigh 90 pounds, write an expression for the payload that can be lifted. Payload is the weight in excess of the weight of the gear, basket, and bag.
9. What volume of helium gas in Exercise 8 is needed to lift a payload of 250 pounds?

10. What payload can be lifted by the balloon in Exercise 8 if the volume of the gas is 10,000 cubic feet?

11. A grocer receives 100 bushels of apples and prices them at $5 a bushel. Toward the end of the day, he cuts the price to $4 a bushel. At the end of the day, all the apples are sold and he has $463. How many bushels did he sell at each price?

12. A hitchhiker starts walking at 4 miles per hour. After a time he catches a ride on a truck at 30 miles per hour. In 10 hours of walking and riding he finds he has covered 157 miles. How long did he walk and how long did he ride?

Problems just for fun

```
         H U N G
U P ⟌ D O P E Y
       D U
       ---
         U P
         U P
         ---
           E Y
           E Y
           ---
```

This is a problem in long division. Each letter represents a number. There are plenty of clues, too! For instance, U must be 1, since U times $U\ P = U\ P$. To help the cause, we will tell you that $D=5$.

4 *Problems Involving Two Unknowns*

■ **Example 3.** A man has $5.25 in nickels and dimes. The number of dimes is three times the number of nickels. How many nickels and dimes does he have?

□ ***Solution***

Let N=number of nickels

Then

$$3N = \text{number of dimes}$$
$$0.05N + 0.10\,(3N) = 5.25$$

or

$$5N + 30N = 525$$
$$N = 15$$

and

$$3N = 45$$

That is, there are 15 nickels and 45 dimes.

☐ ***Check***

$$15(5) + 45(10) = 75 + 450 = 525 = \$5.25$$

5 *Two Equations Involving Two Unknowns*

Many problems that we solve by using only one variable can be solved by using two variables and writing two equations. Let us now solve Example 3 in the following way.

☐ ***Second Solution to Example 3***

Let $N = \text{number of nickels}$
$D = \text{number of dimes}$
$0.05N = \text{value of the nickels}$
$0.10D = \text{value of the dimes}$

Then

$$D = 3N$$

and

$$0.05N + 0.10D = 5.25$$

or

$$5N + 10D = 525$$

which reduces to

$$N + 2D = 105$$

The two equations to be solved are

$$D = 3N \qquad \text{and} \qquad N + 2D = 105$$

Substituting $D = 3N$ into the second equation gives

$$N+2(3N)=105$$
$$7N=105$$
$$N=15$$

and

$$D=3N=45$$

The number of nickels in the collection is 15, and the number of dimes is 45.

The methods of Secs. 4 and 5 are equivalent. Sometimes we find that for a given problem it is easier to write two equations involving two variables than it is to write one equation in one variable. We may on occasion write three equations involving three variables and then solve these three equations.

▲ EXERCISES

Solve the following problems by using one unknown and one equation, or by using two unknowns and writing two equations. *Always check your answers.*

1. The sum of $1.20 is to be divided between two boys. The older boy is to get twice as much as the younger boy. How much does each boy get?
2. A metal pipe 10 feet long is to be divided into two pieces. Find the length of each piece if one is two-thirds as long as the other.
3. A rectangular field is 35 feet longer than it is wide. If the length of the fence around this field is 310 feet, find the dimensions of the field.
4. An estate of $5,628 is to be divided between a mother, two sons, and one daughter. The mother is to receive as much as all the children, and the daughter is to receive one-half as much as each son. How much does each person get?
5. A boy receives $8.64 from the sale of 90 newspapers and magazines. If the newspapers sold for 7 cents and the magazines sold for 25 cents, how many of each did this boy sell?
6. A woman buys 15 pounds of walnuts. Some of the walnuts are 3 pounds for a dollar and the rest are 2 pounds for a dollar. How many pounds of each variety did she buy if she paid $6.00 for the 15 pounds of walnuts?
7. A grocery clerk has some coffee worth $1.04 per pound and some worth 92 cents per pound. How many pounds of each should be used to make a 100-pound mixture worth 98 cents per pound?
8. A foreman assigned a boy the job of unpacking 100 glass articles. He would pay the boy 2 cents for each article he unpacked safely but would charge the boy 7 cents for each article that was broken. If the boy received $1.55, how many articles did he break?

9. A student has quiz grades of 68 and 73. What grade must he achieve on a third quiz to have an average of 79?
10. A student has an average grade of 81 on three quizzes. What grade must he get on a fourth quiz to bring his average up to 82? To lower his average to 75?
11. A boy goes on a spending spree and spends half his money for a movie and one-third of what remains for candy. On the way home he buys a comic book for 15 cents. How much money did he have to start with if he has a nickel left?
12. A child's bank has $11.25 in nickels, dimes, and quarters. There are two more dimes than nickels, and the number of quarters is one-half the number of nickels and dimes together. How many are there of each type of coin?
13. If the sum of two numbers is 32, and four times the smaller exceeds three times the larger by 9, find the numbers.
14. If the sum of two numbers is 42 and their difference is 4, find the numbers.
15. A grocer charged Mrs. Smith $1.90 for 10 pounds of flour and 6 pounds of sugar and charged Mrs. Brown 76 cents for 2 pounds of sugar and 5 pounds of flour. Determine the cost per pound of sugar and flour.
16. A theater charged $1.25 admission for adults and 50 cents for children. How many adults and children were in the audience of 806 if the box-office receipts were $781?

Problems just for fun

What happened to the 10 Cents?

Two small boys each had 60 pieces of homemade candy to sell door to door. One sold his on one street at two for a nickel and brought home $1.50. The other sold his on another street for three for a nickel and brought home $1. The next day they decided to combine their businesses. They put the 120 pieces of candy together and sold them five pieces for a dime (two for a nickel plus three for a nickel). After all the pieces were sold they found they had $2.40 when they expected $2.50. Can you account for the missing dime?

■ **Example 4.** How many cubic centimeters of a solution that is 90% alcohol by volume must be added to 1,000 cubic centimeters of a solution that is 20% alcohol by volume to make a solution that is 45% alcohol?

□ ***Solution***

Let x = number of cubic centimeters of solution to be added

$x + 1{,}000$ = final volume

Number of cubic centimeters of alcohol in original solution $= \frac{20}{100}(1{,}000) = 200$ cubic centimeters

Number of cubic centimeters of alcohol added $= \left(\frac{90}{100}\right)x$

Number of cubic centimeters of alcohol in final solution $= \frac{45}{100}(x + 1{,}000)$

Since the original volume of alcohol plus the volume added must equal the volume of alcohol in the final solution, we get

$$200 + \left(\frac{90}{100}\right)x = \frac{45}{100}(x + 1{,}000)$$
$$20{,}000 + 90x = 45x + 45{,}000$$
$$45x = 25{,}000$$
$$x = \frac{5{,}000}{9} = 555.5 \text{ cubic centimeters}$$

Check this result.

▲ EXERCISES

1. How many cubic centimeters of water must be added to 1,000 cubic centimeters of a 75% sugar solution to make a 60% solution?
2. How many gallons of two liquids, one 90% alcohol and the other 30% alcohol, must be used to make a 30-gallon mixture with 50% alcohol?
3. A creamery desires to obtain 200 pounds of milk that tests 4% butterfat by mixing milk that tests 4.2% butterfat with milk that tests 3.8% butterfat. How much milk of each test is needed?
4. If ice cream is to contain 12% butterfat, how much cream containing 40% butterfat must be added to 100 pounds of a mixture containing 10% butterfat in order that the resulting mixture contain the required 12% butterfat?
5. A man can sell a piece of property for $5,000 and realize a profit of 10% over the purchase price. How much did he pay for the property?
6. A 98% solution of sulfuric acid is to be diluted with distilled water to make a 90% solution. How many gallons of the original solution and

how many gallons of water are necessary to make 10 gallons of the 90% solution?

7. The enrollment in a Scout troop was 32. The troop increased its enrollment by 25%, and later 25% of the boys in the troop moved to another community. How many boys remain in the troop?
8. The sum of two numbers is 65, and their difference is 19. Find the two numbers.
9. Three years from now a mother will be three times as old as her son will be. Two years ago she was four times as old as her son was. Find the present ages of the mother and the boy.
10. A ranch is rectangular in shape, and the long side is twice as long as the short side. The area in square miles is numerically equal to the perimeter in miles. Find the dimensions of the ranch.
11. Two solutions of 5% acid and 25% acid are mixed to produce 40 gallons of a mixture of 10% acid. How much of each solution is used?
12. An industry has a work force of 650 employees, 70% of which are men. How many women workers may be recruited if the total work force is to remain 50% or more men?

■ **Example 5.** A man who can row 3 miles per hour in still water finds that he can row 10 miles upstream in the same time as he can row 20 miles downstream. Find the velocity of the stream.

□ ***Solution***

Distance = average velocity times time

Let V = velocity of the stream in miles per hour

Then

$V+3$ = velocity of the man when rowing downstream

and

$3-V$ = velocity of the man when rowing upstream

$$\text{Time to row 20 miles downstream} = \frac{20}{V+3}$$

$$\text{Time to row 10 miles upstream} = \frac{10}{3-V}$$

These two times are equal; therefore,

$$\frac{20}{V+3} = \frac{10}{3-V}$$

or

$$20(3-V) = 10(V+3)$$

and

Problems just for fun

Two boys shoot marbles. If a boy loses, he gives the other boy one marble. When the boys finish playing, one has won three times, and the other has seven more marbles than when he started. How many games did they play?

$$6-2V=V+3$$
$$3V=3$$
$$V=1 \text{ mile per hour}$$

☐ ***Check***

$$\text{Net upstream velocity}=3-1=2 \text{ miles per hour}$$
$$\text{Net downstream velocity}=3+1=4 \text{ miles per hour}$$
$$\text{Time to row 20 miles downstream}=\frac{20}{4}=5 \text{ hours}$$
$$\text{Time to row 10 miles upstream}=\frac{10}{2}=5 \text{ hours}$$

▲ EXERCISES

1. Two trains start from the same place and run in opposite directions at the rates of 35 miles per hour and 50 miles per hour. When will the distance between the two trains be 170 miles?
2. Work Exercise 1 if the two trains are traveling in the same direction.
3. A man starts walking at the rate of 3 miles per hour. Three hours later another man starts from the same place and rides a bicycle at the rate of 10 miles per hour. How far and how long will the latter have to travel to overtake the first man?
4. At noon a freight train traveling at 25 miles per hour is 90 miles ahead of an express train traveling at 40 miles per hour. When and where will the express train overtake the freight?
5. When will the express train in Exercise 4 be 40 miles behind the freight?
6. A sound made at the end of a steel railroad rail travels in the air at 1,100 feet per second and in the steel at 16,500 feet per second. The two sounds, i.e., the sound traveling through the air and that traveling through the metal, are heard 6 seconds apart by an observer at the other end of the rail. How long is the rail?
7. Two boys start pedaling bicycles toward each other. They are 20 miles apart at the start and meet 2 hours later. If one boy can pedal 1 mile per hour faster than the other, find their speeds.
8. A motorcycle starts west on Interstate 80 at 45 miles per hour. Two hours later an automobile starts from a point 50 miles east of the motorcycle's starting point and drives at an average speed of 60 miles per hour until it overtakes the motorcycle. How long does it take for the car to overtake the motorcycle?
9. A motorboat that has a speed of 20 miles per hour in still water requires 3 hours to make a trip upstream and 2 hours to return to the original starting position. What was the velocity of the stream?
10. A motorboat travels 8 miles downstream in 20 minutes and makes the return trip in 30 minutes. What is the speed of this boat in still water and the rate of the current?
11. A salesman drove to Chicago in 6 hours and returned home over the same route in 6 hours and 45 minutes. His average speed going was 5

miles more than his average speed returning. Find how far he lives from Chicago and his average speed each way.

12. An airplane makes a flight of 1,000 miles in 5 hours. For the first 3 hours it has a tail wind of 30 miles an hour and for the remaining 2 hours it has a head wind of 10 miles an hour. How fast does the airplane fly in still air?

A "Weighty" Problem

Using a balance, Eager Beaver discovers that 3 tacks plus 1 file will exactly balance 12 nails. Also, 1 file will exactly balance 1 tack plus 8 nails. How many tacks will 1 file balance?

■ **Example 6.** A tank can be filled by one pump in 20 hours and by a second pump in 50 hours. How long will it take to fill the tank using both pumps?

□ ***Solution***

Let $x=$ number of hours required to fill the tank if both pumps are used

$\frac{1}{20}=$ part of tank filled by first pump in 1 hour

$\frac{1}{50}=$ part of tank filled by second pump in 1 hour

so that

$x\cdot\frac{1}{20}=$ part of tank filled by first pump in x hours

and

$x\cdot\frac{1}{50}=$ part of tank filled by second pump in x hours

But in x hours the tank is exactly filled, hence

$$\frac{x}{20}+\frac{x}{50}=1$$

where we write 1 to indicate 1 tankful. Multiplying both members of this equation by 100, we get

$$5x + 2x = 100$$

$$7x = 100 \qquad \text{and} \qquad x = \tfrac{100}{7} = 14\tfrac{2}{7}$$

Therefore, the two pumps working together will fill the tank in $14\frac{2}{7}$ hours.

☐ ***Check***

$$\frac{100}{7} \cdot \frac{1}{20} + \frac{100}{7} \cdot \frac{1}{50} = \frac{5}{7} + \frac{2}{7} = \frac{7}{7} = 1$$

▲ EXERCISES

1. A tank can be filled by one pipe in 5 hours and emptied by another pipe in 8 hours. When the tank is half full, both pipes are opened. How long will it take to fill the tank from this half-full condition?
2. One pipe can empty a tank in 2 hours and another pipe can empty it in 4 hours. A third pipe can fill the tank in 3 hours. If the tank is full and all three pipes are opened, how long will it take to empty the tank?
3. A printing press can print the daily quota of papers in 4 hours. After operating $2\frac{1}{2}$ hours, this press breaks down and is replaced by another press that prints the daily quota in 7 hours. How long will it take the second press to finish printing the daily quota?
4. A pipe A can fill a tank in one-half the time that it takes a second pipe B to fill the tank. Pipe B can fill the tank in two-thirds the time it takes a third pipe C. If it takes 18 hours to fill the tank when all three pipes are turned on, how long will it take each pipe working alone to fill the tank?
5. Two pumps pumping together can fill a reservoir in 10 days. At the end of 7 days one pump breaks down and the other pump finishes the job in 5 days. How long would it take each pump working alone to fill the reservoir?
6. One farmer operating a tractor can plow a field of 200 acres in 10 days. After he has been plowing for 3 days, a neighbor using another tractor helps the farmer to finish plowing the 200 acres in 4 days. How long would it have taken the neighbor to plow the field if he had worked alone?
7. One machine can cap 1,800 bottles in 1 hour, and another machine can cap 1,500 bottles in an hour. If there are 20,000 bottles to be capped and the slower machine is started 2 hours after the faster machine, how long will it take to cap the 20,000 bottles?
8. After 40 quarts of beer was drawn from one of two equal barrels and 120 quarts of beer was drawn from the other, the first barrel contained twice as much as the other. If each barrel was full at the start, how many quarts did they hold?
9. Sally spent two-thirds of her money for a dress and one-fifth of her money for a hat. How much did she spend for each article if she had $3 left?

10. When John opened his savings bank, he found that he had 28 coins whose value was \$3.40. If there were only dimes and quarters in this bank, how many of each were there?
11. A radiator which has a 16-quart capacity is filled with a 10% alcohol solution. If it requires a 30% alcohol solution to protect the radiator, how much 95% alcohol must be added to the original 10% solution to make a 30% alcohol solution?
Hint: Some of the 10% solution will have to be drained before more can be added.
12. There are three pipes attached to an empty tank. The first fills it in 3 hours. The second fills it in 4 hours. The third empties it in 1 hour. They are opened in that order at 1 P.M., 2 P.M., and 3 P.M. When will the tank be empty?

Vital Statistics

In one year during the gold-rush days, 30% of the female population of Nome, Alaska, got married. But during the same year only 1.7% of the male population got married. Assuming that bigamy was outlawed, calculate the ratio of men to women in Nome.

Review Exercises

1. A pile of nickels, dimes, and quarters has a total value of \$5.30. There are twice as many quarters as nickels and one more dime than nickels. How many coins are there of each denomination?
2. A tank can be filled by one pipe in 5 hours and emptied by another pipe in 8 hours. When the tank is half full, both pipes are turned on. How long will it take to fill the tank?
3. In a group of baby chicks, it was found that there were 120 more female chicks than male chicks. If the ratio of female to male chicks is $\frac{7}{2}$, how many chicks were there in the group?
4. A farmer with a team of horses can plow a field in 28 days. His son using a tractor can plow the same field in 10 days. If the farmer plows 9 days and then quits, how many days will it take for the son to finish?

5. One alloy is 45% silver and another alloy is 25% silver. How many ounces of these alloys should be melted together to form 100 ounces of a new alloy which will be 40% silver?
6. If the difference between two numbers is 16 and one number is three-fifths of the other, what are the numbers?
7. A train leaves Milwaukee at noon for Seattle and travels 90 miles per hour. At 3 P.M. a plane leaves Milwaukee following the same route as the train and travels 210 miles per hour. At what time will the plane overtake the train, and how far from Milwaukee will they both be at this time?
8. A merchant bought some card tables for $3 each. If he wishes to make a profit of 40% on the selling price, at what price should he sell the tables?
9. A clothing dealer plans to buy suits at $40 each. He estimates that three-fourths of them can be sold at $70 each and the rest at $30 each. How many suits should he buy if he expects to realize a profit of $1,200?
10. A farmer persuaded Sam to try to work for 30 days on a job. Sam would be paid $8 a day for each day he worked. But, Sam would have to pay the farmer $2 a day for each day he failed to work. At the end of 30 days neither the farmer nor Sam owed the other anything, which convinced Sam of the folly of work. How many days did Sam work?
11. A retailer cannot stock more than 10 units of item A and 5 units of item B. Tell the retailer what possibilities he has if he is to have at least one of each item in stock.
12. A merchant stocks two items A and B which can be as many as 4 items each. But he discovers that for some reason the number of items A is never greater than B.

    a. Write the necessary relations that describe this situation.
    b. Tell the merchant what possibilities he has.
    c. What is the maximum number of each item that he has in stock at any one time?

13. A merchant has space to stock 4 items. If the number of items A is always greater than the number of items B

    a. Write the necessary relations that describe this situation.
    b. Tell the merchant what possibilities he has.
    c. What is the maximum number of each item that he has in stock at any one time?

14. In working with some data a person found that two quantities A and B which he knew had to be positive also had to satisfy the two conditions

    $$2y+x \geqq 10 \qquad \text{and} \qquad x+y \leqq 10$$

    Help this person solve his problem.
15. If the two inequalities had been $2y + x \leqq 10$ and $x + y \leqq 10$, what can you tell this person about his work?

CHAPTER SIX

Quadratic Equations and Relations

It is often said that an equation contains only what has been put into it.... But there is something more: analysis, by the simple play of its symbols, may suggest generalizations far beyond the original limits. E. PICARD

1 *The Graph of the Equation* $y = kx^2$

The statement that one quantity y varies as the square of another quantity x is encountered in many problems of a practical nature. The standard equation for this variation is given by $y = kx^2$. In order to visualize this variation, we will plot the graph of this equation for $k = 1, 4, \frac{1}{4}, -1, -4$, and $-\frac{1}{4}$. As before, we form a table of values for each equation and plot the corresponding pairs of values.

When we examine the curves in Fig. 13, we see that they all pass

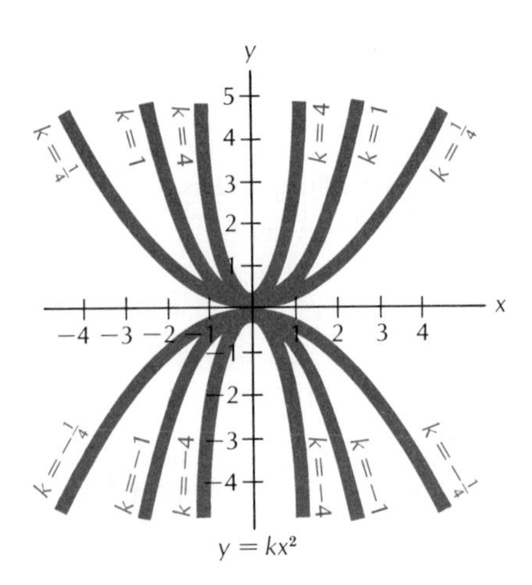

FIG. 13

through the point (0,0). If k is positive, the curves open upward and do not extend below the x axis. If k is negative, the curves open downward and do not extend above the x axis. The origin is called the *vertex* of these curves. The steepness of the curves depends upon the value of k.

We also notice that the left-hand side of each curve is the mirror image of the right-hand side. We call this property *symmetry* and say that each curve is symmetrical with respect to the y axis. This axis of symmetry is called the *axis of the curve.*

If x is proportional to y^2, $x = ky^2$. We plot this equation for $k = 1$, 4, $\frac{1}{4}$, $-\frac{1}{4}$, -1, and -4 (see Fig. 14).

We see that these curves are symmetrical with respect to the x axis, and in this case, the x axis is the axis of symmetry for these curves. The curve opens to the left if k is negative and does not extend to the right of the y axis. The curve opens to the right if k is positive and does not extend to the left of the y axis. The origin is the vertex for these curves.

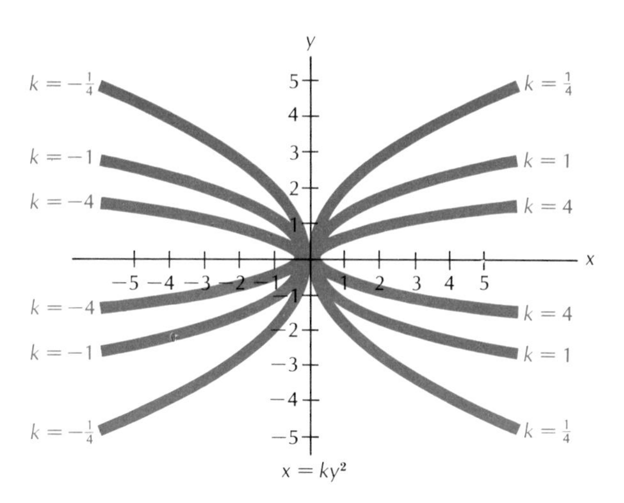

FIG. 14

The Parabola

The curves whose equations are $y = kx^2$ or $x = ky^2$ are called *parabolas.* The axis of symmetry is called the *axis of the parabola,* and the origin is the *vertex of the parabola.*

The parabola is a much-used curve and is frequently encountered in the world about us. The orbits of some of the comets and the path of a projectile in a vacuum are parabolas. The arches of a bridge and the high ceilings in churches are often in parabolic form. The cable of a suspension bridge which is loaded uniformly per horizontal foot will hang in the shape of a parabola. The surface of a rotating liquid and the reflecting surfaces used in headlights, searchlights, and telescopes are parabolic in form. By this we mean that any cross section through the axis of symmetry is a parabola.

A parabolic mirror has the property that parallel rays of light coming from a distant object are reflected from the mirror in such a way that all pass through a point. This point is called the *focus*. Likewise, if a source of light is placed at the focus of the mirror, the light rays will reflect from the mirror as parallel rays. Such parabolic reflectors are used when it is desired to deliver an intense beam of light, for example, the automobile headlight. A parabolic reflector may also be used to direct sound waves. The sound created at the focus of a parabolic band shell is reflected back to the audience in the same manner as light rays would be reflected (Fig. 15). This eliminates "dead spots" in hearing conditions.

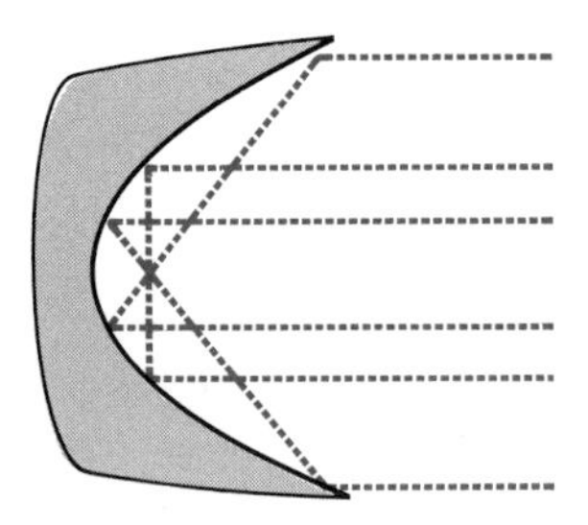

FIG. 15

▲ EXERCISES

Plot the curves for the following equations:

1. $y=2x^2$
2. $y=-2x^2$
3. $x=3y^2$
4. $x=-3y^2$
5. $x=-2y^2$
6. $y=-3x^2$
7. $y^2+3x=0$
8. $2y=5x^2$

Problems just for fun

The Duchess owns a necklace which has 21 diamonds. The middle diamond is the largest and most expensive. The diamonds are arranged so that starting from one end of the necklace, each successive diamond is worth \$100 more than the preceding one. This is true right up to and including the middle diamond. Starting from the other end, each diamond up to and including the large one is worth \$150 more than the preceding one. If the necklace is worth \$32,450, what is the value of the large diamond?

3 *The Equation for a Functional Relation Exhibited by a Table*

We recall that the curve in Fig. 4 (page 36) for the functional relation given by Table 2 (page 36) has a shape similar to the upper portion of the curves for $x = ky^2$ in Fig. 14. We wonder if the equation giving the functional relation between the distance D and the height h is $h = kD^2$. In order to determine the value of k, we select a pair of values from the table, for example, $D = 12.3$ miles when $h = 100$ feet, substitute these values into the equation, and solve for k.

$$100 = k(12.3)^2$$

or

$$k = \frac{100}{(12.3)^2}$$

Then

$$h = \frac{100}{(12.3)^2} D^2$$

or

$$100D^2 = (12.3)^2 h$$

If this equation represents the relation given by Table 2, we can check it by substituting another value for h and solving for D. When $h=400$ feet,

$$100D^2=(12.3)^2(400)$$

and

$$D^2=4(12.3)^2$$
$$D=\pm 2(12.3)=\pm 24.6$$

The negative value for D has no meaning in this problem. We may check other values also and conclude that we have the true equation of the functional relation.

▲ EXERCISES

Plot the tabulated data for the following problems. If the curve appears to have the shape of a parabola, find and check the equation for the functional relation between the variables. Do the equations have meaning for all values of the dependent variable? Discuss.

1. Neglecting friction, the distance s in feet that a body falls from rest is a function of the time of fall t in seconds. This relation is given in Table 21.

TABLE 21

t	0	0.5	1	2	3	4
s	0	4.025	16.1	64.4	144.9	257.6

2. For a fixed opening, the length of exposure t in seconds necessary to photograph a given subject by photoflood lamp is a function of the distance s in feet of the subject from the source of light. The relation between t and s is given in Table 22.

TABLE 22

s	0	4	6	8	9
t	0	0.89	2	3.55	4.5

3. If the source of illumination is constant, the exposure time t in seconds is a function of the lens "f opening." Table 23 expresses the relation between f and t. What exposure time is needed for an opening of f 11? f 5.6?

TABLE 23

f	2	4	8	12
t	0.000625	0.0025	0.0100	0.0225

4. The heat H in calories generated in a resistance element per second depends upon the amount of electric current I in amperes sent through the wire. Table 24 gives the relation between H and I.

TABLE 24

I	0	1	2	3	4
H	0	2.4	9.6	21.6	38.4

5. The surface area S of a solid depends upon a dimension d. Table 25 gives the relation between S and d.

TABLE 25

d	0	$\frac{1}{4}$	$\frac{1}{3}$	$\frac{3}{4}$	1.0
S	0	0.786	1.40	7.07	12.57

6. Neglecting friction, the distance s in feet that a ball rolls down an inclined plane is a function of the time t in seconds. Table 26 shows the relation between s and t for a certain inclined plane.

TABLE 26

t	0	1	2	2.5	3
s	0	8	32	50	72

Problems just for fun

Here we have a window which is square, with inside dimensions of 1 foot. The window is divided by narrow bars (width negligible) into four lights measuring $\frac{1}{2}$ foot on every side. Make another window each of whose four sides shall be 1 foot, but divided by narrow bars into eight lights whose sides shall all be $\frac{1}{2}$ foot.

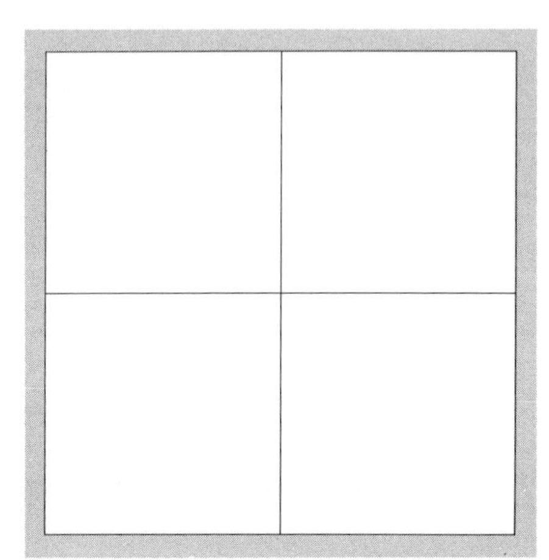

4 *Equations of the Second Degree in One Unknown*

If the highest power of the variable in an expression is 2, the expression is said to be *quadratic,* or of the *second degree,* in this variable. The expression $ax^2 + bx + c$ is quadratic in x; the expression $dx^2 + exy + fy^2$ is quadratic in both x and y. Any equation of the second degree in one unknown x may be written as

$$ax^2 + bx + c = 0$$

where a, b, and c are constants. This equation states that for some value or values of x (to be determined), the left-hand side $ax^2 + bx + c$ is equal to the right-hand side, namely, 0. To find these solutions, we might set the right-hand side equal to y and plot the equation $y = ax^2 + bx + c$. The solutions would be the points where $y = 0$, that is, the points at which the curve crosses the x axis.

To illustrate this method, let us consider the following problem:

■ **Example.** A gardener has a rectangular plot of ground 30 feet by 20 feet, on which he plans to build a flower bed surrounded by a grass strip. If he wants 400 square feet of grass, how wide should the strip be?

□ ***Solution (Fig. 16)***

$$\begin{aligned} \text{Let } x &= \text{width of strip in feet} \\ 20-2x &= \text{width of flower bed} \\ 30-2x &= \text{length of flower bed} \\ (20)(30) &= 600 \text{ square feet} = \text{area of plot} \\ (20-2x)(30-2x) &= \text{area of flower bed} \\ 600-(20-2x)(30-2x) &= \text{number of square feet in grass} \end{aligned}$$

Hence

$$600-(30-2x)(20-2x)=400$$

This equation reduces to

$$4x^2-100x+400=0$$

or

$$x^2-25x+100=0$$

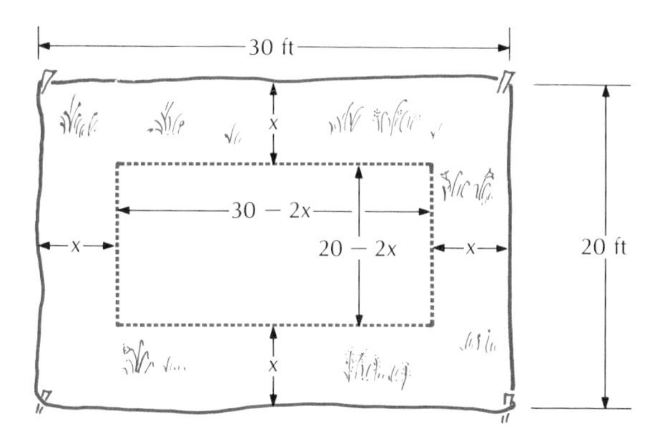

FIG. 16

In order to find a solution of this equation graphically, set the left-hand member equal to y. The solution will be the value or values of x for which y is zero. In order to plot the equation $x^2-25x+100=y$, we form a table of values (Table 27) and plot the curve for $y=x^2-25x+100$.

TABLE 27

x	0	5	10	15	20	25
y	100	0	−50	−50	0	100

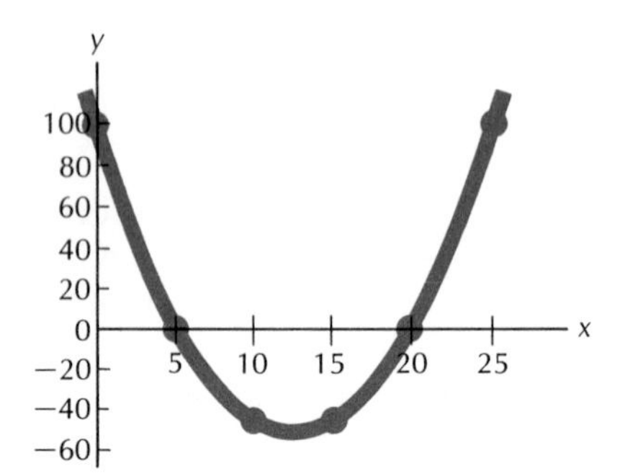

FIG. 17

From Fig. 17 we conclude that $y = 0$ when $x = 5$ and 20. Both $x = 5$ and $x = 20$ satisfy the equation $x^2 - 25x + 100 = 0$.

We find that we have two values of x which formally satisfy the equation $x^2 - 25x + 100 = 0$. We wish to decide whether both these solutions satisfy the given problem. If we take $x = 5$ and put it back into the original statement of the problem, we find that

$$\text{Area of flower bed} = 10(20) = 200 \text{ square feet}$$
$$\text{Area of plot} = 20(30) = 600 \text{ square feet}$$
$$\text{Area in grass} = 600 - 200 = 400 \text{ square feet}$$

which is exactly the value demanded by the problem (Fig. 18).

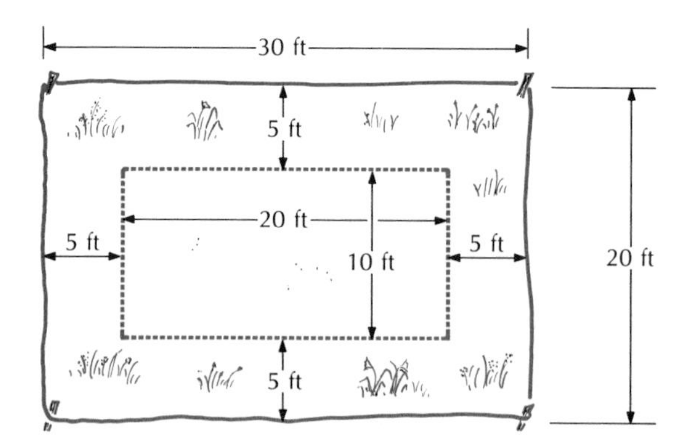

FIG. 18

If we take $x = 20$ feet and try to put it back into the statement of the problem, we find at once that we cannot satisfy the conditions of the problem:

$$\text{Width of flower bed} = 20 - 2x = 20 - 2(20) = -20$$
$$\text{Length of flower bed} = 30 - 2x = 30 - 2(20) = -10$$

We get two negative values, which when multiplied together give a positive value, which is a correct mathematical solution of the equation, although the negative values have no meanings as dimensions of a flower bed and a grass strip. It is always necessary to check our answers to see if they make sense.

We were able to find a solution to our problem using graphical methods. But it was a long and tedious process, because we had to construct the table of values and plot this table.

The Quadratic Formula

To avoid the tedious graphical method, we generally use the quadratic formula. The quadratic equation of the form

$$ax^2 + bx + c = 0$$

has as its solutions

$$x = \frac{-b \pm \sqrt{b^2 - 4ac}}{2a}$$

This is called the *quadratic formula.* This formula may be derived by a method called *completing the square.* The two solutions of the quadratic equation are

$$x = \frac{-b + \sqrt{b^2 - 4ac}}{2a} \qquad \text{and} \qquad x = \frac{-b - \sqrt{b^2 - 4ac}}{2a}$$

These two solutions are formal solutions of the quadratic equation and, when substituted into the left-hand side of the equation, will reduce that side to zero. Let us try this out below.

$$a\left(\frac{-b + \sqrt{b^2 - 4ac}}{2a}\right)^2 + b\frac{-b + \sqrt{b^2 - 4ac}}{2a} + c =$$

$$\frac{b^2 - 2b\sqrt{b^2 - 4ac} + b^2 - 4ac}{4a} + \frac{-b^2 + b\sqrt{b^2 - 4ac}}{2a} + c =$$

$$\frac{2b^2 - 2b\sqrt{b^2 - 4ac} - 4ac}{4a} + \frac{-2b^2 + 2b\sqrt{b^2 - 4ac}}{4a} + \frac{4ac}{4a} = 0$$

All terms cancel, and we see that

$$x = \frac{-b + \sqrt{b^2 - 4ac}}{2a}$$

does, indeed, reduce the left-hand side to zero. Similarly we may see that the second solution with the minus sign before the square root is also a solution of the original general equation.

If we have a practical problem to solve (for example, the flower-bed problem solved by the graphical method) we may use the quadratic formula to solve it. The equation in the flower-bed problem was

$$x^2 - 25x + 100 = 0$$

In this equation, $a = 1$, $b = -25$, and $c = 100$. Substituting into the quadratic formula, we obtain

$$x = \frac{-(-25) \pm \sqrt{(-25)^2 - 4(1)(100)}}{2(1)} = \frac{25 \pm \sqrt{625 - 400}}{2}$$

$$= \frac{25 \pm \sqrt{225}}{2} = \frac{25 \pm 15}{2} = 20 \text{ or } 5$$

However, we remember that the answers may be mathematically correct but physically impossible. In this case, $x = 20$ gives a flower bed with negative length and width, so only $x = 5$ has physical meaning.

▲ EXERCISES

1. Find two consecutive integers whose product is 462.
   Hint: N = one number, $N + 1$ = the other number.
2. Find two consecutive even numbers whose product is 1,224.
   Hint: $2N$ = one number and $2N + 2$ = the other number, where N is an integer.
3. The length of a rectangle is 14 feet greater than the width. Find the length and width of the rectangle if the area is 240 square feet.
4. The length of a rectangle is 4 feet greater than the width. Find the dimensions of the rectangle if the area is 96 square feet.
5. The perimeter of a rectangle is 30 feet. Find the length and width of this rectangle if the area is 56 square feet.
6. The hypotenuse of a right triangle is 17. Find the legs of the triangle if one leg is one unit less than twice the other leg.
7. If the number of feet in the perimeter of a square is equal to the number of square feet in the area, find the length of the side of the square.
8. The area of a square after adding 2 feet to each side is 576 square feet. What were the original dimensions of the square?
9. Find a positive number which when increased by 20 is equal to 69 times the reciprocal of the number.
10. A picture which is 9 inches wide and 12 inches long is surrounded by a frame which has 162 square inches. What is the width of the frame?
11. A merchant bought a shipment of vases for $100. He sold all but three of them which were broken by careless clerks. If he makes a profit of $4 on each vase sold and a total profit of $98, how many vases were in the shipment?
12. Two ships start from the same point at noon. One sails north at a speed of 6 knots, and the other sails east at a speed of 8 knots. When will the

ships be 100 sea miles apart? (A knot is nautical language for 1 sea mile per hour.)

13. A box contains 600 cubic inches. The length is 12 inches, and the width is 5 inches greater than the height. Find the dimensions of the box.
14. A rectangular piece of metal is 3 inches longer than it is wide. From each corner a 1-inch square is cut out and the sides are turned up to form an open box which contains 130 cubic inches. Find the dimensions of the piece of metal.
15. One pipe alone can fill a tank in 4 hours less than it takes a second pipe by itself to fill the tank. If both pipes working together can fill the tank in 2 hours, how long does it take each pipe to fill the tank?

Problems just for fun

Six little kittens sittin' on a fence,
One weighs more because he is more dense.
Can you separate him from the other five,
Which all have the same weight, dead or alive?
Put them on a balance, but only twice,
To find the one so fat and nice.

6 *Relations between the Coefficients of and the Solution to the Quadratic Equation*

In Sec. 5 we used the quadratic formula to solve the equation

$$ax^2 + bx + c = 0$$

We now establish two interesting facts:
We had

$$x_1 = \frac{-b + \sqrt{b^2 - 4ac}}{2a} \qquad \text{and} \qquad x_2 = \frac{-b - \sqrt{b^2 - 4ac}}{2a}$$

as the solutions to the equation $ax^2 + bx + c = 0$. If we add them, we get

$$x_1+x_2=\frac{-b+\sqrt{b^2-4ac}}{2a}+\frac{-b-\sqrt{b^2-4ac}}{2a}=-\frac{b}{a}$$

If we multiply them together, we get

$$x_1x_2=\frac{-b+\sqrt{b^2-4ac}}{2a}\ \frac{-b-\sqrt{b^2-4ac}}{2a}=\frac{c}{a}$$

If we rewrite the equation $ax^2+bx+c=0$ in the form

$$x^2+\left(\frac{b}{a}\right)x+\frac{c}{a}=0$$

then the sum of the two solutions x_1+x_2 must be equal to $-\frac{b}{a}$, the negative of the coefficient of x, and the product x_1x_2 must be equal to $\frac{c}{a}$, the constant term.

The two facts that we have just established are very useful in writing our quadratic equation if we know its two solutions or in checking our work.

■ **Example 1.** The equation $x^2-21x-46=0$ cannot have $x=23$ and $x=2$ as solutions since $x_1+x_2=23+2=25\neq 21$ and $x_1x_2=+46$ instead of -46.

■ **Example 2.** Write the quadratic equation which has roots $\frac{1}{2}$ and -3.

$$x_1=\frac{1}{2} \qquad \text{and} \qquad x_2=-3$$

$$x_1+x_2=\frac{1}{2}-3=\frac{-5}{2}=-\frac{b}{a} \qquad \text{and} \qquad x_1x_2=\frac{-3}{2}=\frac{c}{a}$$

Therefore, we write

$$x^2+\frac{b}{a}x+\frac{c}{a}=x^2+\frac{5}{2}x-\frac{3}{2}=0 \qquad \text{or} \qquad 2x^2+5x-3=0$$

Which is the required equation.

■ **Example 3.** If one solution of the equation $2x^2+5x-12=0$ is $-\frac{3}{2}$, find the other solution. If $x_1=-\frac{3}{2}$, x_2 is the other root.

$$x_1+x_2=-\frac{3}{2}+x_2=-\frac{5}{2}$$

or

$$x_1x_2 = -\frac{3}{2}x_2 = \frac{12}{2} = 6$$

We find that $x_2 = -4$.

▲ EXERCISES

1. Form the quadratic equation whose solutions shall be

   a. $2, 5$ b. $a-1, a+1$
   c. $\frac{3}{5}, -\frac{2}{3}$ d. $-\frac{2}{7}, \frac{1}{5}$

2. A quadratic equation has solutions 2 and 3. Write the quadratic equation that has solutions that are twice these values.
3. One solution of $5x^2 - 26x + 5 = 0$ is 5. Find the other solution.
4. One solution of $8x^2 = 15x + 2$ is $-\frac{1}{8}$. Find the other solution.
5. Find the value of a in the equation $ax^2 - x - 1 = 0$ if the two solutions are $-\frac{1}{3}$ and $\frac{1}{2}$.
6. Find the value of b in the equation $2x^2 + bx - 6 = 0$ if the sum of the two solutions is equal to twice the product of the two solutions.
7. For what value of a would the equation $ax^2 - 6x + 4 = 0$ have equal solutions? What are the two solutions?
8. Prove that one solution to the equation $ax^2 + bx + c = 0$ is the negative of the other if and only if $b = 0$.

What is the value of the color area shown in the two overlapping squares?

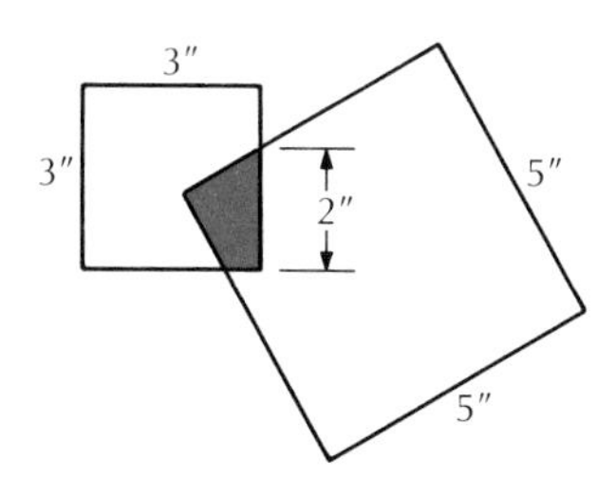

Another Equation for the Parabola

We already know that $y = kx^2$, or $x = ky^2$, is the equation for the parabola with vertex at the origin. The equation

$$y = kx^2 + bx + c \qquad \text{or} \qquad x = ky^2 + dy + e \qquad (k \neq 0)$$

when plotted, gives a curve which is a parabola. If the constants b and c or d and e are equal to zero, the vertex of the parabola is at the origin. The position of the vertex of the parabola depends upon the value of the constants b and c or d and e.

Let us plot the curve defined by the equation $y = x^2 + x - 6$. In drawing the graph of the equation, it is helpful to find the points at which the curve cuts the two axes. Set $y = 0$ and get the equation $x^2 + x - 6 = 0$, which has $x = 2$ and $x = -3$ as solutions. Hence our curve crosses the x axis at $x = 2$ and -3. Also, when $x = 0$, $y = -6$, and the curve crosses the y axis at the point $y = -6$. Since we know that the equation represents a parabola, we have some knowledge of the shape of the curve. We mark these axis crossings on the graph paper and find the coordinates of a few other points, for example, $(1, -4)$ and $(-2, -4)$, to help us draw the curve. The vertex of the parabola $y = x^2 + x - 6$ is at the point $(-0.5, -6.25)$ (Fig. 19).

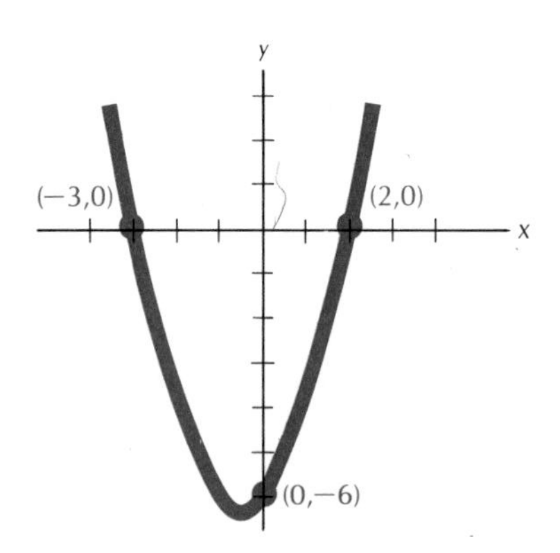

FIG. 19

▲ EXERCISES

Plot the curves for the following equations:

1. $y = 2x^2 + 3$
2. $y = -4 - x^2$
3. $x = y^2 - 4$
4. $9x = y^2$
5. $y = 4x - x^2$
6. $x = 9y - y^2$
7. $y = (x - 5)(x + 1)$
8. $y = 2(x - 3)(x - 1)$

A tree is 4 feet in circumference. A squirrel, in climbing this tree, goes once around the tree for each vertical 6 feet of climb. If the squirrel climbs to a height of 18 feet, how far does he climb?

8 *Finding the Exact Representation for a Relation Given by a Table*

We have previously found the exact formula for functional relations exhibited by a table of values when the graph was a straight line or a parabola with the vertex at the origin. If the graph has the shape of a parabola with the vertex at a point different from the origin, we can still find the equation. It will be of the form

$$y = kx^2 + ax + b \qquad \text{or} \qquad x = ky^2 + cy + d$$

We need to find the value of the three constants. If the axis of the parabola is parallel to the y axis, we use the first form; and the second when the axis is parallel to the x axis.

■ **Example.** The height h in feet of a cable of a suspension bridge above the deck of the bridge is a function of the horizontal distance D in feet from the center of the bridge. Table 28 gives the relation between h and D. Let us find the equation of the curve of the cable and the length of a support needed to reach vertically from the cable to the floor of the bridge at a point 30 feet from the center.

TABLE 28

D	0	±10	±20	±40	±50	±100
h	10	10.4	11.3	16.4	20	50

□ ***Solution.*** Since the axis of the parabola is parallel to the y

axis, we assume that the relation between h and D is given by the equation

$$h = kD^2 + aD + b$$

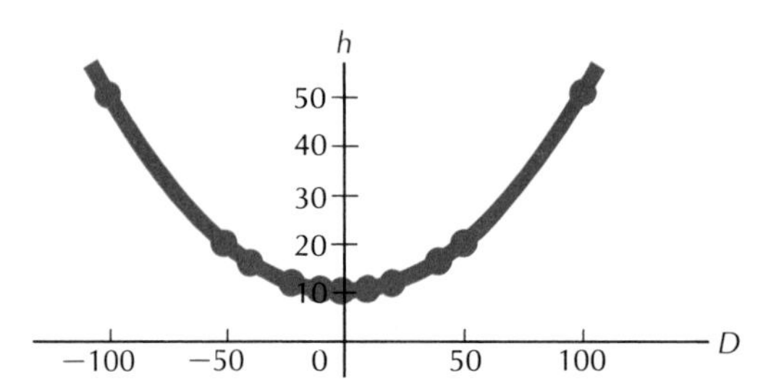

FIG. 20

In order to determine the three constants k, a, and b, we select three points on the curve, for example, (0,10), (50,20), and (100,50) in Fig. 20, and substitute these values into the equation. We get

$$10 = b$$
$$20 = k(50)^2 + a(50) + b$$
$$50 = k(100)^2 + a(100) + b$$

Solving these equations, we find that $b = 10$, $a = 0$, and $k = \frac{1}{250}$. The equation for the parabola is

$$h = \frac{D^2}{250} + 10$$

When $D = 30$ feet,

$$h = \frac{(30)^2}{250} + 10 = 13.6 \text{ feet}$$

This is the length of a support needed to reach from the cable to the deck of the bridge at a distance of 30 feet from the center of the bridge.

▲ EXERCISES

1. Neglecting air resistance, a stone thrown into the air rises for a time and then falls back to the ground. Table 29 gives the height h in feet of the stone above the ground at various times t in seconds. Determine the equation for this relation. What is the maximum height reached by the stone? At what time is the stone 20 feet above the ground? Can t take on negative values? Can t have values greater than 4? Explain.

TABLE 29

t	0	1	2	3	4
h	0	48	64	48	0

2. The height h in feet of a baseball when thrown upward at an angle of 45° with the horizontal is a function of the horizontal distance D. The relation between D and h is given in Table 30. Find the equation of the path. How high will the ball rise? What is the horizontal distance it will travel? At what horizontal distance will the ball be 18 feet above the ground?

TABLE 30

D	0	10	20	50	75	100
h	0	9	16	25	18.75	0

3. The height h in feet of an arch in a church above the floor is a function of the horizontal distance D in feet from the center of the floor. The relation between D and h is given in Table 31. Find the equation of the arch. How far is it above the floor at a point 30 feet from the center of the floor?

TABLE 31

D	0	± 10	± 20	± 40
h	70	$68\frac{3}{4}$	65	50

4. A shot-putter 6 feet tall puts his shot so that it reaches a maximum height of 16 feet when it is 20 feet away horizontally and a height of 6 feet when it is 40 feet away. The relation between h and D is parabolic and is given in Table 32. Find the equation of the shot. What is the distance of the put?

TABLE 32

D	0	20	40
h	6	16	6

5. Find the equation of the parabola which has its axis parallel to the y axis and also passes through the points (1,3), (4,0), and (0,0).

6. Find the equation of the parabola which has its axis parallel to the x axis and passes through the points (0,2), (0,–2), and (3,1).
7. The cross section of the mirror of a reflecting telescope is a parabola. If the vertex is placed at the point (0,0) on a coordinate system and the axis lies along the y axis, the equation of the parabola is $x^2 = 100y$. How deep is the mirror if the diameter of the mirror is 5 feet?

A child weighs four-fifths of its weight and 10 pounds. What is its weight?

9 *The Graph of the Equation* $xy = k$

Let us now determine the curve that represents the relation given by the statement that y is inversely proportional to x; that is, $y = k/x$, or $xy = k$. This type of variation is one of the common types and is frequently encountered in the sciences.

We will graph $xy = k$ for $k = 1, 2, -1$, and -2. Notice that if we set x or y equal to zero we are unable to solve for the other variable.

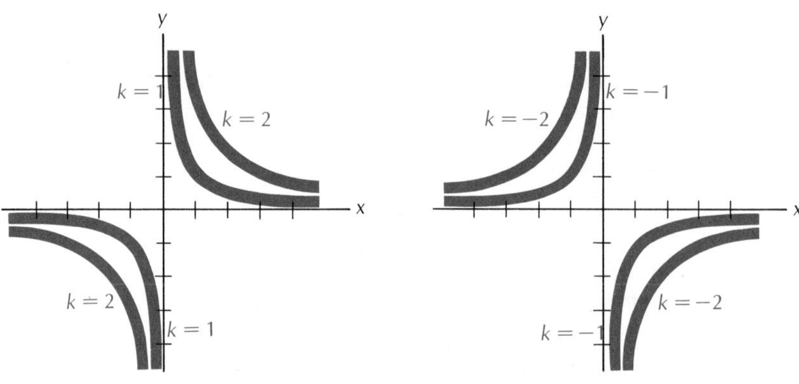

FIG. 21a FIG. 21b

This means that the curve does not cross the x axis or y axis. Likewise, since $y = k/x$, y will increase as x decreases. Also, y will decrease as x becomes larger.

We find that the curve for $xy = k$ has two branches for each value of k. If k is positive, the two branches lie in the first and third quadrants (Fig. 21a). If k is negative, the two branches lie in the second and fourth quadrants (Fig. 21b). The curves do not cross the x or y axes.

Curves represented by the equation $xy = k$ have this characteristic shape, and they are called *rectangular hyperbolas.*

In many physical problems in which the variation is expressed by the equation $xy = k$, only one branch of the hyperbola has any physical significance. That is, the independent variable does not take on both positive and negative numbers. For example, if the temperature is kept constant, the volume of an enclosed gas is inversely proportional to the pressure, and we have $PV = k$. Since negative values of pressure and volume are never possible, only that portion of the hyperbola lying in the first quadrant has meaning.

▲ EXERCISES

Plot the curves for the following equations:

1. $PV = 6$
2. $xy = -5$
3. $yx = 4$
4. $yx^2 = 4$
5. $y^2x = 9$
6. $yx^2 = -4$

10 *Inequalities Involving Squares*

In Chap. 4 we worked with relations which involved inequalities between two quantities. We call these relations linear inequalities because the highest power of either of the variables is 1. However, we have just seen that we need to work with relations between quantities which are not linear; that is, the power of one or more of the variables is 2 or a larger number. These relations may also be expressed in terms of inequalities. For example

$$y > x^2 \qquad \text{or} \qquad y \leqq x^2 \qquad \text{or} \qquad y^2 + x^2 > 1 \qquad \text{etc.}$$

We expect the solution to this type of inequality to be an infinite number of pairs of values (x,y) or points. The easiest way to solve these inequalities is by the graphical method. First change the inequality to an equality and plot. Next select the set of points (x,y) for which the inequality is true.

Problems just for fun

From the main tracks M, two shunting tracks A and B branch off. Both tracks A and B lead to a short track S, which is just long enough to accommodate one car. There is a coal car standing on track A, and a tank car standing on track B. The engineer of a switch engine has orders to move the coal car onto track B and the tank car onto track A. How does the engineer accomplish this change?

■ **Example.** $y > x^2$
The graph of $y = x^2$ is given in Fig. 22*a*. The points in the shaded area (not on the curve $y = x^2$) will satisfy $y > x^2$.

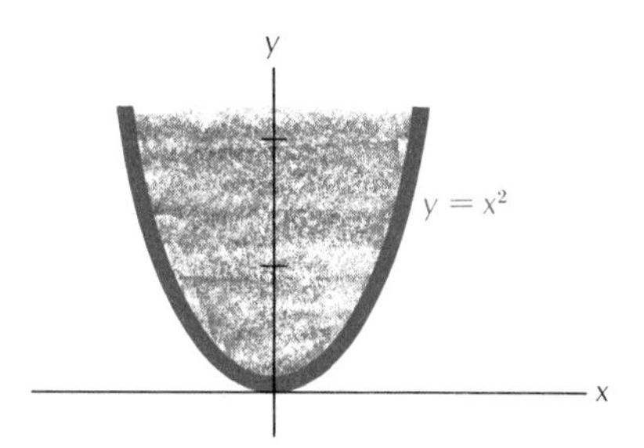

FIG. 22a

■ **Example.** $y \leqq x^2$. Looking at Fig. 22*a* we see that the points on the curve $y = x^2$ and in the unshaded area will satisfy the inequality $y \leqq x^2$.

We may encounter more than one inequality involving two variables and ask for the set of values (x,y) or points that will satisfy both inequalities at the same time.

■ **Example.** Find the pairs of values (x,y) or points that will satisfy $y > x^2$ and $y \leqq 2 - x^2$.

We plot the curve for $y = 2 - x^2$ (Fig. 22*b*). The points (x,y) in the shaded area and on the curve $y = 2 - x^2$ will satisfy the inequality

$$y \leqq 2 - x^2$$

Since we have the solution to $y > x^2$ in Fig. 22*a*, we superimpose the two figures to get Fig. 22*c*. The overlap of the two shaded areas will contain the points that satisfy the two inequalities. This is the area

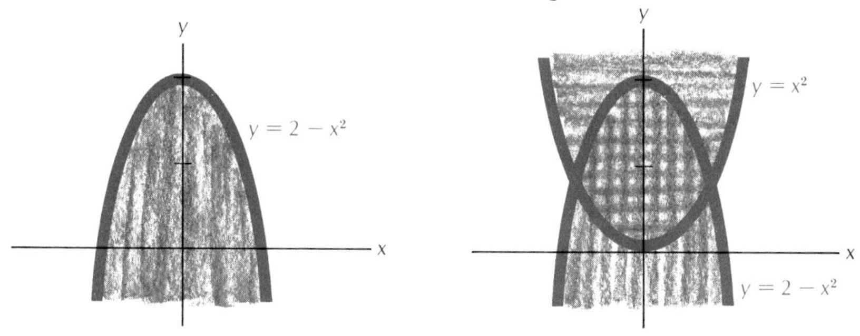

FIG. 22b FIG. 22c

above the curve $y = x^2$ and below and on the curve $y = 2 - x^2$. We can see that the largest value of y is 2, but the smallest value $y = 0$ is not included in the possible solutions. If we had $y \geqq x^2$ and $y \leqq 2 - x^2$, then we could read the maximum and minimum values for x and y.

▲ EXERCISES

1. Plot the points that will satisfy:

   a. $y \geqq 1 - x^2$
   b. $y < 1 - x^2$
   c. $x > -y^2$
   d. $x < -y^2$
   e. $x^2 + y^2 = 1$
   f. $x^2 + y^2 \leqq 1$
   g. $x^2 + y^2 \geqq 1$

2. Find the pairs of numbers (x,y) or points that will satisfy

   a. $xy = 1$
   b. $xy < 1$
   c. $xy > 1$

3. Find the pairs of numbers (x,y) or points that will satisfy the following pairs or systems of inequalities

   a. $y \geqq x^2$, $y \leqq x$
   b. $y \geqq x^2$, $y \geqq 1$
   c. $y^2 + x^2 \leqq 1$, $y \leqq x$
   d. $y^2 + x^2 \geqq 1$, $y \geqq x$
   e. $y^2 + x^2 \leqq 1$, $y < 2$
   f. $y^2 + x^2 \leqq 1$, $y \leqq -1$
   g. $y^2 + x^2 \leqq 1$, $y < -1$
   h. $y^2 + x^2 \leqq 1$, $y \leqq 1 - x^2$
   i. $y^2 + x^2 \leqq 1$, $y \geqq 1 - x^2$
   j. $x \geqq 0$, $xy < 1$, $y \leqq x$
   k. $x \geqq 0$, $xy \leqq 1$, $y \geqq x$

CHAPTER SEVEN

Geometry of Triangles

Let no one ignorant of Geometry enter my door. PLATO

1 *Plane Triangles*

In geometry we learned that a triangle is characterized by six elements —three sides and three angles. The triangle is completely determined and can be constructed provided we know the size of:

1. Two angles and one side, or
2. Two sides and the angle between, or
3. Three sides

If the triangle is a right triangle, one of the angles, namely, the right angle, is known. In order to construct a right triangle we need to know only the size of:

1. One side and one acute angle, or
2. Any two sides

We will follow the notation used in plane geometry. The vertices of the triangle will be denoted by the capital letters A, B, and C, and the sides opposite the vertices will be denoted by the lowercase letters

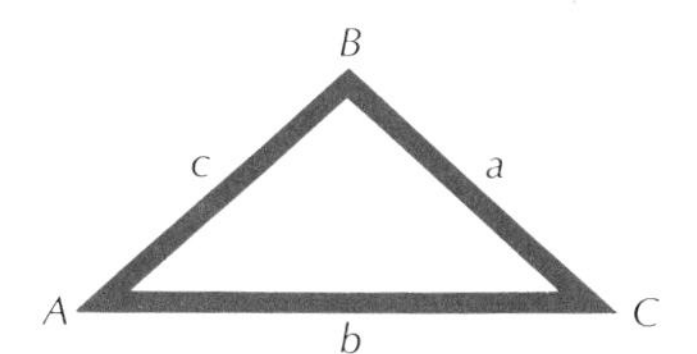

FIG. 23

a, b, and c. As shown in the triangle in Fig. 23, a is the side opposite the vertex A, b is the side opposite the vertex B, and c is the side opposite the vertex C.

2 *Right Angles*

The peoples of the ancient world defined the right angle as the angle that a plumb line makes with the water level or the horizon. These people had a very simple rule for making an angle of 90°. They knew that a triangle with sides three, four, and five units in length is a right triangle, and they staked out such a triangle whenever they needed to construct a right angle.

3 *Construction of Triangles. Scale Drawings. Similar Triangles*

In plane geometry we learned that if we know three parts of a triangle (one of which must be a side), we can construct the triangle, using a ruler and a compass. The next step is to find the size of the remaining sides and angles. With a ruler, we can measure the length of the sides of the triangle. Using the protractor, we can measure an angle of any size. The accuracy with which we can measure the length of the sides and the angles of the triangle depends upon the accuracy with which we can read the ruler and the protractor we are using. Surveyors have devices by which they can measure a distance or an angle with great accuracy. In general, we are not concerned with such great accuracy here.

If the given dimensions of length are too great for the triangle to fit onto our paper, we choose a smaller length to represent the original length. That is, we choose a suitable scale and draw the triangle to scale. The lengths are multiplied by convenient fractions *but the angles are not changed.* From such scale drawings we can determine the size of the unknown parts of the triangle by measuring these parts with the ruler and protractor.

When we make a scale drawing, we are assuming that the figure is similar to the original or desired figure, and we apply all the

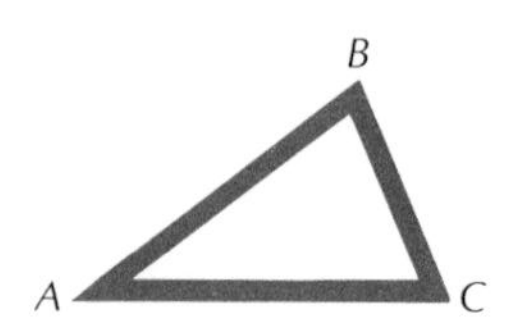

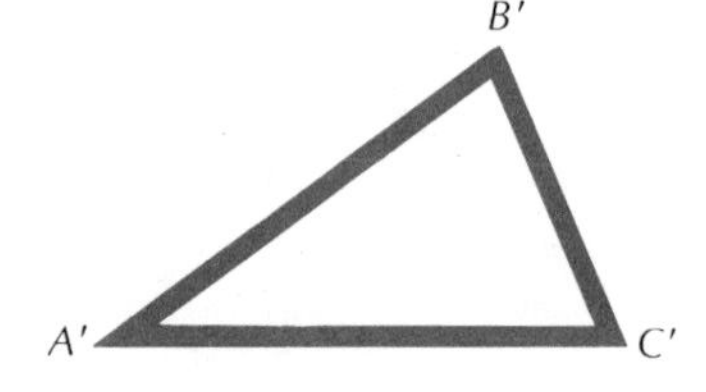

FIG. 24

properties about similar geometric figures that we learned in plane geometry. Two triangles are similar if their angles are equal.

If the triangles ABC and $A'B'C'$ (Fig. 24) are similar, the ratio of the corresponding sides is a constant. That is,

$$\frac{AB}{A'B'} = \frac{AC}{A'C'} = \frac{BC}{B'C'} = \text{constant}$$

Two triangles are congruent if their angles and sides are equal (Fig. 25). In this case, the ratio of the corresponding sides is equal to 1.

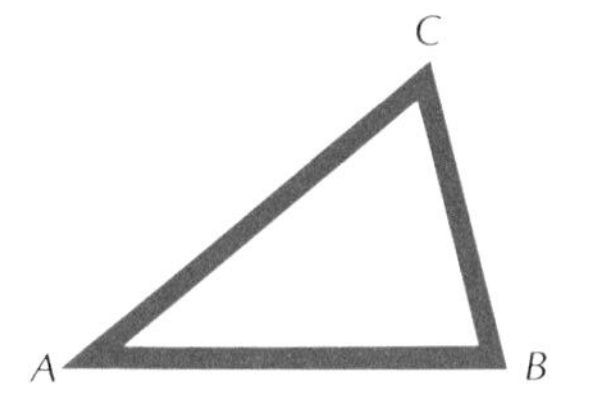

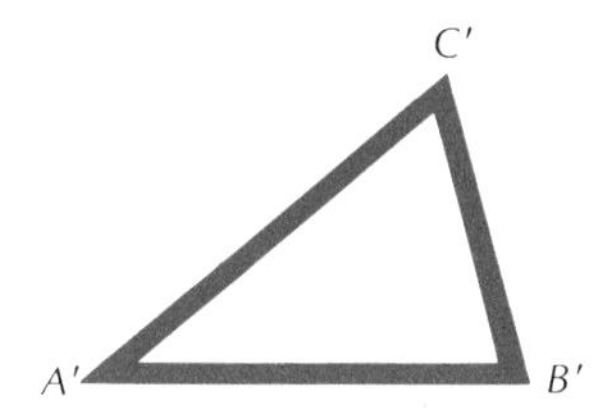

FIG. 25

We can use scale drawings to solve practical problems involving triangles. From scale drawings we can often obtain dimensions or measurements of distances which might be difficult or impossible to measure directly. Scale drawings are also very useful:

1. In estimating results
2. As an aid in visualizing the conditions of the problem
3. As a check on the results obtained by computation

■ **Example.** In order to find the distance between two buildings A and B, which are separated by a high hill, a point C was chosen from which both A and B could be seen. The angle between CA and CB and the two distances CA and CB were measured and found to be $70°30'$, 2,000 feet, and 3,500 feet. Find the distance AB.

□ ***Solution.*** In order to solve this problem we select a suitable scale and construct the triangle ABC, which is similar to the triangle formed by the two buildings and the point C (Fig. 26).

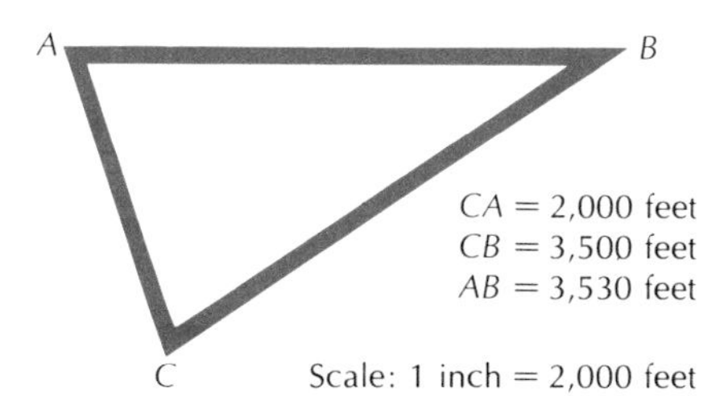

FIG. 26

Thus to find the distance between the two buildings A and B, we need to measure the side AB of the triangle. The accuracy of the work depends upon the size of the drawing and the care with which the triangle is constructed.

Some Definitions

Before proceeding to solve verbal problems, we need to define three special angles.

Angles of elevation and depression. When an observer views an object O, the line OE joining O to E (the eye of the observer) is called the *line of sight* (Fig. 27). The angle between the line OE and a horizontal line EH is called the *angle of elevation*, or *angle of depression*, of O, according as the object O is higher or lower than the eye E.

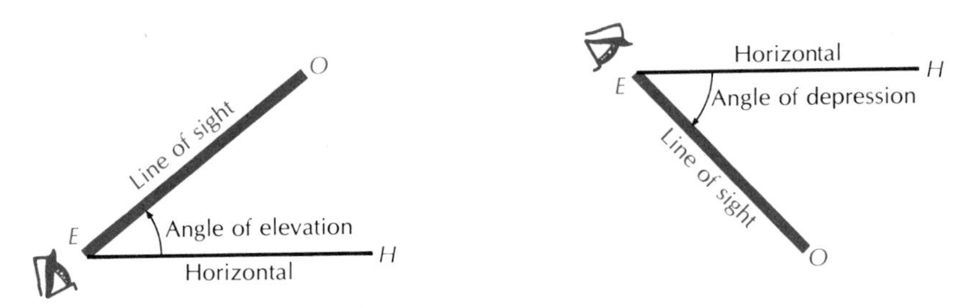

FIG. 27

Angle of inclination. The angle between a line and the horizontal is called the *angle of inclination*. Both the angle of elevation and the angle of depression are special cases of the angle of inclination.

▲ EXERCISES

Use the graphical method to solve the following problems:

1. The angle of elevation of the sun is 40°, and the length of a man's shadow is 7 feet. Find the height of the man.
2. The inclination of a hill is 20°. If a boy walks 1 mile up the hill, how many feet has he risen? How many feet has he advanced in the horizontal direction?
3. A chord of a circle is 20 feet. The angle between the two lines joining the end points of the chord to the center of the circle is 60°. Find the radius of the circle.
4. In order to measure the width of a river (Fig. 28), the distance between two points A and B, close to and parallel to the river bank, was measured and found to be 400 feet. A tree T on the opposite bank was used as a

sighting point. The angle between AT and AB was measured by a transit and found to be 40°. The angle between BT and AB was found to be 90°. Find the width of the river.

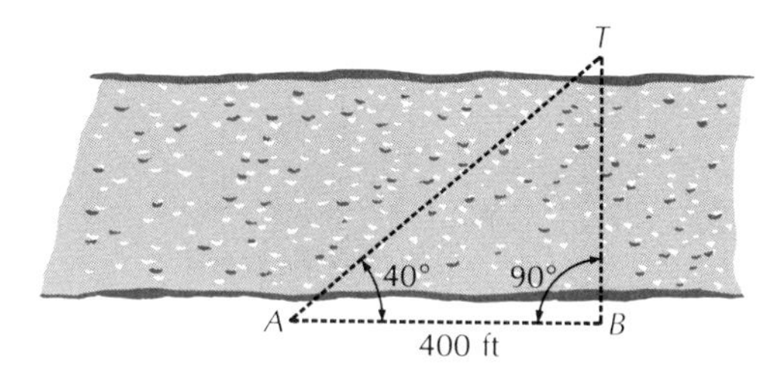

FIG. 28

5. A tree standing on the bank of a stream is known to be 60 feet high. From the opposite bank the angle of elevation of the top of the tree, measured from a point 4 feet above the foot of the tree, is found to be 20°. Find the width of the stream.
6. Mount Hood is 51 miles from Portland, Ore., in a direction 14° south of east. How far south and how far east is the mountain from the city?
7. From the top of a 40-foot bank, a man observes that a sailboat makes an angle of depression of 10°. How far from the bank is the sailboat?
8. A swimmer is 200 yards west of a beach which runs north and south. How far will he swim if he approaches the beach by swimming northeast?
9. A shed 30 feet wide has a sloping roof. If the roof is 35 feet wide, what angle does it make with the horizontal?
10. From a road intersection one motorist travels due north at 30 miles per hour. A second motorist leaves the intersection 30 minutes later at 40 miles per hour on a straight road leading northeast. How far are the two cars apart when the second motorist has traveled 1 hour?

Problems just for fun

Two men start walking toward each other from two points 10 miles distant. One man walks at the rate of 6 miles per hour and the other at the rate of 4 miles per hour. A bee leaves the nose of the first man as he starts and flies back and forth between the two at the rate of 25 miles per hour. When the two meet, how far has the bee flown?

That's a lot of mathematics to exterminate one bee.

I think it would be simpler if we used some DDT.

Shadow Reckoning and Some History

We can often use similar triangles to solve practical problems. For example, we can determine the height of a tree or a building by

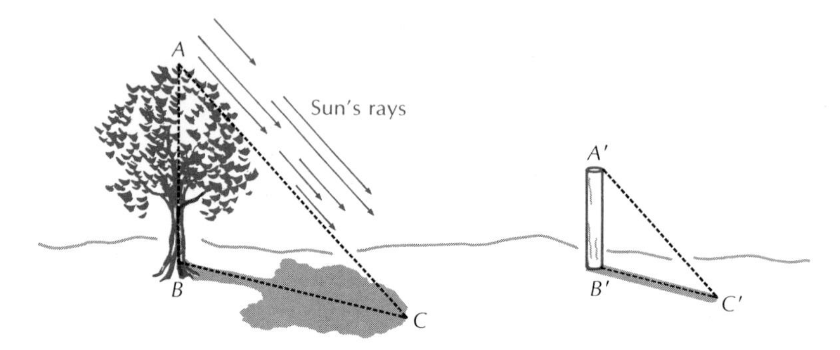

FIG. 29

measuring its shadow. To find the height of a tree we compare the triangle formed by the top and foot of the tree and the tip of its shadow with the triangle formed by the top, bottom, and tip of the shadow of a vertical pole (Fig. 29). The two triangles ABC and $A'B'C'$ are similar, and we can write

$$\frac{AB}{A'B'}=\frac{BC}{B'C'}$$

Since we are able to measure the three distances BC, $A'B'$, and $B'C'$, we can easily calculate the inaccessible height AB of the tree.

Thales (640-546 B.C.), who was one of the founders of Greek geometry, used this method to determine the height of the Egyptian pyramid of Cheops. To be sure, the Egyptians had measured the pyramid before Thales showed them his method. The Egyptians also used the shadow method to measure the height of their pyramids, but they used a very special right triangle, namely, the right triangle in which the two legs are equal (Fig. 30). They constructed

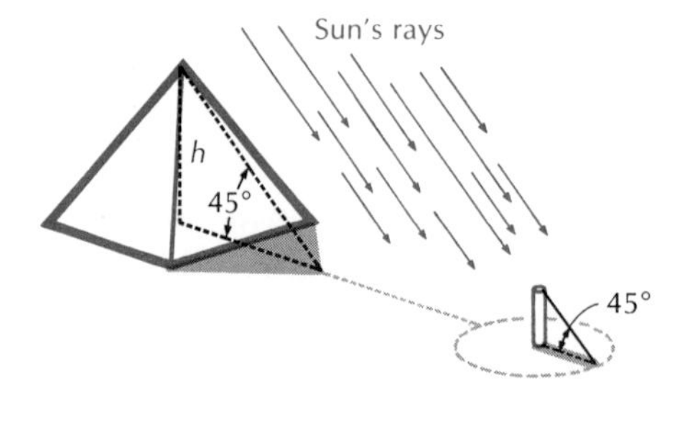

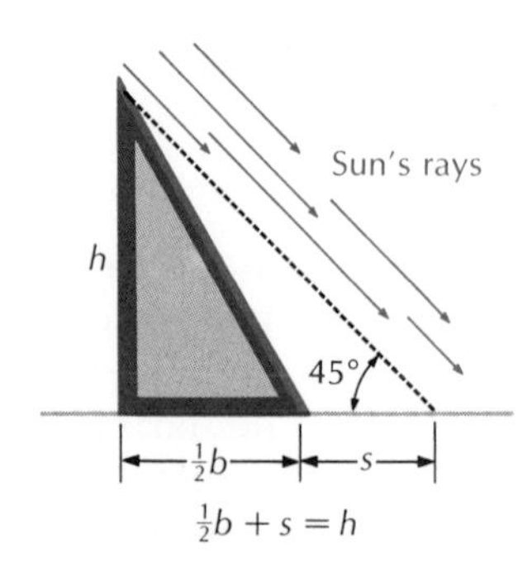

FIG. 30

the perpendicular bisector of the base of the pyramid, and on this bisector they erected a pole and constructed in the sand a circle with radius equal to the height of the pole and centered at the center of the pole. When the shadow of the pole just touched the circle of the perpendicular bisector, they knew that the length of the shadow of the pyramid plus half the base of the pyramid was equal to the height of the pyramid. The Egyptians knew that there were only two days each year on which they could use this method to measure the heights of the pyramids. Thales' method did not depend upon the position of the sun, and he could determine the height on any day that the sun shone.

▲ EXERCISES

Solve the following problems, making use of the properties of similar triangles:

1. A man 6 feet tall is standing at the foot of a tree. If the shadow of the tree and of the man are 40 feet and 4 feet in length, respectively, find the height of the tree.
2. A 1-foot pole casts a shadow of 2 feet when the shadow of a cliff is 30 feet. What is the height of the cliff?
3. A boy 5 feet tall stands in an attic the roof of which slopes down to the floor from a center ridge. When the boy stands 4 feet from the edge of the floor, his head touches the roof. Find the angle that the roof makes with the floor and the height of the roof at the mid-point, if the attic is 20 feet wide.
4. A 15-foot ladder leans against a wall with its foot 5 feet from the wall. A man stands on a rung which is 8 feet from the bottom of the ladder. How far is the man from the wall and from the ground?
5. A silver dollar has a diameter of $1\frac{1}{2}$ inches and when placed 4.6 yards from the eye will just block out the disk of the moon. If the diameter of the moon is 2,160 miles, find the distance of the moon from the earth.
6. A tree 64 feet high casts a shadow 4 feet tall on a wall 20 feet away. How long would the shadow of the tree be if the wall were missing?
7. A man 6 feet tall standing on a ladder 8 feet tall casts a 4-foot shadow. What is the length of the shadow of the ladder?
8. An equilateral triangle has sides three times as long as the sides of another equilateral triangle. What is the ratio of their areas?

More Historical Notes

The word "geometry" comes from two Greek words: *ge*, which means "the earth," and *metron*, which means "to measure." The ancient peoples found it necessary to measure the space they lived in, and geometry had its early roots in the determination of the size and shape, distances and angles, areas and volumes of this space. Much of the early geometric knowledge was discovered by Egyptians, who were practical workmen who built the great pyramids, surveyed the land along the Nile River, and achieved prominence as astronomers. The Egyptians combined their knowledge of astronomy and architecture to orient their temples and pyramids so that rays of certain stars could penetrate deep into the buildings by shining through specially planned hallways and arches.

Eratosthenes (born about 284 B.C.) combined the knowledge of geometry, geography, and astronomy to determine the size of the earth. The method he used to measure the circumference of the earth is as follows:

When Eratosthenes was at Syene, which was 500 miles south of Alexandria, he noticed that at noon the sun shone straight down into a deep well, and its reflection from the water at the bottom struck his eye as he looked, which proved that the sun was directly overhead. On the same day at Alexandria it was observed from the shadow of a tall pillar that the sun was $7\frac{1}{2}°$ south from the vertical at noon. Since the rays from the sun are practically parallel, we can draw the diagram shown in Fig. 31. Erastosthenes reasoned that

$$\frac{360°}{7.5°} = \frac{\text{circumference of the earth in miles}}{500 \text{ miles}}$$

Since 360° is approximately fifty times $7\frac{1}{2}°$, we find that the value of the circumference of the earth is approximately 25,000 miles.

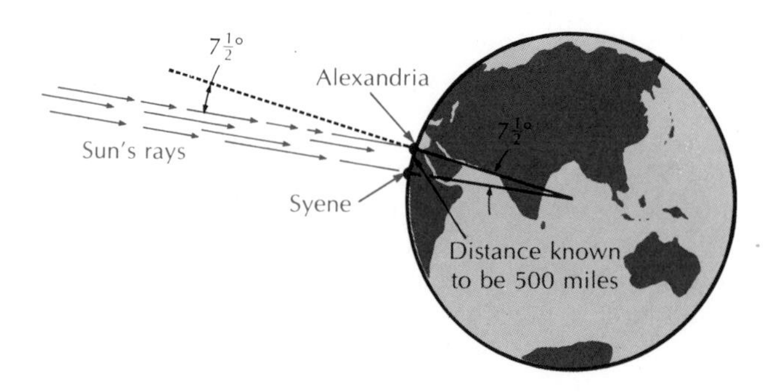

FIG. 31

The Egyptians used special right triangles, usually one in which the acute angles were 45° or one in which the sides had ratios of 3:4:5, to solve their practical problems. Also they frequently solved their problems by making scale drawings.

The Greek scholars observed the use that the Egyptians made of geometry and gathered and systematized the Egyptian geometrical formulas, which were often only practical approximations. The most important among these scholars were Thales (640-546 B.C.), Pythagoras (580?-500 B.C.), Plato (429-348 B.C.), and Euclid (born about 365 B.C.). They idealized the practical geometrical ideas of the Egyptians, and the concepts of points, lines, areas, and volumes became abstractions in the minds of the Greeks. Since the Greeks were by nature fond of logic, they applied this logic to the formulas of the Egyptians to see why these statements should be true. This is the way that the mensuration facts of the Egyptians were organized into the systematic, deductive study which we call geometry. The geometry written by Euclid reads much the same as our textbooks on geometry today. The outstanding contribution made by the Greek scholars to the science of mathematics was the establishing of geometry as a deductive system.

Problems just for fun

Consider the following equations:

$$3x + y = 5$$
$$x = 2 - (\tfrac{1}{3})y$$

Substituting the value of x from the second equation into the first, we obtain

$$6 - y + y = 5$$

or $$6 = 5$$

What is wrong?

CHAPTER EIGHT

Definition of the Trigonometric Functions

There is perhaps nothing which so occupies,
as it were, the middle position of mathematics,
as trigonometry. J. F. HERBART

1 *Ratio of the Sides of Similar Triangles*

We have seen that it is possible to solve problems involving triangles by the graphical method. But this method of solving problems is much too crude for a modern science. The mathematician prefers to do less measuring and more figuring, and tries to find ways by which he can solve these same problems analytically.

We already know some facts that will enable us to solve for the unknown parts of a triangle (Fig. 32). If two of the angles are given,

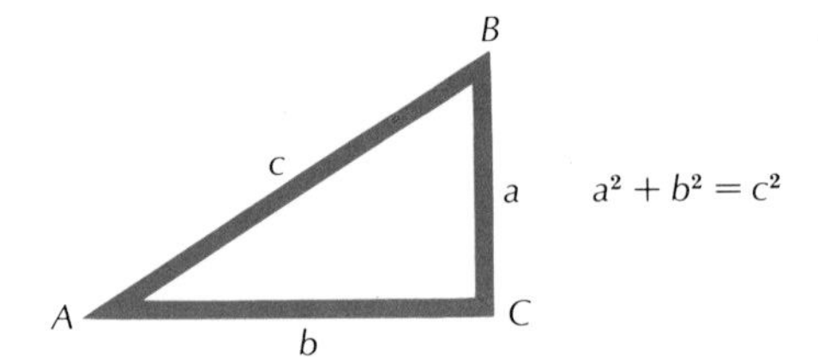

$a^2 + b^2 = c^2$

FIG. 32

we can find the value of the third angle, because the sum of the three angles of a triangle is 180°. If the triangle is a right triangle, one acute angle is sufficient to enable us to find the other acute angle, because the sum of the two acute angles in a right triangle is 90°. If two sides of a right triangle are given, the other side may be found by using the Pythagorean theorem.

Earlier we were able to determine the height of a tree by measuring the length of its shadow and finding the ratio of two sides of a right triangle $A'B'C'$ which had the same acute angle $B'C'A'$ as the angle BCA in the given triangle (Fig. 33). We should become very weary if every time we wished to determine an inaccessible side of a right triangle, we had to construct a similar right triangle whose sides we could measure in order to determine their ratio.

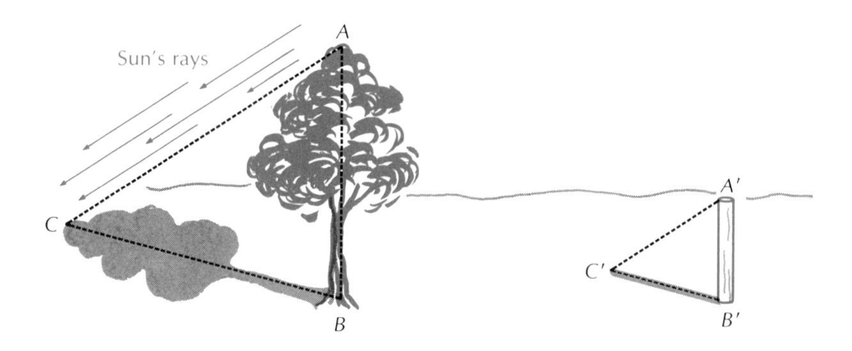

FIG. 33

It is natural for us to wonder whether there is some special magic in these ratios of the sides of these triangles having an acute angle equal to angle BCA. Let us investigate these ratios of the sides of triangles.

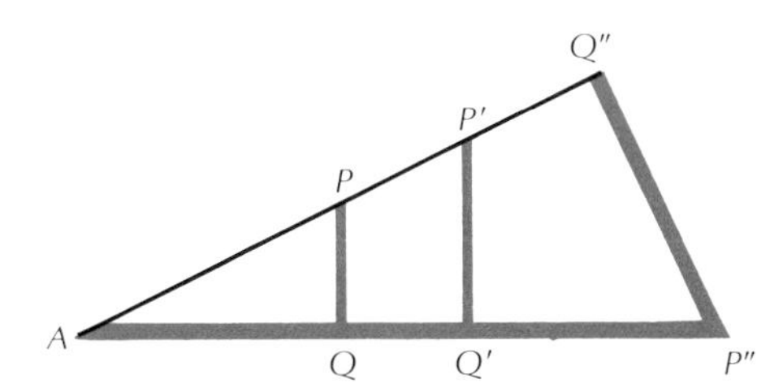

FIG. 34

Given an acute angle A (Fig. 34), from a point P on one side we drop a perpendicular PQ to the other side. In this way a right triangle AQP is formed, in which the line AP is the hypotenuse, PQ is the side opposite angle A, and AQ is the side adjacent to angle A. Then the first ratio used in the example above in this case is

$$\frac{PQ}{AQ} = \frac{\text{side opposite}}{\text{side adjacent}}$$

If we take another point P' on the same side of angle A and drop another perpendicular $P'Q'$ to the other side, we form another right triangle. For this triangle we may write

$$\frac{\text{Side opposite}}{\text{Side adjacent}} = \frac{P'Q'}{AQ'}$$

If we take a point P'' on the other side of angle A and drop a perpendicular $P''Q''$ to the other side, we form a third right triangle. For this triangle we may write

$$\frac{\text{Side opposite}}{\text{Side adjacent}} = \frac{P''Q''}{AQ''}$$

Since angle A is common to the three right triangles, the three triangles are similar, and the corresponding sides are proportional. Then

$$\frac{PQ}{AQ} = \frac{P'Q'}{AQ'} = \frac{P''Q''}{AQ''} = \text{constant} = \frac{\text{side opposite angle } A}{\text{side adjacent angle } A}$$

That is, keeping angle A fixed, the ratio $\frac{\text{side opposite}}{\text{side adjacent}}$ is equal to a constant and does not depend upon the size of the sides.

Further Properties of These Ratios

Let us see what effect changing the size of angle A has upon the ratio of the side opposite an acute angle to the side adjacent in a right

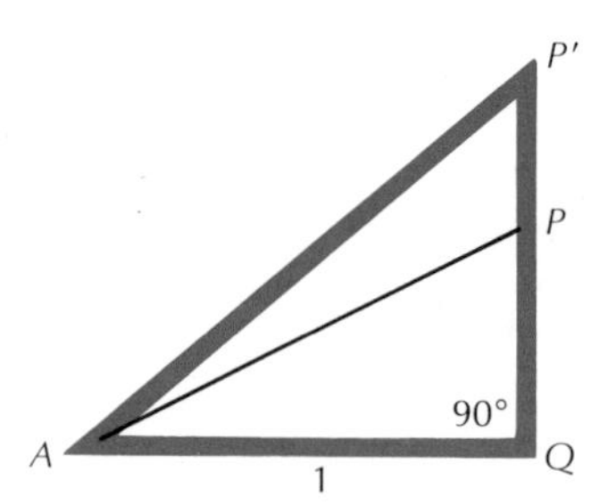

FIG. 35

triangle. To do this, we construct a right angle (Fig. 35) and measure off a distance 1 on one of the legs. At the point A we construct two different acute angles PAQ and $P'AQ$. Since the side AQ is equal to 1,

$$\frac{\text{Side opposite angle } QAP}{\text{Side adjacent angle } QAP} = QP$$

and

$$\frac{\text{Side opposite angle } QAP'}{\text{Side adjacent angle } QAP'} = QP'$$

We conclude that this ratio does depend upon the size of the angle, and the larger angle gives the larger ratio.

We conclude that in a right triangle the ratio of the side opposite an acute angle to the side adjacent is a function of the angle only and for a fixed angle does not depend upon whether the triangle is large or small. For a given angle this ratio is constant and needs to be determined only once. This ratio plays such an important role in the solution of right triangles that it has been given the name *tangent of the angle*, which we define

$$\text{Tangent of } A = \tan A = \frac{\text{side opposite angle } A}{\text{side adjacent angle } A}$$

■ **Example.** Let us illustrate the usefulness of this property of the angle. At a point 30 feet from a steep cliff (Fig. 36), the angle of elevation of the cliff was measured and found to be 20°. If for an

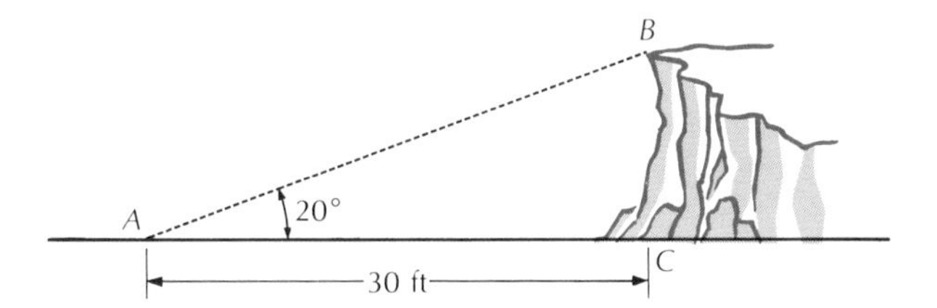

FIG. 36

angle of 20° the ratio of the side opposite to the side adjacent, which is the tan 20°, is known to be 0.3640, find the height of the cliff.

□ ***Solution.*** Applying the definition, we write

$$\tan 20° = \frac{BC}{30}$$

$$0.3640 = \frac{BC}{30}$$

and we get

$$BC = 30(0.3640) = 10.92 \text{ feet}$$

Because we knew the value of the ratio, we did not need to construct a similar triangle to solve this problem.

3 *The Six Trigonometric Functions*

Looking at Fig. 37, we see that there are six different ratios of the sides, namely, $\frac{a}{b}$, $\frac{a}{c}$, $\frac{b}{c}$, $\frac{b}{a}$, $\frac{c}{a}$, and $\frac{c}{b}$. Just as we did in the case of the

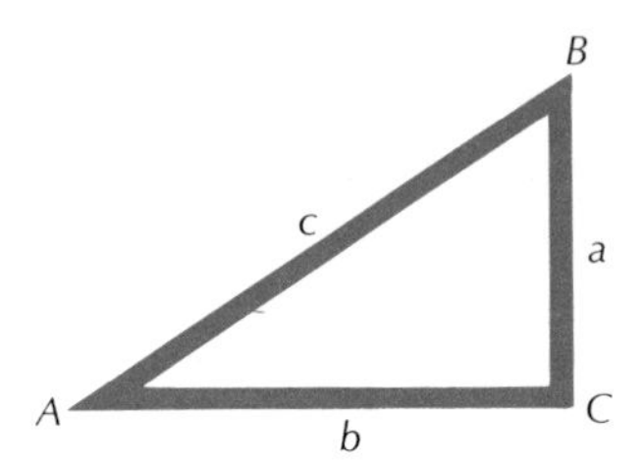

FIG. 37

tangent, we can show that all these ratios are functions of the angle A only and do not depend on the magnitude of the sides of the triangles. These ratios have all been given names. They are called tangent, sine (pronounced *sign*), cosine (pronounced *co-sign*), cotangent, cosecant, and secant, respectively. Since we have seen that these ratios do not depend on the size of the right triangle formed, we may compute them directly from the sides of any one triangle (Fig. 37). We may write

$$\text{Tangent of } A = \tan A = \frac{\text{side opposite angle } A}{\text{side adjacent angle } A} = \frac{a}{b}$$

$$\text{Sine of } A = \sin A = \frac{\text{side opposite angle } A}{\text{hypotenuse}} = \frac{a}{c}$$

$$\text{Cosine of } A = \cos A = \frac{\text{side adjacent angle } A}{\text{hypotenuse}} = \frac{b}{c}$$

$$\text{Cotangent of } A = \cot A = \frac{\text{side adjacent angle } A}{\text{side opposite angle } A} = \frac{b}{a}$$

$$\text{Cosecant of } A = \csc A = \frac{\text{hypotenuse}}{\text{side opposite angle } A} = \frac{c}{a}$$

$$\text{Secant of } A = \sec A = \frac{\text{hypotenuse}}{\text{side adjacent angle } A} = \frac{c}{b}$$

We notice that cotangent, cosecant, and secant are reciprocals of tangent, sine, and cosine, respectively. For this reason they are used less often than the first three functions. In our work we will use only the tangent, sine, and cosine.

We must remember that these functions are defined for an acute angle by means of right triangles. The angles of a triangle which is not a right triangle also have sines, cosines, and tangents, but we cannot compute them directly from the sides of the triangle. We must drop perpendiculars and create right triangles before the preceding definitions apply.

▲ EXERCISES

Write out the value of the sine, cosine, and tangent for the following:

1. Angle B in the triangle ABC given in Fig. 37.
2. Angles A and B in the triangle in Fig. 38.

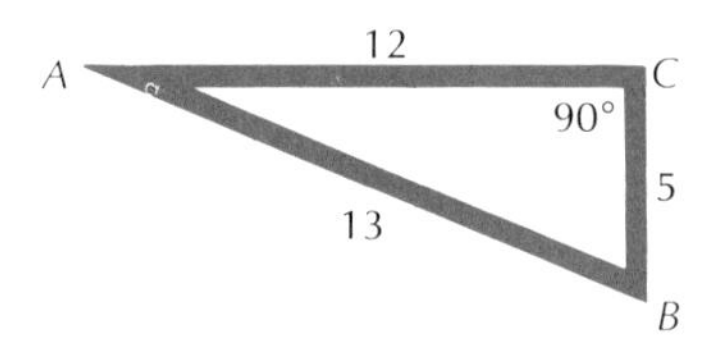

FIG. 38

3. Angles M and N in the triangle in Fig. 39.

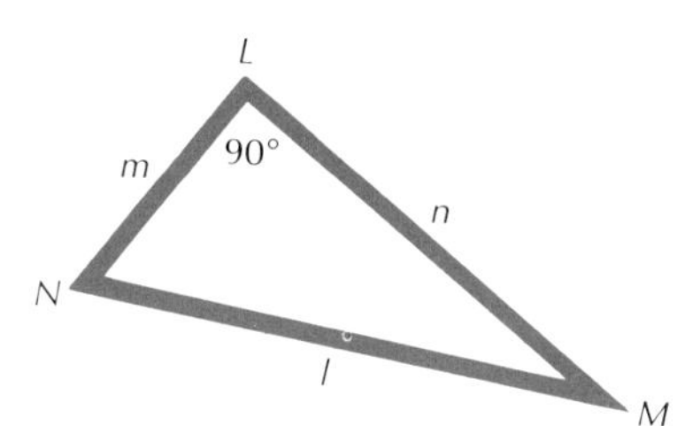

FIG. 39

4. Angles P and R in the triangle in Fig. 40.

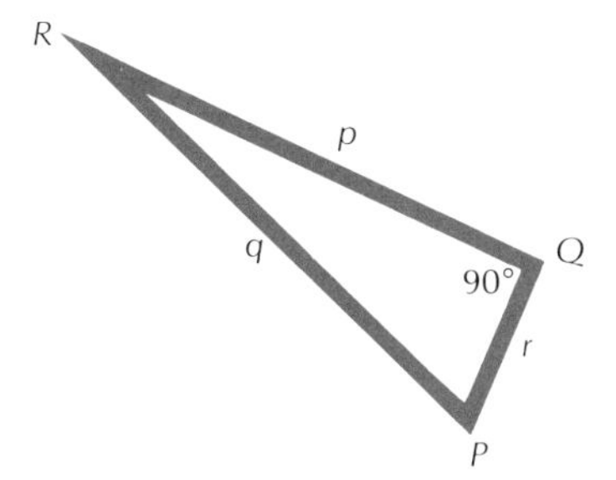

FIG. 40

5. Angles A and C in the triangle in Fig. 41.

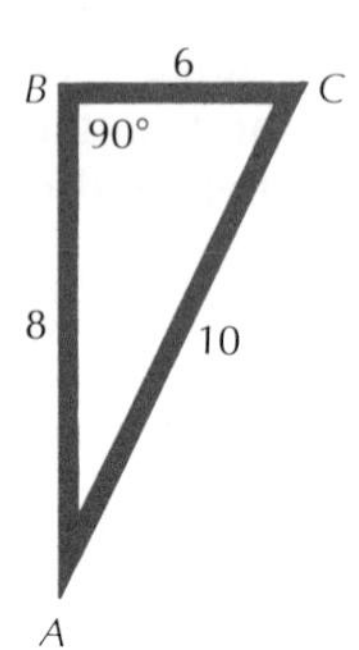

FIG. 41

Problems just for fun

Six coins may be changed from the arrangement shown at the left to the circle shown at the right in four moves. A proper move consists of sliding one coin along the table (no coins may be picked up) to a new position where it touches two other coins. Can you do it?

CHAPTER NINE

General Right Triangles

The utmost care must be taken to avoid errors, and that it is taken is proved by the wonderful accuracy with which the headings driven from opposite ends of the Musconetcony tunnel meet. The tunnel is about 5,000 feet long. When the headings met, the error in alignment was found to be only half an inch, and the error in level only about one-sixth of an inch. A. WILLIAMS

1 *Tables and Trigonometry*

We must find the value of sine, cosine, and tangent for all values of the angle between 0° and 90°. If we have such a complete table we shall be able to solve completely any right triangle. If we are given one side and one acute angle of a right triangle, we can find the other two sides and the other acute angle. If we are given two sides of a right triangle, we can find the third side and the ratios of the sides. From the ratio of the sides we can find in the table the value of the angle for which this ratio corresponds. We shall discuss this idea in more detail shortly.

With such a table we have the means of solving any right triangle by numerical methods. The name of this mathematics is *trigonometry.* The word "trigonometry" comes from Greek words *trigonon,* which means "triangle," and *metron,* which means "to measure." From the meaning of the word, we see that trigonometry studies the relations between the sides and angles of the triangle and gives methods of deducing from the given parts the unknown sides and angles of a triangle. The early use of trigonometry was to indirectly measure inaccessible distances. We can use trigonometry to measure the heights of buildings and inaccessible mountains such as Mount Everest, to determine the position of ships at sea, to navigate an airplane by the aid of stars, or to measure the length of a tunnel through a mountain or under a river.

Plane geometry enables us to *construct* the triangle, given three parts; while trigonometry enables us to *compute* these unknown parts of the triangle. From plane geometry we know that lengths of 3, 4, and 5 feet form a right triangle. But how large are the acute angles so formed? Geometry has no answer. We must turn to trigonometry for the solution.

While trigonometric functions were developed for and first used in the solutions of triangles, these functions have many uses in other branches of mathematics and fields of science. For example, any study of alternating-current machinery is full of trigonometric functions.

Constructing a Table

Before we proceed further, we must construct a table of sines, cosines, and tangents of all the angles between 0° and 90°. To make such a table, we shall use the method of Hipparchus (about 140 B.C.). We draw a right triangle in which one of the acute angles A is the desired angle. We measure the three sides and calculate the value of the three functions, sine A, cosine A, and tangent A, from the definition of these trigonometric functions. But we realize that the accuracy of our results will depend upon the care with which we draw the triangle and how accurately we can measure the length of the sides of the triangle.

Since we have at our disposal accurately constructed graph paper, we shall use it to aid us in measuring the length of the sides. Select a piece of graph paper. Using the lower left-hand corner as

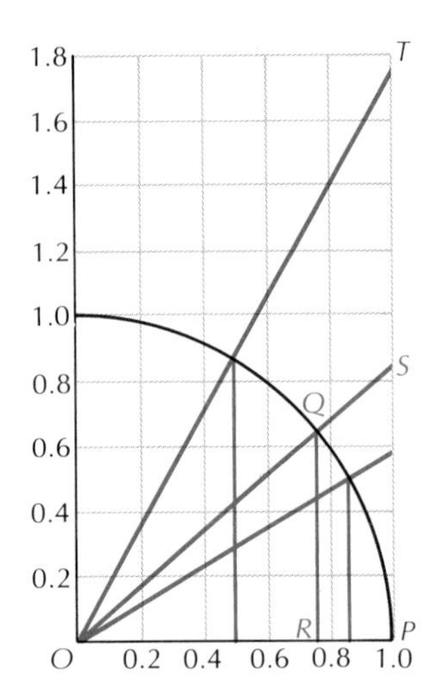

FIG. 42

origin, carefully draw a circle with radius equal to 10 of the large divisions (see Fig. 42). If we consider each one of these divisions equal to $\frac{1}{10}$, the radius of the circle is equal to 1. We wish to find the value of the three trigonometric functions for an angle of 40°. We lay off this angle at 0 as carefully as possible and complete the right triangle QOR by drawing the vertical line QR. Triangle QOR is the right triangle with angle QOR equal to 40° and the hypotenuse OQ equal to 1. By definition,

$$\sin 40° = \frac{QR}{OQ} = \text{length of } QR$$

$$\cos 40° = \frac{OR}{OQ} = \text{length of } OR$$

Thus we see that in order to find the value of the sine and cosine for any acute angle, we draw with the aid of a protractor the angle POQ on the graph. The height of the point Q gives the value of the sine of the angle, and we see that sin 40° = 0.64. The x coordinate of the point Q gives the value of the cosine of the angle, and we read from the graph, cos 40° = 0.77.

At the point P draw the vertical line PT. Extend the line OQ to S, thus forming the right triangle POS in which the side PO is equal to 1. Then, using the right triangle POS, we may write

$$\tan 40° = \frac{PS}{OP} = PS = y \text{ coordinate of the point } S$$

From the graph we read tan 40° = 0.84. Of course, we could find the tangent of the angle by dividing the ordinate of Q by the abscissa of Q, but it is easier to use a triangle so that the denominator is equal to 1.

▲ EXERCISES

1. Use the method outlined above to find the value of sine, cosine, and tangent for the angles 10°, 20°, 30°, 40°, 50°, and 60°, as accurately as possible.
2. What can you say about the size of the trigonometric functions for angles close to 0° and 90°?
3. What value would you assign to the following functions: sin 0°, cos 0°, tan 0°, sin 90°, and cos 90°?

Use the results of Exercise 1 to solve the following problems:

4. When the sun's angle of elevation is 20°, a tree casts a shadow 132 feet long. How high is the tree?
5. From the top of a water tower, the angle of depression of a water hy-

drant is 50°. How high is the water tower if the hydrant is 300 feet from the base of the tower?

6. The diagonal of a rectangle makes an angle of 40° with the length of the rectangle. If the rectangle is 13.2 feet long, find its width and the length of the diagonal.
7. The top of a ladder makes an angle of 20° with a wall. If the base of the ladder rests on the level ground 6 feet from the wall, how long is the ladder and how high up the wall does it reach?

Mr. Stupid did a division problem this way $\frac{1\not{6}}{\not{6}4}=\frac{1}{4}$. The answer is correct, but Teacher thought the method just suited Mr. Stupid. Can you find two other numbers which Mr. Stupid can divide and get the correct answer?

3 *Functions of 0° and 90°*

The values of the sine, cosine, and tangent of angles close to 0° and 90° are easily obtained from the definitions and Fig. 42. Now, if we make the angle ROS smaller and smaller until it is nearly zero, the point R is very near P. At $\cos ROS = OR$, we see that the cosine of a very small angle is nearly 1 and becomes closer as the angle gets smaller. Thus it seems natural to write

$$\cos 0° = 1$$

Similarly, as the angle approaches zero the distances QR and PS approach zero, and we define quite naturally

$$\sin 0° = 0 \qquad \cos 0° = 1 \qquad \tan 0° = 0$$

Likewise, if we make the angle larger and larger until it is almost 90°, the y coordinate of the point Q approaches 1, and the x coordinate of the point Q approaches zero, and we take

$$\sin 90° = 1 \qquad \text{and} \qquad \cos 90° = 0$$

Let us try to find a value for tan 90°. From Fig. 42 we see that the height of the point S as the angle gets close to 90° becomes larger and larger. In fact, when the angle becomes 90°, the line OQ is parallel to the line PT, and it is impossible to find the point of intersection of the two lines OQ and PT. We are not able to determine the distance PS. We conclude that $\tan 90° = PS$ does not exist. That is, we are not able to assign any definite value to tan 90°.

4 *Functions of 30°, 45°, and 60°*

Using two theorems from plane geometry, we can find the value of three special angles without tables.

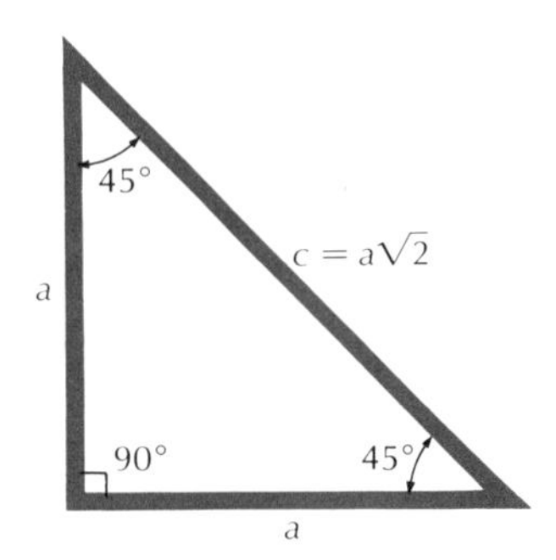

FIG. 43

We construct a right triangle in which the two acute angles are equal to 45° (Fig. 43). We know that the triangle is isosceles and that two sides are equal since the angles are each 45°. We may find the hypotenuse c by the Pythagorean theorem

$$c^2 = a^2 + a^2 = 2a^2$$
$$c = \sqrt{2}\, a$$

Now we write the values of the three functions from Fig. 43.

$$\tan 45° = \frac{\text{side opposite}}{\text{side adjacent}} = \frac{a}{a} = 1$$

$$\sin 45° = \frac{\text{side opposite}}{\text{hypotenuse}} = \frac{a}{a\sqrt{2}} = \frac{1}{\sqrt{2}} = 0.707$$

$$\cos 45° = \frac{\text{side adjacent}}{\text{hypotenuse}} = \frac{a}{a\sqrt{2}} = \frac{1}{\sqrt{2}} = 0.707$$

Next we construct a right triangle with angles 30° and 60° (Fig. 44). One theorem of plane geometry states that, if one acute angle

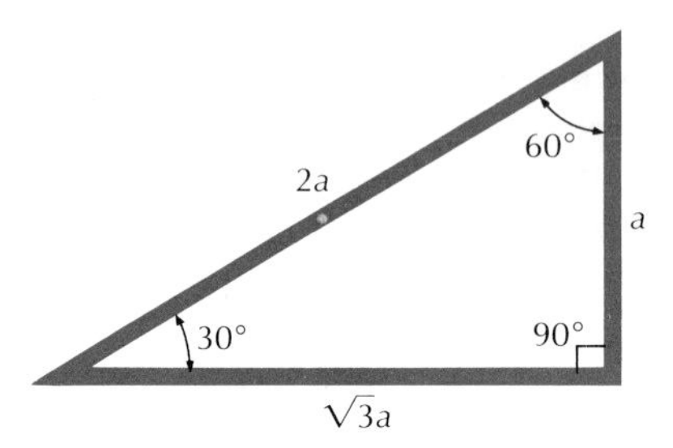

FIG. 44

is half the other, the hypotenuse is twice the shortest side. By the Pythagorean theorem, we see that the third side is $\sqrt{3}\,a$.

Now we write the values of the trigonometric functions from Fig. 44.

$$\sin 30° = \frac{\text{side opposite}}{\text{hypotenuse}} = \frac{a}{2a} = \frac{1}{2} = 0.500$$

$$\sin 60° = \frac{\text{side opposite}}{\text{hypotenuse}} = \frac{\sqrt{3}\,a}{2a} = \frac{\sqrt{3}}{2} = 0.866$$

$$\cos 30° = \frac{\text{side adjacent}}{\text{hypotenuse}} = \frac{\sqrt{3}\,a}{2a} = \frac{\sqrt{3}}{2} = 0.866$$

$$\cos 60° = \frac{\text{side adjacent}}{\text{hypotenuse}} = \frac{a}{2a} = \frac{1}{2} = 0.500$$

$$\tan 30° = \frac{\text{side opposite}}{\text{side adjacent}} = \frac{a}{\sqrt{3}\,a} = \frac{1}{\sqrt{3}} = 0.577$$

$$\tan 60° = \frac{\text{side opposite}}{\text{side adjacent}} = \frac{\sqrt{3}\,a}{a} = \sqrt{3} = 1.732$$

Functions of Complementary Angles

For the right triangle ABC shown in Fig. 45 we may write

$$\sin A = \frac{a}{c} \qquad \sin B = \frac{b}{c}$$

$$\cos A = \frac{b}{c} \qquad \cos B = \frac{a}{c}$$

$$\tan A = \frac{a}{b} \qquad \tan B = \frac{b}{a}$$

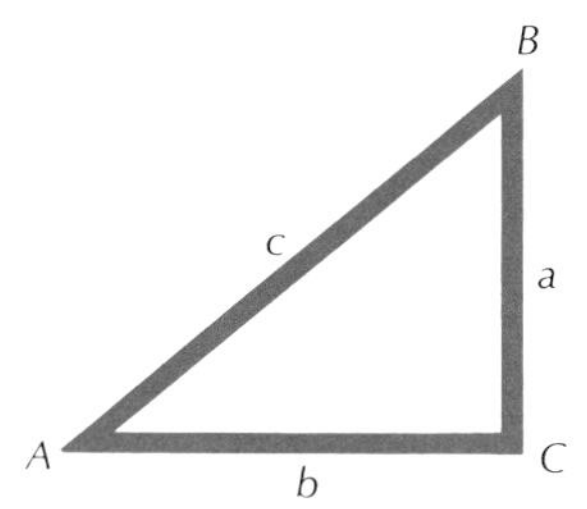

FIG. 45

We observe that $\sin A = \cos B$ and $\cos A = \sin B$. Also

$$\tan A = \frac{\sin A}{\cos A} \qquad \text{and} \qquad \tan A = \frac{1}{\tan B}$$

6 *Four-place Tables of Trigonometric Functions*

The early scholars at the University of Alexandria in Egypt started to make tables of the ratios of the sides of right triangles. Their values for these ratios were not too accurate, because they did not have an accurate table of square roots and a decimal system of fractions. We can say, however, that their development of trigonometry brought back measurement to geometry. As a need for greater accuracy developed, the mathematicians improved their tables to keep pace with the other sciences.

The values of the sines, cosines, and tangents of angles between 0° and 90° were computed first for every degree, then for every 10 minutes, later for every minute, and at present we have tables for every second to four, eight, and even twelve decimal places. These tables were computed by the use of special infinite series.

For our purpose, trigonometric tables giving the values of sine, cosine, tangent, and cotangent for every 10 minutes between 0° and 90° to four places will be accurate enough. Such a table will be found at the back of the book (Table II). Some time should be devoted to a study of this table in order to learn how to use it correctly and rapidly.

Looking at the table, we note that we can read from the top downward to get the sine, tangent, cotangent, or cosine of angles between 0° and 45°. Likewise, we can read from the bottom up to get the value of the same functions from 45° to 90°. We encounter no difficulty in finding that $\sin 28°20' = 0.4746$ and $\cos 13°30' = 0.9724$. Check these results for yourself.

▲ EXERCISES

Use the table to find the sine, cosine, and tangent of each of the following angles:

1. 20°40′
2. 57°10′
3. 86°30′
4. 8°50′
5. 65°20′
6. 72°
7. 13°10′
8. 42°20′
9. 72°40′
10. 7°

Problems just for fun

A rabbit is 60 leaps ahead of a dog which is chasing it. The rabbit can take 3 leaps while the dog is taking 2, but 3 of the dog's leaps equal 7 of the rabbit's. How many leaps will the dog take before it overtakes the rabbit?

To Find the Angle from the Table

Sometimes we know the value of the trigonometric function and need to find the angle which has this value. For example, to find the value of angle A if $\sin A = 0.5495$, we look in the sine column of the table until we find the value 0.5495. We find that $\sin 33°20' = 0.5495$. Then angle A is equal to 33°20′. In many cases we will not find the exact value in the table. For example, if we want the value of angle A for which $\tan A = 1.7596$, we look in the tangent column of the table and do not find the number 1.7596. However, we see that $\tan 60°20' = 1.7556$ and $\tan 60°30' = 1.7675$, and our value lies between these two values. We take 60°20′ for the value of A because 1.7596 is closer to 1.7556 than to 1.7675. There are methods by which we can determine more closely the value of the angle. We will not discuss these methods here, because usually the accuracy of our measurements does not warrant the determination of an angle closer than 10 minutes.

▲ EXERCISES

Find the angle to the nearest 10 minutes when the value of the trigonometric function is given.

1. $\sin A = 0.1132$
2. $\cos A = 0.9727$
3. $\sin A = 0.5195$
4. $\sin A = 0.9920$
5. $\tan A = 0.4210$
6. $\tan A = 1.3934$
7. $\cos A = 0.2720$
8. $\tan A = 2.7750$
9. $\cos A = 0.6614$

8 A Practical Problem

Let us use the facts that we have learned about trigonometry to solve a practical problem.

■ **Example.** From the top of a lighthouse 100 feet above the water, the angles of depression of two ships in the same straight line with the lighthouse are 15°30′ and 37°50′, respectively. What is the distance between the two ships, and how far are they from the point directly below the lighthouse?

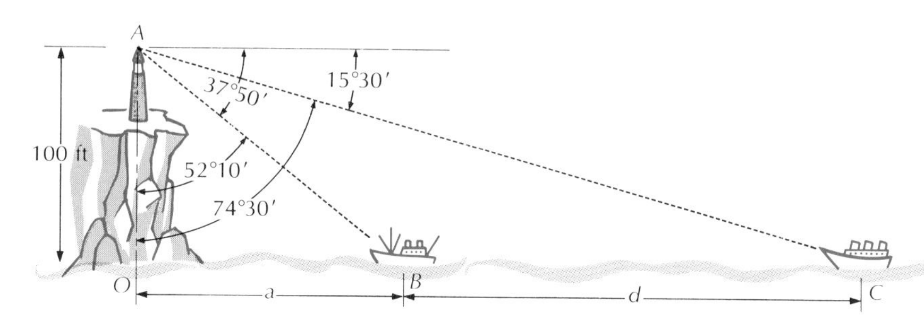

FIG. 46

□ ***Solution.*** First let us draw a diagram (Fig. 46). From triangles OAB and OAC we get

$$\tan 52°10' = \frac{a}{100} \qquad \text{and} \qquad \tan 74°30' = \frac{a+d}{100}$$

From the first equation we find that

$$a = 100 \tan 52°10' = 100(1.2876) = 129 \text{ feet}$$

From the second equation we get

$$a + d = 100 \tan 74°30' = 100(3.6059) = 361 \text{ feet}$$

Then

$$d = 361 - a = 361 - 129 = 232 \text{ feet}$$

Problems just for fun

Three boats going upstream meet three boats coming downstream in a river with a channel so narrow that two boats cannot pass. However, at the point where they meet there is a side basin just large enough to hold one boat. Can you figure out how the six boats manage to pass one another?

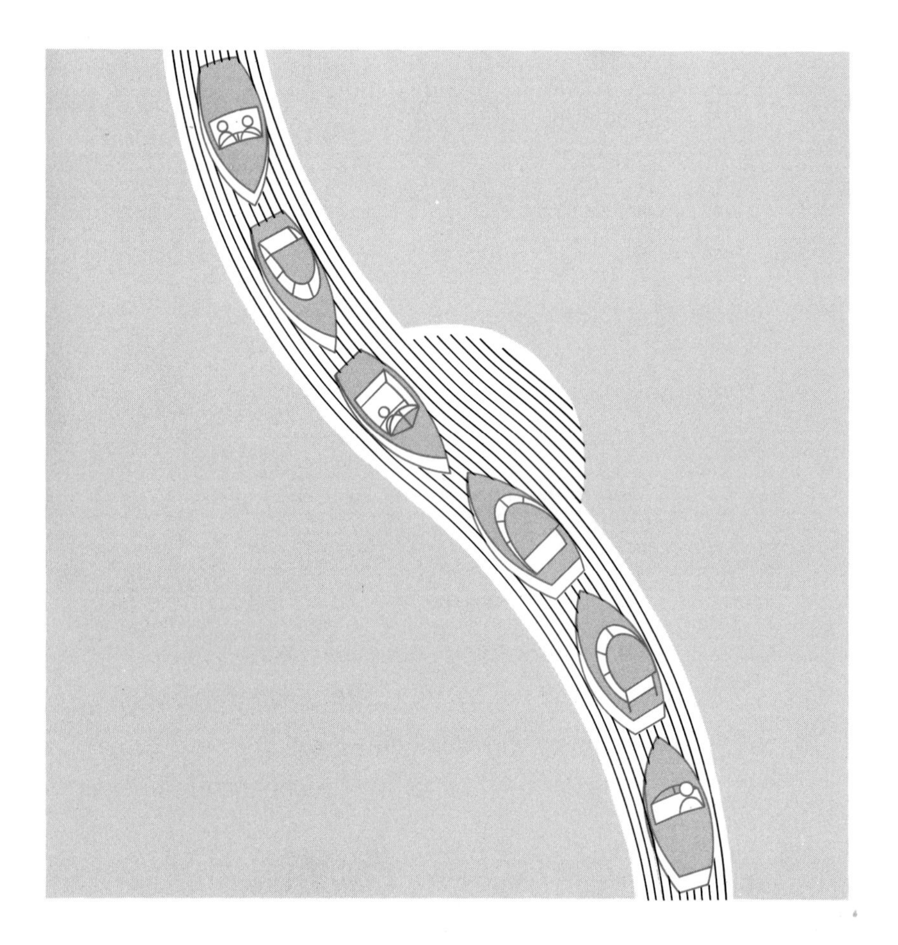

▲ EXERCISES

Use the trigonometric tables to solve the following problems:

1. A tree 60 feet high casts a shadow of 70 feet. Find the angle of elevation of the sun.
2. How far is a surveyor from a building 700 feet high if the angle of elevation of the top of the building is 53°10′?
3. If a road rises steadily a distance of 210 feet in a horizontal distance of 1 mile, find the angle the road makes with the horizontal.
4. A boy flying a kite estimates that the angle of elevation of the kite is 60° when the kite is directly above a telephone pole which is 200 feet away. How high is the kite, and how many feet of string has the boy played out?
5. An iron wedge used for splitting trees has a base 3 inches wide and a vertex angle of 15°. Find the length of the edges of the wedge.
6. From a forest lookout tower perched on a cliff 200 feet above a lake, the angle of depression of a small blaze across the lake is 14°30′. How far is the blaze from the foot of the cliff? How far is the blaze from the ranger?
7. Find the area of a parallelogram if the angle between sides 8 inches and 12 inches long is 37°50′.
8. The pitch of a roof gable is the ratio of its height to its entire width, i.e., CD/AB. If the pitch of the roof is $\frac{2}{3}$, what is the size of angle A? See Fig. 47.

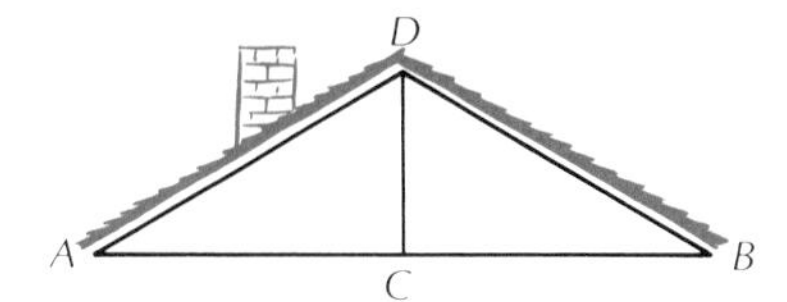

FIG. 47

9. Two ranger lookout stations are on hills which are the same height and 4.8 miles apart. When an airplane is directly above one station, the pilot finds the angle of depression of the second station to be 28°40′. How high is the airplane above the ranger stations?
10. A field-artillery gun has a range of 18 miles. If the gun is fired in the direction which is 20° north of west, how many miles north and how many miles west does the artillery shell go?
11. Lots are laid out by lines perpendicular to Main Street and running through to First Avenue as shown in Fig. 48. If the angle between the streets is 30°, find the frontage on First Avenue and the depth of each lot.
12. An airplane climbs 820 feet while traveling 1 mile west, then climbs 500 feet while traveling a second mile in the same direction. What angle of climb could the pilot have used in a steady climb to have gained the same total height of 1,320 feet in 2 miles?

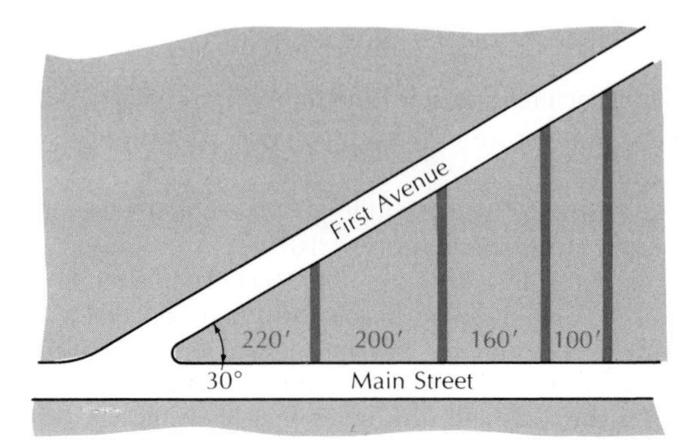

FIG. 48

13. A 12-foot ladder rests against a wall at a point 10 feet above the ground. Find the angles the ladder makes with the wall and the ground.
14. If the ladder in Exercise 13 slips so that the top is 8 feet above the ground, find the angle of inclination of the ladder and the distance from the foot of the ladder to the wall.
15. How far must the foot of the ladder in Exercise 13 be pulled out if the angle of inclination of the ladder is to be 36°?
16. The summit of a mountain is seen at an angle of elevation of 23°20′ from a camp whose altitude is 5,280 feet. If the height of the mountain is 11,450 feet, what is the air-line distance from the camp to the summit of the mountain?
17. A ship sails a distance of 50 miles in a direction 20° north of east and then sails 60 miles straight north. What is the distance of the ship from its starting point?
18. Two guy wires are fastened to a vertical pole at a height of 40 feet. The longer wire is 80 feet long, and the shorter wire is 60 feet. If both wires are anchored in the ground due west of the pole, find the distance between the lower ends of the wires.
19. Two church steeples are 150 feet and 120 feet high and are 500 feet apart. If a line were drawn joining the tops of the steeples, what would be the angle of inclination of this line, and how far from the two churches would it meet the ground?
20. Two city streets meet at an angle of 30°. The fire department arrives at a fire which is 200 feet from the intersection and discovers that the only fire hydrant is around the corner on the other street 100 feet from the intersection. What is the shortest length of hose needed, assuming that there are no buildings so that the hose can cut across lots?

Problems just for fun

Pete Smith, Sr., who was anxious to encourage Pete Smith, Jr., in this course, agreed to award him \$6 for every problem he solved correctly but fined him \$4 for every incorrect or unsolved problem. Junior did the preceding group of problems and found he had just broken even. How many did Junior solve correctly?

9 *Distance of the Moon from the Earth*

Aristarchus (310–250 B.C.) made the first estimate of the relative distances of the moon and the sun from the earth. Hipparchus (about 150 B.C.) made a table of sines and determined the distance of the moon from the earth. Hipparchus also made maps of the stars in latitude and longitude.

The method used by Hipparchus to calculate the distance of the moon from the earth is essentially as follows:

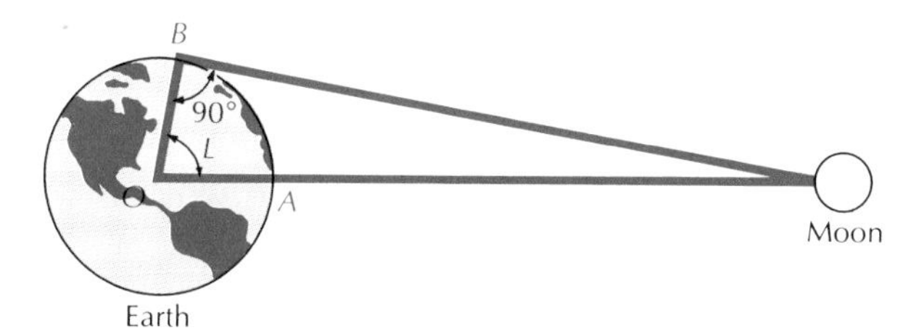

FIG. 49

An observer at A (Fig. 49) observes the moon when it is directly overhead, and at exactly the same instant an observer at B observes the moon when it rises on the horizon. If A and B are in the same latitude but are separated by L degrees of longitude, we may write

$$\cos L = \frac{OB}{OM} = \frac{\text{radius of the earth}}{\text{distance to the moon}}$$

If we know that $L = 89\frac{1}{16}°$ and find the value of $\cos 89\frac{1}{16}°$ from a table, we can solve for OM for the distance to the moon, which is approximately 245,000 miles.*

Now, knowing the moon's distance, it is easy to determine its diameter and circumference. We need only measure the angle sub-

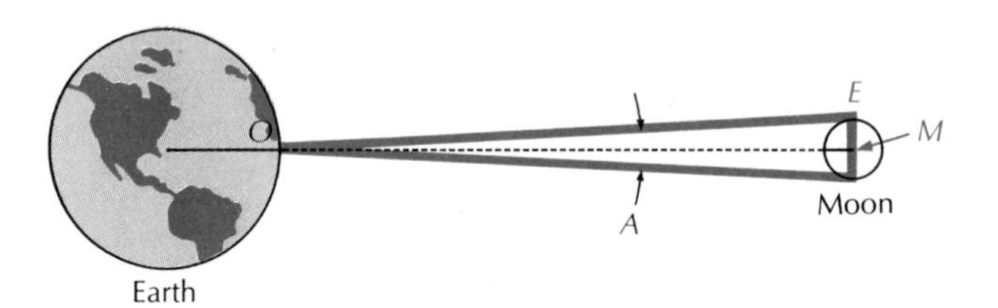

FIG. 50

tended by opposite spots on the boundary of a full moon when it is directly overhead (Fig. 50). We can write

$$\tan \frac{A}{2} = \frac{EM}{OM} = \frac{\text{radius of the moon}}{\text{moon's distance from the earth}}$$

If $A = \frac{1}{2}°$ and $\tan \frac{1}{4}° = 0.0044$, using $OM = 245{,}000$ miles, we find that $EM = 1{,}078$ miles. The best modern measurements of the moon's radius give 1,081 miles. We can compute the circumference of the moon as 2π times the radius.

It is now apparent that if we wish to measure astronomical distances, we are going to need tables of trigonometric functions for very small angles. Various formulas have been devised to obtain very accurately the values of sines, cosines, and tangents of angles close to zero.

Archimedes' Method for Finding the Value of π

But how much did the ancients know about the value of π, which is the symbol in mathematics for the ratio between the lengths of the circumference and diameter of a circle? More than two centuries before Christ, the Greek scientist Archimedes (287–212 B.C.) developed a method for calculating the value of π to any desired accuracy.

Consider a circle of radius 1 with a regular polygon circumscribed about it (Fig. 51). From a triangle RQO we can write

$$\tan A = \frac{RQ}{RO} = RQ$$

*Lancelot Hogben, "Mathematics for the Million," p. 247, W. W. Norton & Company, Inc., New York, 1937.

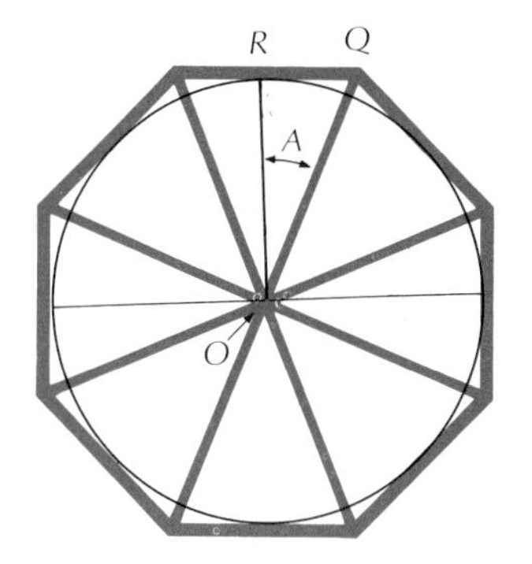

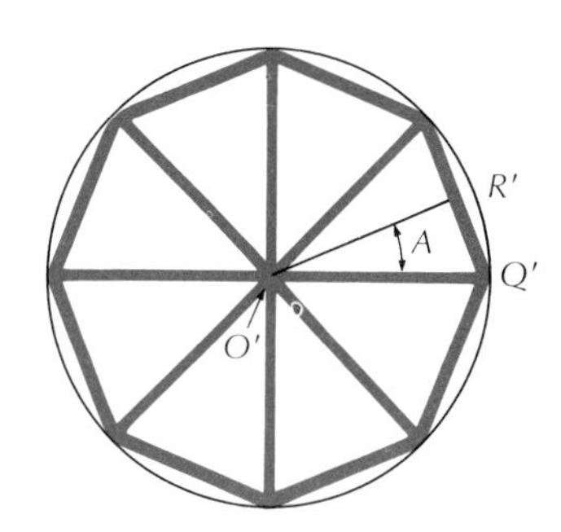

FIG. 51

because the radius of the circle is equal to 1. If the polygon has n sides, the angle $A = \frac{360°}{2n}$ and

$$RQ = \tan \frac{360°}{2n}$$

The perimeter P of this circumscribed regular polygon becomes

$$P = 2n \tan \frac{360°}{2n}$$

Let us consider the regular polygon of n sides inscribed inside this circle with unit radius. For this case, we may write

$$\sin A = \frac{R'Q'}{O'Q'} = R'Q'$$

and $A = \frac{360°}{2n}$ so that

$$R'Q' = \sin \frac{360°}{2n}$$

In this case the perimeter P' of the inscribed regular polygon becomes

$$P' = 2n \sin \frac{360°}{2n}$$

It seems obvious that the circumference of the circle is less than P, the perimeter of the circumscribed polygon, and greater than P', the perimeter of the inscribed polygon. That is,

$$P > 2\pi \text{ radius} > P'$$

$$2n \tan \frac{360°}{2n} > 2\pi > 2n \sin \frac{360°}{2n}$$

or

$$n \tan \frac{360°}{2n} > \pi > n \sin \frac{360°}{2n}$$

If the polygon has six sides, this last inequality becomes

$$6 \tan 30° > \pi > 6 \sin 30°$$
$$6(0.577) > \pi > 6(\tfrac{1}{2})$$
$$3.462 > \pi > 3$$

We find that the value of π must be between 3.462 and 3. If we take the average of 3.462 and 3, we may write that the value is $\pi = 3.231$, with an error of at most 0.231.

It is clear that, if we increase n, the number of sides of the inscribed and circumscribed polygons, we may find the value of π as accurately as we wish. Archimedes used for π a number lying between $3\frac{1}{7}$ and $3\frac{10}{71}$. To get π more accurately, we see that Archimedes needed a table of sines and tangents. No record has been found that such a table existed before the tables compiled by Hipparchus.*

By modern methods the value of π has been computed to hundreds of decimal places. Below we give the value to 30 places, together with a little rhyme which has been used to remember these figures.

$\pi = 3.141\ 592\ 653\ 589\ 793\ 238\ 462\ 643\ 383\ 279\ldots$

3 1 4 1 5 9
Now I, even I, would celebrate
2 6 5 3 5
In rhymes inapt, the great
8 9 7 9
Immortal Syracusan, rivaled nevermore,
3 2 3 8 4
Who in his wondrous lore,
6 2 6
Passed on before,
4 3 3 8 3 2 7 9
Left men his guidance how to circles mensurate.

A. C. Orr

Review Exercises

1. A triangle with sides 5 feet, 12 feet, and 13 feet is a right triangle. Why? Find the sine, cosine, and tangent of the angle opposite the 12-foot side.

*For a table of values of π used by the Babylonians, Hebrews, Chinese, Egyptians, Hindus, Arabs, Japanese, and Europeans, see Hogben, *op. cit.*, p. 261.

2. Find the sine, cosine, and tangent of the angle opposite the 5-foot side.
3. Show that the area of a right triangle (Fig. 52) is given by the formula

$$\text{Area} = \tfrac{1}{2}c^2 \tan B$$

4. A guy wire 300 feet long extends from the top of a radio tower to an anchor on level ground. If the wire makes an angle of 45° with the ground, how high is the tower?
5. A motorboat travels 10 miles an hour due east for 2 hours and then goes due north for 3 hours at 15 miles per hour. What will be its direction from its starting point?

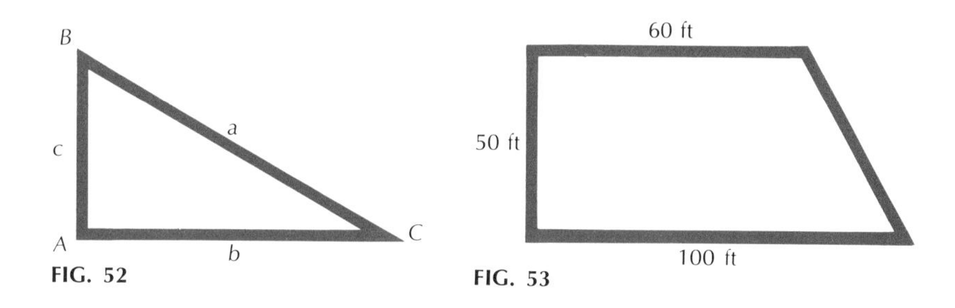

FIG. 52 FIG. 53

6. A city lot is in the shape of a right trapezoid (Fig. 53). If the width is 50 feet and the two parallel sides are 60 and 100 feet, find the length of the fourth side and the angles which it makes with the two parallel sides.
7. From a window on one building, an ornament on another building makes an angle of elevation of 30° and the roof of the building makes an angle of elevation of 45°. If the ornament is known to be 12 feet below the roof, what is the distance between the buildings?
8. Using the Archimedes method for finding the value of π, find the two numbers between which π lies when we use a polygon of 20 sides.

CHAPTER TEN

General Triangles

The student of mathematics often finds it hard to throw off the uncomfortable feeling that his science, in the person of his pencil, surpasses him in intelligence—an impression which the great Euler confessed he often could not get rid of. ERNST MACH

1 *A General Triangle*

If we reflect a bit, we realize that in the preceding chapters we have been solving problems that involved right triangles only. Of course, many problems involve triangles in which none of the angles are right angles. So we must roll up our sleeves and face this situation.

Suppose we have the following problem: Two towers at points A and B are viewed from a point C which is 100 feet from tower A and 400 feet from tower B. The angle ACB is 66°. How far are the towers apart? We make a rough sketch (Fig. 54), and we find that the tri-

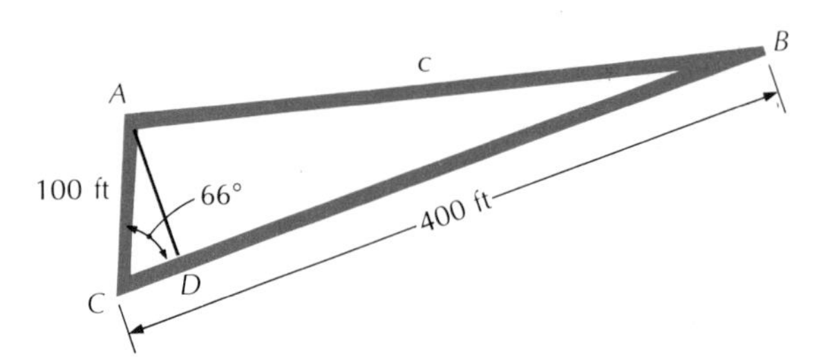

FIG. 54

angle ABC is not a right triangle. Of course, we could make our drawing carefully and use the graphical method to determine the distance, but this is time-consuming and not very accurate.

Since we have been working with right triangles, it is natural to make two right triangles *CAD* and *ADB* by dropping the perpendicular *AD* upon the side *CB*. From the right triangle *ACD* we can easily determine the distances *AD* and *CD*. Subtracting the length of *CD* from 400 will give us *BD*, and we can find the length *AB* in the right triangle *ABD* by the *square of the hypotenuse* method. But this is also tedious and we hope to find a way or a formula that combines in a general result the steps taken in the preceding example.

The Law of Cosines

Let us consider the triangle *ABC*, where angles *A* and *C* are acute (Fig. 55). From the vertex *B*, drop the perpendicular *BD* upon the side *AC*,

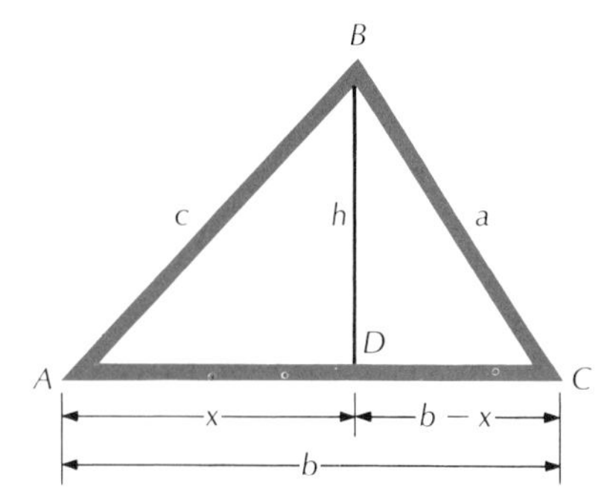

FIG. 55

forming the two right triangles *ABD* and *BCD*. Let $BD = h$ and $AD = x$. Then $DC = b - x$. From triangle *ABD*, we may write

$$h^2 = c^2 - x^2$$

From triangle *BCD* we may write

$$h^2 = a^2 - (b - x)^2 = a^2 - b^2 + 2bx - x^2$$

Equating these two values for h^2, we obtain

$$c^2 - x^2 = a^2 - b^2 + 2bx - x^2$$

Solving for a^2, we get

$$a^2 = b^2 + c^2 - 2bx$$

In the triangle *ABD* we see

$$\cos A = \frac{x}{c} \qquad \text{or} \qquad x = c \cos A$$

If we replace x by this value, we obtain one of the equations which is known as the *law of cosines,*

$$a^2 = b^2 + c^2 - 2bc \cos A$$

The other two equations are

$$b^2 = a^2 + c^2 - 2ac \cos B$$

and

$$c^2 = a^2 + b^2 - 2ab \cos C$$

These are obtained in a similar manner if the perpendicular h is dropped from angles A and C, respectively.

These three equations are easy to remember in words as follows: *The square of one side of a triangle is equal to the sum of the squares of the other two sides minus twice the product of the two sides times the cosine of the included angle.*

We can use one of these equations to determine the unknown side of a triangle if we know the length of two sides and the included angle. This is exactly the situation in the preceding tower problem, and we can find the distance c between the two towers as follows:

$$\begin{aligned} c^2 &= 100^2 + 400^2 - 2(100)(400) \cos 66^\circ \\ &= 10{,}000 + 160{,}000 - 80{,}000(0.4067) \\ &= 137{,}464 \\ c &= \sqrt{137{,}464} = 371 \text{ feet} \end{aligned}$$

Thus far we have not defined the cosine of an angle greater than 90°. This will be done in the next chapter. With this extended definition the law of cosines holds for triangles containing an obtuse angle.

If the angle A is a right angle, $\cos A = \cos 90^\circ = 0$; then the last term of the law of cosines disappears, leaving

$$a^2 = b^2 + c^2$$

which is our familiar Pythagorean theorem. It seems natural to think of the law of cosines as being the Pythagorean theorem for general triangles.

3 *The Law of Sines*

Returning to the triangle ABC of Fig. 55, we shall develop another useful formula. Using triangle ABD, we may write

$$\sin A = \frac{h}{c} \qquad \text{or} \qquad h = c \sin A$$

From triangle BCD, we may write

$$\sin C = \frac{h}{a} \qquad \text{or} \qquad h = a \sin C$$

Equating these two values of h, we get

$$c \sin A = a \sin C$$

or

$$\frac{a}{\sin A} = \frac{c}{\sin C}$$

In exactly the same way, we may drop a perpendicular line from A to the side BC, forming two triangles. We follow the same line of reasoning and obtain

$$\frac{c}{\sin C} = \frac{b}{\sin B}$$

We combine these two equations into one statement, which is known as the *law of sines,*

$$\frac{a}{\sin A} = \frac{b}{\sin B} = \frac{c}{\sin C}$$

This statement is also easy to remember, for it says that *the ratio of a side of a triangle to the sine of the angle opposite that side is the same for all three sides.*

Like the law of cosines, the law of sines is valid for any triangle. Notice that it reduces to the usual form for the sine of an angle in a right triangle if $A = 90°$. It is used when we know the angles and one side and need to find the other sides of the triangle. Also, if we know two sides and an angle opposite one of them, the law of sines does the job for us.

Suppose we want to know the angle at tower B in Fig. 54. We write

$$\frac{c}{\sin C} = \frac{AC}{\sin B}$$

or

$$\frac{371}{\sin 66°} = \frac{100}{\sin B}$$

$$\sin B = \frac{100 \sin 66°}{371} = 0.2462$$

and

$$B = 14°20'$$

▲ EXERCISES

Use the laws of sines and cosines to solve the following problems:

1. When 200 feet of string has been played out, the angle of elevation of a kite is 43°. Find the height of the kite above the ground.
2. Find the length of the shorter diagonal of a parallelogram if the angle between two sides is 43° and the sides are 8 and 12 feet.
3. A triangular lot has sides 100 feet, 80 feet, and 60 feet. Find the angles at the corners and the area of the lot.
4. Two piers A and B are on opposite sides of a lake. The point C on land is taken a distance of 120 feet from A and 190 feet from B, and the angle between BC and AC is found to be 57°. Find the distance between the two piers.
5. A 10-foot pole placed vertically on a hillside casts a shadow 15 feet straight down the hill. The angle formed by the hill and the line joining the top of the pole with the tip of the shadow is 25°. Find the angle of elevation of the sun. What is the angle between the horizontal and the hill?
6. An observer in a lighthouse notices that one ship is 20 miles in a direction 67° north of west and another ship is 26 miles in a direction 33° north of east. How far apart are the ships?
7. In what direction must an observer on the second ship in Exercise 6 look to see the first ship?
8. Two angles of a triangular lot are 60° and 63°30′. If the largest side is 102 feet, find the length of the other two sides.
9. A man walks 10 miles in a direction 20° north of west and then 5 miles in a direction 30° east of north. How far is he from the starting point?
10. In what direction should the man in Exercise 9 have walked to have reached the same point without changing his course?
11. A man walks due north 300 feet. He turns to his right and walks 250 feet in a straight line. What direction should he walk to return to his starting point, which is 410 feet away?
12. Two angles of a triangle are 97° and 32°. The radius of the inscribed circle is 12 inches. Find the length of the three sides of the triangle. *Hint:* The line from the center of the circle to each vertex bisects the angle.
13. On level ground a man sees the top of a building at an angle of elevation of 8°30′. He walks 100 feet nearer the building and again measures the angle of elevation at 19°. How high is the building?
14. A man crosses the Mississippi River at a point where it is exactly 1 mile wide. If the river flows 3 miles per hour and the motorboat can make 7 miles per hour in still water, at what angle must he steer up stream in order to land exactly opposite? How long will it take to cross?
15. Intelligence reports to the U.S. Fleet Admiral that an enemy task force is located 120 nautical miles 42° east of north and is steaming 39°30′ east of south at 25 knots (1 knot = 1 nautical mile per hour). A destroyer group capable of 40 knots is dispatched to intercept the enemy. In what

direction should they steam and how long will it take to intercept the enemy task force?
Hint: Use the law of sines.

Eight coins lie in a straight row. In any one move we may jump a coin over *exactly* two others, stacked or not, onto the top of the next one. How do you make the moves so that you end with four stacked pairs?

CHAPTER ELEVEN

General Angles

No more impressive warning can be given to those who would confine knowledge and research to what is apparently useful, than the reflection that conic sections were studied for eighteen hundred years merely as an abstract science, without regard to any utility other than to satisfy the craving for knowledge on the part of mathematicians, and that then at the end of this long period of abstract study, they were found to be the necessary key with which to attain the knowledge of the most important laws of nature.

A. N. WHITEHEAD

1 Functions of General Angles

Suppose a man walked 10 miles in a direction 20° north of west and then 5 miles in a direction 40° north of west. How far is he from the starting point? If we make a sketch (Fig. 56), we see that the desired

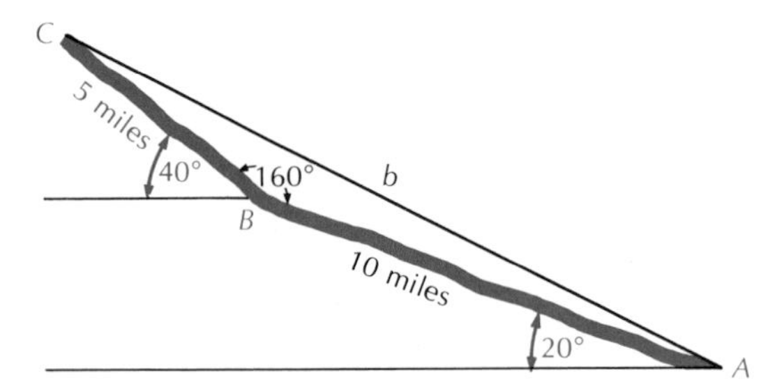

FIG. 56

distance $CA = b$ from the starting point A is opposite an angle of 160°. Using the appropriate form of the law of cosines derived in Sec. 2 of the previous chapter, we write

$$b^2 = 10^2 + 5^2 - 2(5)(10) \cos 160°$$

We can easily find the value of b if we are able to find the value of cos

160°. If we look in our tables, we find that they stop at 90°. We do not want to resort to the graphical method or to break the triangle ABC into two right triangles by dropping a perpendicular from the vertex B upon AC. The simplest thing to do is to find the value of the cosine of 160°.

Angles of Any Magnitude

First, we must extend our idea of angles. Let us draw a set of rectangular axes. We will consider an angle as being in standard position if its vertex is placed at the origin, with one side OP on the positive x axis, the other side OQ falling as shown in Fig. 57.

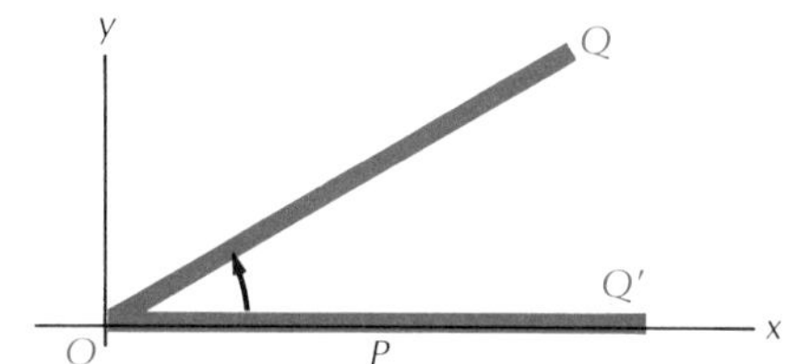

FIG. 57

Now let a line OQ' start coincident with the line OP and rotate about O, coming to rest coincident with OQ. This line generates an angle whose magnitude is determined by the amount of rotation. If the line rotates counterclockwise, the angle is taken as positive; if the line rotates clockwise, the angle is taken to be negative. To obtain positive angles of 70°, 150°, and 250°, start with the line OQ' along the x axis and rotate it counterclockwise 70°, 150°, and 250°,

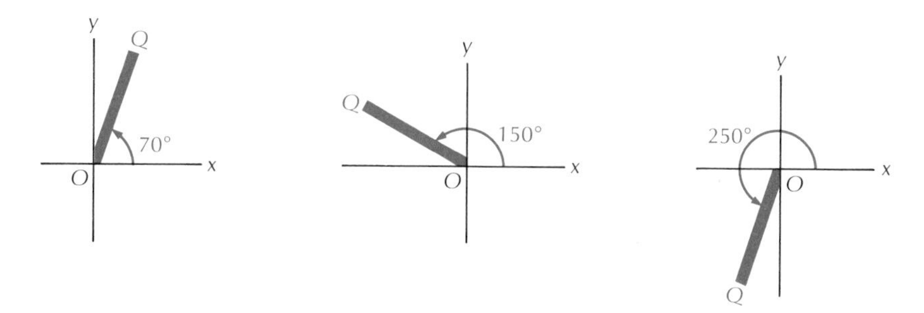

FIG. 58

respectively (Fig. 58). To obtain negative angles of 70°, 150°, and 250°, rotate the line OQ' in the clockwise direction. An arc is usually drawn connecting the initial and terminal sides of the angle. An arrow is placed on the arc to denote the direction of rotation.

We have replaced the idea of an angle being an opening between two lines by an angle being an "amount of rotation." As the line OQ' rotates about O, the angle will increase up to 360° in one revolution. We need not stop with one revolution. If the line rotates through two revolutions, the angle generated is 720°, or twice 360°. In this way we can produce an angle of any size (Fig. 59).

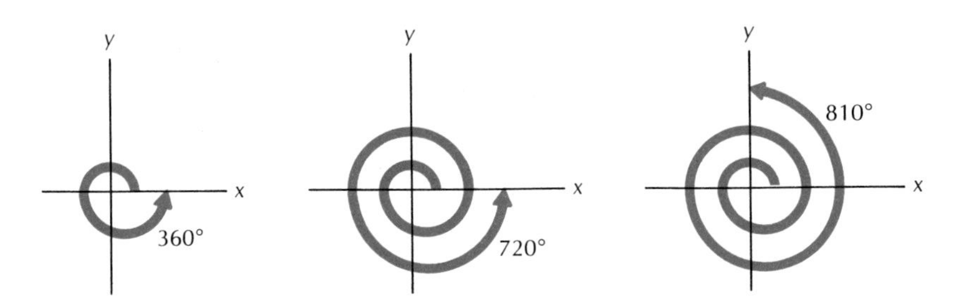

FIG. 59

Such large angles may appear to be unnecessary, but our modern machinery contains many rotating parts (airplane propellers, flywheels, electric fans, etc.), and we often need to know the angle that a part turns through in a minute, which may be many revolutions.

We recall that the values of the trigonometric functions sine, cosine, and tangent for angles between 0° and 90° were found by evaluating the ratios $\frac{\text{side opposite}}{\text{hypotenuse}}$, $\frac{\text{side adjacent}}{\text{hypotenuse}}$, and $\frac{\text{side opposite}}{\text{side adjacent}}$, respectively. In order to extend these definitions to angles greater than 90°, we shall have to decide what is meant by hypotenuse, side opposite, and side adjacent for such angles. We shall agree arbitrarily to

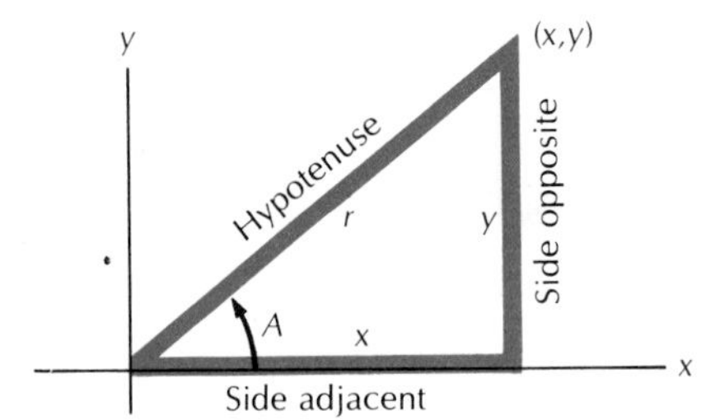

FIG. 60

call x, y, and r (see Fig. 60) the side adjacent, the side opposite, and the hypotenuse. We shall always regard the distance r as positive, but we see from the figures that the distances x and y may be positive or negative, depending upon the position of the terminal side of the angle.

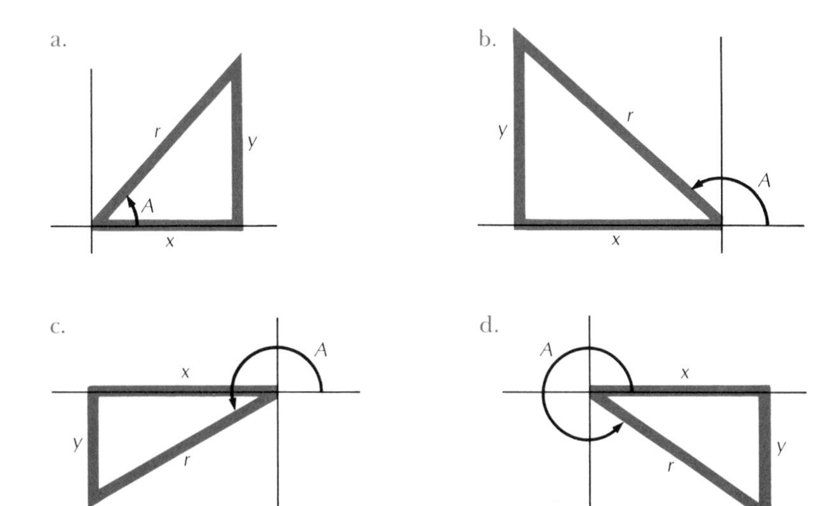

FIG. 61

In order to find the value of the trigonometric functions for angles greater than 90°, we draw the angle carefully and measure its sides x, y, and r (Fig. 61). Then

$$\sin A = \frac{\text{side opposite}}{\text{hypotenuse}} = \frac{y}{r}$$

$$\cos A = \frac{\text{side adjacent}}{\text{hypotenuse}} = \frac{x}{r}$$

$$\tan A = \frac{\text{side opposite}}{\text{side adjacent}} = \frac{y}{x}$$

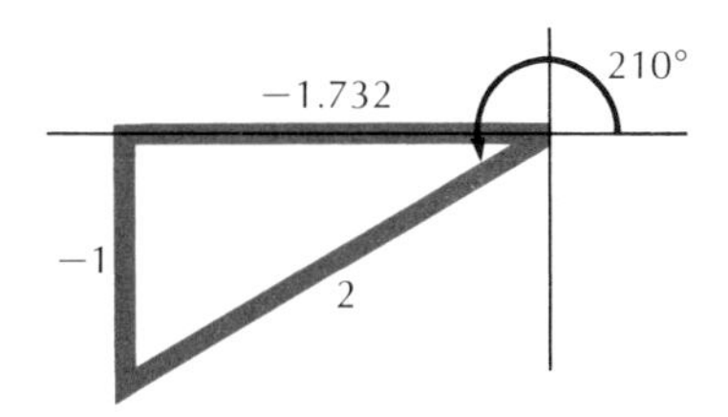

FIG. 62

For example, if angle $A = 210°$, its terminal side lies in the third quadrant, and we draw the diagram as in Fig. 62 and measure the sides x, y, and r. Using the values from the figure and the definitions of the trigonometric functions, we write

$$\sin 210° = -\frac{1}{2} = -0.500$$

$$\cos 210° = \frac{-1.732}{2} = -0.866$$

$$\tan 210° = \frac{-1}{-1.732} = 0.577$$

▲ EXERCISES

1. Write out the value of sine, cosine, and tangent for the angles shown in Fig. 63. Be careful about signs.

a.

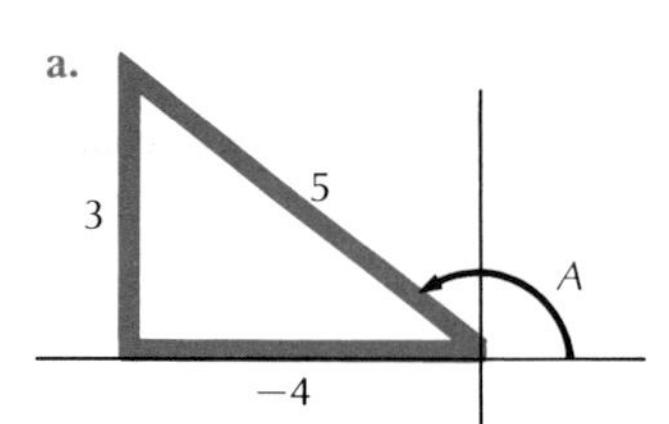

b.

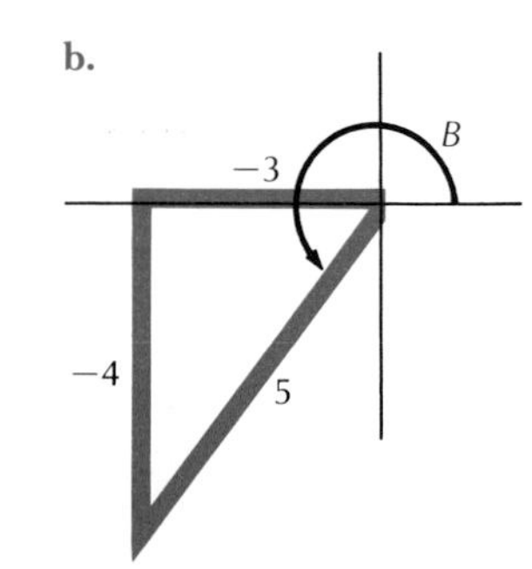

c.

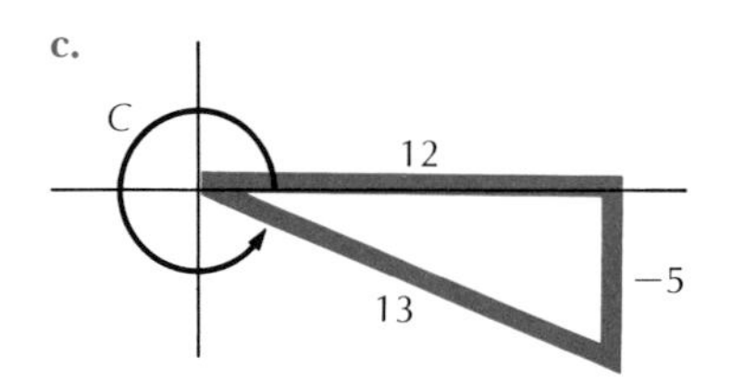

d.

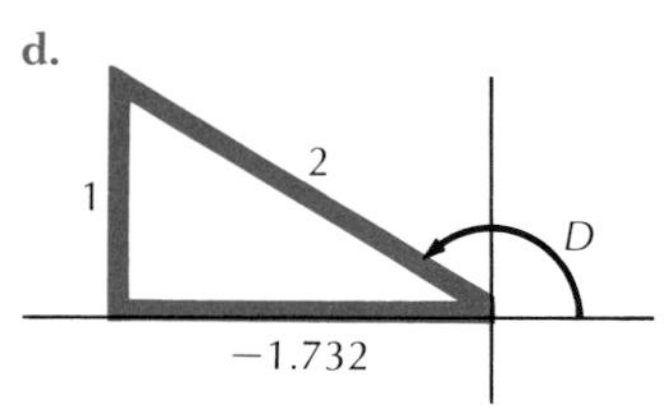

FIG. 63

2. Given each of the following functions, find all possible angles A between −360° and +360°.

a. $\sin A = \frac{1}{2}$ b. $\sin A = -\frac{1}{2}$
c. $\cos A = \frac{1}{2}$ d. $\cos A = -\frac{1}{2}$
e. $\tan A = 1$ f. $\tan A = -1$
g. $\cos A = 1$ h. $\cos A = -1$
i. $\sin A = 1$ j. $\sin A = -1$

3. Given each of the following functions, draw each possible angle which is less than 360°, and write by inspection the value of the two remaining trigonometric functions, sine, cosine, or tangent.

a. $\cos A = \frac{12}{13}$ b. $\sin A = -\frac{1}{3}$
c. $\sin A = \frac{4}{5}$ d. $\cos A = \frac{1}{2}$
e. $\tan A = -\frac{3}{2}$ f. $\tan A = \frac{4}{3}$

4. If A, B, C, and D are the angles shown in Exercise 1, find the value of

a. $\sin A + \cos B$
b. $\sin A - \sin B$
c. $\sin B \tan C$
d. $\tan B + \tan C$
e. $\cos A + \sin B + \sin D$

5. For each of the angles A, B, C shown in Exercise 1, show that the two following equations are true.

 a. $\sin^2 A + \cos^2 A = 1$
 b. $\cos^2 A(1 + \tan^2 A) = 1$

Problems just for fun

Mike and Pat between them won one glass of beer from Jake on a bet. Jake paid off but insisted that Mike and Pat divide the one glass of beer equally without the aid of any measuring device. How was this division accomplished, assuming that the glass was clear, with perfectly vertical sides?

3 *More about General Angles*

We could calculate a table of the trigonometric functions for all angles between 90° and 360° in the same way that we calculated the values of functions for angles between 0° and 90°. This would be too much work. Since the values of the trigonometric functions have not been tabulated for angles greater than 90°, there must be some simple way to get the value of these functions for large angles.

Perhaps we have overlooked some clues that might be hidden in the above examples. Let us look carefully at the angle of 210° (Fig. 64). When we measured the sides x, y, and r, we found that $r = 2$, $y = -1$, and $x = -1.732$. Since the hypotenuse is twice the opposite side y, a 30° angle must be involved. Inspection of the figure confirms our suspicions, since angle ROQ is 30°. If we draw a 30° angle in standard position, we obtain the triangle in Fig. 65. By comparing these two figures, we find that the hypotenuse, side opposite, and side adjacent of the two angles 30° and 210° are equal except for sign. Thus

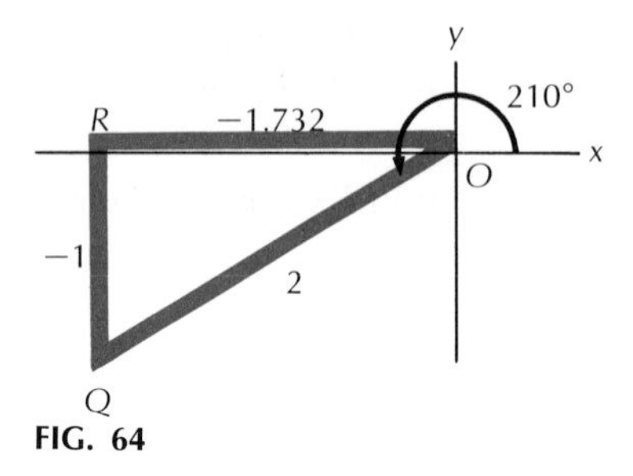

FIG. 64

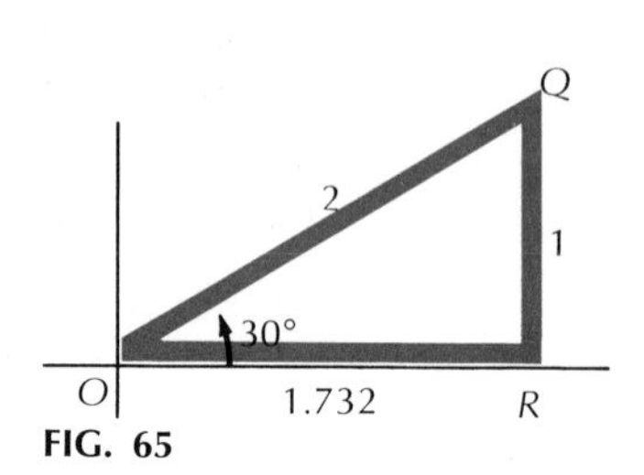

FIG. 65

$$\sin 210° = -\frac{1}{2} \qquad \sin 30° = \frac{1}{2}$$

$$\cos 210° = \frac{-1.732}{2} = -0.866 \qquad \cos 30° = \frac{1.732}{2} = 0.866$$

$$\tan 210° = \frac{-1}{-1.732} = 0.577 \qquad \tan 30° = \frac{1}{1.732} = 0.577$$

We see that the values of sin 210°, cos 210°, and tan 210° are numerically the same as sin 30°, cos 30°, and tan 30°, respectively. Since the terminal side of 210° lies in the third quadrant, we expect the sine and cosine to be negative. We conclude that in order to find the sine, cosine, and tangent of 210°, we look up the sine, cosine, and tangent of 30° and assign the proper + or − sign.

If we have any angle A whose terminal side lies in the third quadrant (Fig. 66), we determine the size of the acute angle ROQ and find the value of its trigonometric functions from the table. The sine, cosine, or tangent of angle A is then obtained by assigning the proper sign + or − to the sine, cosine, or tangent of the acute angle ROQ.

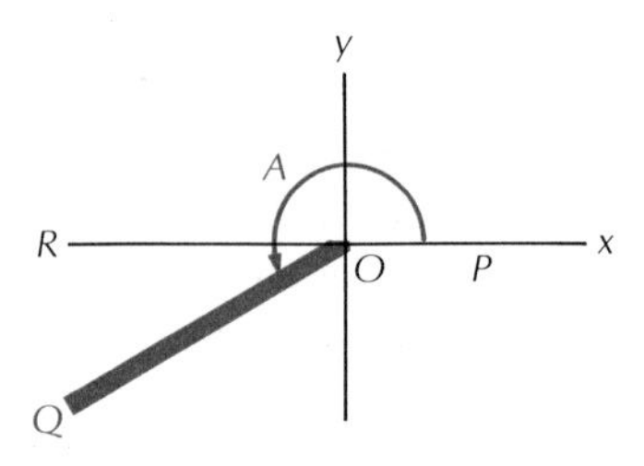

FIG. 66

If the terminal side of angle A lies in the second or fourth quadrants, the acute angle ROQ is as shown in Fig. 67. The sine, cosine, and tangent of angle A have the same numerical value as the sine, cosine, and tangent of the acute angle ROQ. All we need to do is attach the proper + or − sign. We can determine the proper sign by picturing the angle mentally and noticing whether the sides x and y are negative or positive.

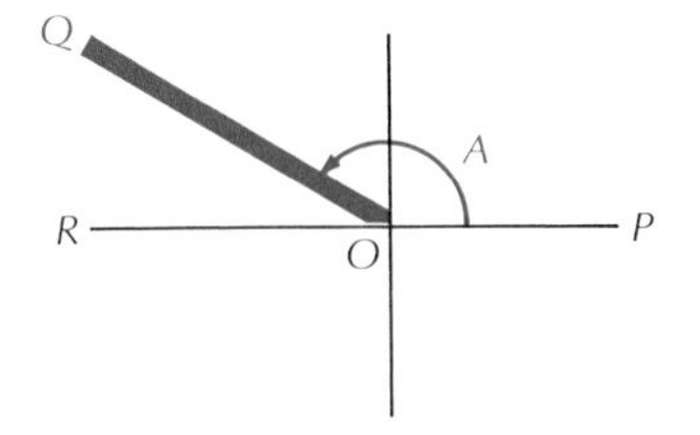

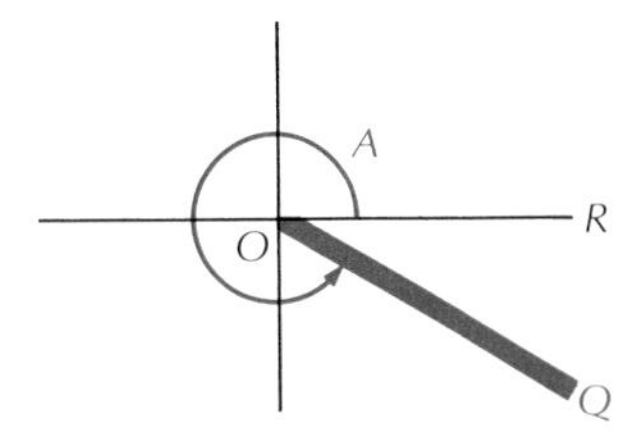

FIG. 67

▲ EXERCISES

1. Determine the value for each of the following functions:

   a. $\sin 150°$
   b. $\sin 300°$
   c. $\cos 160°$
   d. $\cos 225°$
   e. $\cos 210°$
   f. $\sin 248°$
   g. $\tan 330°$
   h. $\tan 110°$
   i. $\tan 200°$

2. For each of the following functions, draw each possible angle which is less than 360° and determine the value of the angle:

   a. $\cos\theta = -\frac{3}{4}$
   b. $\tan\theta = -\frac{4}{10}$
   c. $\tan\theta = \frac{1}{2}$
   d. $\cos\theta = \frac{2}{3}$
   e. $\sin\theta = -\frac{7}{10}$
   f. $\sin\theta = \frac{1}{3}$

3. Determine the value for each of the following relations:

   a. $\sin 120° \cos 240°$
   b. $\sin 240° + \cos 60°$
   c. $\tan 300° \tan 120°$
   d. $\tan 210° \cos 150°$
   e. $\cos 135° \tan 300°$

Problems just for fun

At the bottom of a well which is 45 feet deep is a snail which starts crawling toward the top in a vertical line. The snail climbs 3 feet each day but slides back 2 feet every night. How many days does it take the snail to get out of the well?

Unfinished Business from Sec. 1

We are now in a position to finish the problem mentioned at the beginning of the chapter. In order to find the value of cos 160°, we look up the value for cos 20° and, since 160° is in the second quadrant, we find that $\cos 160° = -0.9397$. Then

$$\begin{aligned} b^2 &= 10^2 + 5^2 - (2)(5)(10)(-0.9397) \\ &= 100 + 25 + 93.97 = 218.97 \\ b &= \sqrt{218.97} = 14.8 \text{ miles} \end{aligned}$$

In fact, we can now use the law of sines or cosines to solve for unknown parts of a triangle even if the triangle has an angle greater than 90°. In the example above we used the law of cosines.

Two cabins A and B are 350 feet apart on one side of a lake. Cabin C is due east across an arm of the lake from A. The owner of A looks 40° north of east when observing B. The owner of C looks 20° north of west when observing B. How far apart are the two cabins A and C? We draw the figure (Fig. 68).

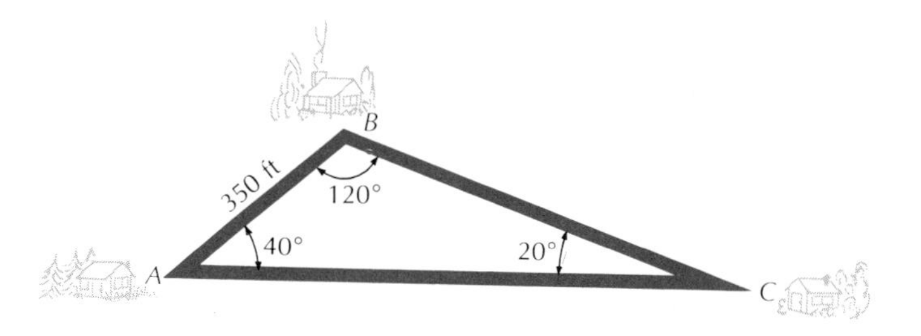

FIG. 68

We will use the law of sines. Let the distance $AC = x$. Then

$$\frac{x}{\sin 120°} = \frac{350}{\sin 20°}$$

and

$$x = \frac{350 \sin 120°}{\sin 20°} = \frac{350 \times 0.8660}{0.3420} = 886.2 \text{ feet}$$

since $\sin 120° = 0.8660$.

▲ EXERCISES

1. A road runs up a hill at an angle of 18°. When a hiker has climbed 800 feet from the bottom of the hill, he observes another man who is approaching the hill. If the angle of depression is 8°, how far is the man from the base of the hill?

2. One man walks 5 miles directly east, while another man walks 10 miles in a direction 35° north of west. How far apart are the men if they started at the same point?
3. A telephone pole 100 feet high was knocked out of a vertical position during a windstorm. A guy wire 150 feet long attached at the top of the pole and at a point 95 feet from the base kept the pole from falling. What angle did the pole make with the horizontal?
4. A tree which stands erect upon a hillside which has an inclination of 20° is viewed from two points A and B which are farther down the hill, 400 feet apart, and in line with the tree. If the angles subtended by the tree at A and B are 8° and 14°, respectively, find the height of the tree.
5. When the sun is at an angle of 24° with the ground, the Leaning Tower of Pisa casts a shadow 385 feet long on the level ground opposite to the direction the tower is leaning. If the tower is 180 feet tall, what angle does it make with the vertical?
6. At 3 P.M. a radio station is 110 miles in a direction 32° north of east of a car moving due west at 45 miles per hour. If the effective range of the radio station is 250 miles, at what time will the program begin to fade on the car radio?
7. At a point 140 feet up the slope of a river bank, the river subtends an angle of 9° and the far edge of the river is seen at an angle of depression of 5°. How wide is the river?
8. A center fielder catches a fly ball 350 feet from home plate and directly in line with second base. If he throws to third base, how long is his throw? (A baseball diamond is a square, 90 feet on a side.)
9. An airplane has a cruising speed of 210 miles per hour and is heading due north. As a result of a 25 mile per hour wind, its ground speed is 224 miles per hour. Find the direction of the wind and the direction the plane is traveling.
10. Find the radius of the largest circular target which can be cut from a triangular piece of plywood with sides 42 inches, 70 inches, and 62 inches.

Problems just for fun

A Great Lakes ore ship carries rock ballast. The ballast is thrown overboard, and this tends to raise the water level of the lake. On the other hand, the ship, being empty, now displaces less water. Will the level of the lake rise or fall as a result of the unloading of the rock ballast?

CHAPTER TWELVE

Graphs of the Trigonometric Functions

Trigonometry contains the science of continually undulating magnitude: meaning magnitude which becomes alternately greater and less, without any termination to succession of increase and decrease.

A. DE MORGAN

1 Graph of the Sine Function

In the circle with radius 1, we see from Fig. 69 that the side QP opposite angle A increases from 0 to 1 as A changes from 0° to 90°;

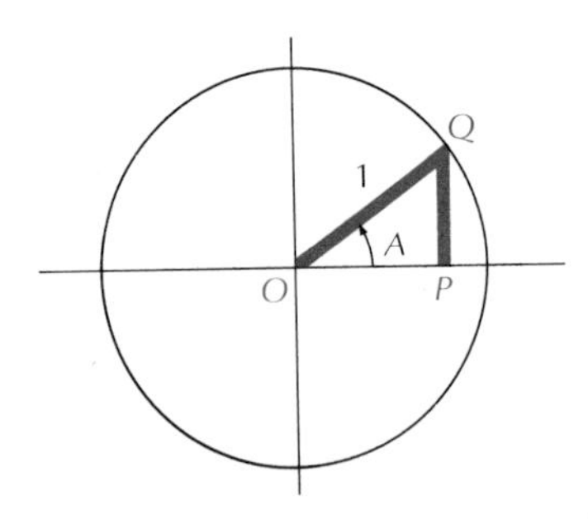

FIG. 69

decreases from 1 to 0 as A changes from 90° to 180°; decreases from 0 to −1 as A changes from 180° to 270°; and increases from −1 to 0 as A changes from 270° to 360°. Since the value of sin A starts at 0 for $A = 0°$ and

$$\sin A = \frac{\text{opposite side}}{\text{hypotenuse} = 1} = \text{opposite side} = QP$$

we see that sin A builds up to a maximum value of 1 at $A = 90°$, then its value decreases from 1 to 0 at $A = 180°$. As A changes from 180° to 270°, the value of sin A changes from 0 to a minimum value of -1 at $A = 270°$. Then, as A changes from 270° to 360°, the value of the sin A increases from -1 to 0.

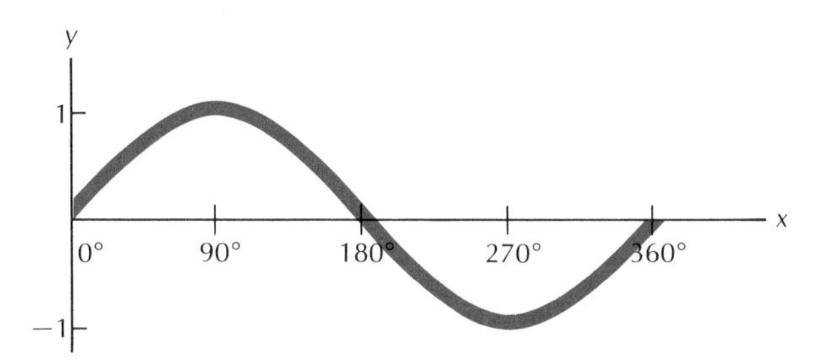

FIG. 70

If we plot the values of sin A for angles between 0° and 360°, we get the graph in Fig. 70. As angle A increases from 360° to 720°, the values for sin A will repeat themselves, and we get the curve shown in Fig. 71. Also, we can extend this curve to the left of the origin, because $\sin(-A) = -\sin A$. Why?

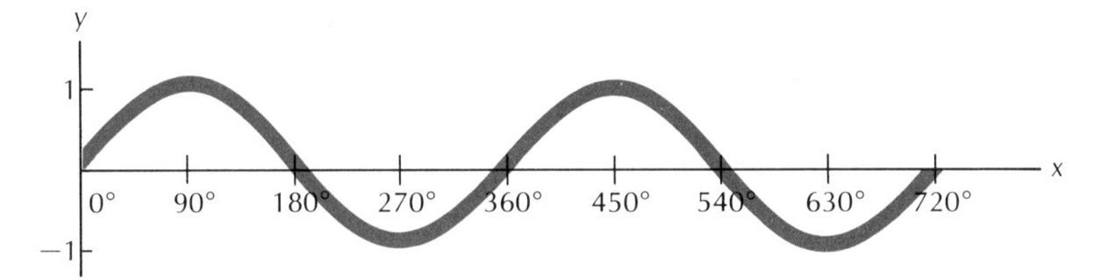

FIG. 71

We see (Fig. 72) that the function sin A, when plotted, yields a wavy curve which has a maximum value of 1 at 90° and a minimum value of -1 at 270°. The curve repeats itself every 360°. That is, sin x takes on the same value when x is increased by 360° or multiples of 360° and is said to be periodic, with a period of 360°. Because this curve represents sin A, it is called a *sine curve.*

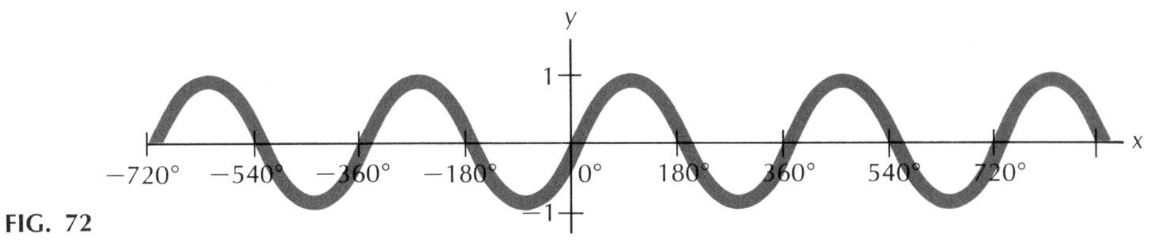

FIG. 72

If we study this curve, we can readily understand why it is necessary only to tabulate the values of sin A for angles between 0° and

90°. If we become familiar with that portion of the sine curve for angles between 0° and 360°, we will have no trouble determining the sign to use. Sin A is positive for angles between 0° and 180° and negative for those between 180° and 360°.

▲ EXERCISES

1. Prove that $\sin(-A) = -\sin A$.
2. Discuss the values of $\cos A$ for

   a. $-180° \leqq A \leqq 180°$
   b. $-540° \leqq A \leqq 720°$

3. Plot the values of $\cos A$ for

   a. $-180° \leqq A \leqq 180°$
   b. $-540° \leqq A \leqq 720°$

4. a. Is the cosine curve periodic?
   b. If periodic, what is the period?

5. Prove that $\cos(-A) = \cos A$.

6. On the same axes plot the curves for $y = \sin x$ and $y = \cos x$; $-360° \leqq x \leqq 360°$. Compare these two curves and answer the following questions.

   a. Is there a way to change the sine curve into the cosine curve?
   b. How would you accomplish this?
   c. Did you expect to find this relation?

7. Using the curves for $\sin x$ and $\cos x$ show that

   a. $\sin(x + 90°) = \cos x$
   b. $\sin(x + 180°) = -\sin x$
   c. $\cos(x + 90°) = -\sin x$
   d. $\cos(x + 180°) = -\cos x$

Problems just for fun

A truck and a car meet one foggy night on a bridge that is so narrow they cannot pass each other or turn. The car is twice as far on the bridge as the truck. But it has taken the truck twice as long as the car to reach this meeting point. Both car and truck can back at only half their forward speed. Which of them should back up to allow both to cross the bridge in the shortest time?

2 *More General Sine Curves*

When data from certain physical measurements are plotted, a curve of the sine or cosine type is obtained, and the function which the data represent is said to be a *sine or cosine function.* Just knowing that the function is a sine function is very informative, because we know that such a function starts with a zero value, builds up to a maximum value, falls to zero again, and then falls to a minimum value before coming back to zero. This process is repeated again and again, and we can reduce the study of the whole process to a study of this one period, knowing that whatever is true for this one period will also be true for any other period.

When a light jump rope is attached to a post and the end moved up and down in a vertical path with the right frequency, the rope assumes the shape shown in Fig. 73. We notice that the portion of the curve between the points AB, BC, and CD is the same and that this curve looks like the sine curve. The vertical displacement y at any point x is given by the equation $y = A \sin kx$.

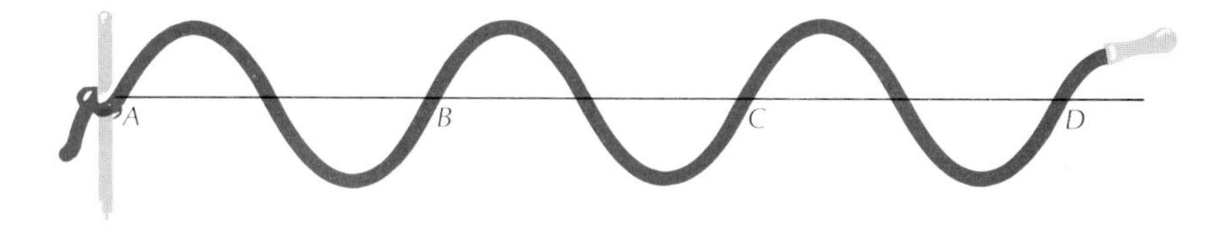

FIG. 73

Pictures of vibrating strings have been taken. The shape assumed by the string is similar to that of the rope, and we say that the vertical displacement is a sine function of the horizontal distance along the string.

Since sound and light are propagated by wave motion, the physicist uses the sine function to help explain and understand the phenomena. In fact, the physicist loses no opportunity for using the trigonometric function of sine or cosine to describe mathematically a motion which is periodic.

The alternating current used to light our homes starts with a value of zero, builds up to a maximum value, then decreases to zero. After this, the current increases up to a maximum value in the opposite direction and then falls to zero again. The whole process takes a very short time, $\frac{1}{60}$ second, and then is repeated again and again. This alternating current as a function of time is represented by the sine function,

$$I = A \sin kt$$

The above uses of the trigonometric function of sine give some insight into the important role played by such functions in the solution of problems in fields other than mathematics. Because of this wide use, we shall spend some time studying this more general sine function.

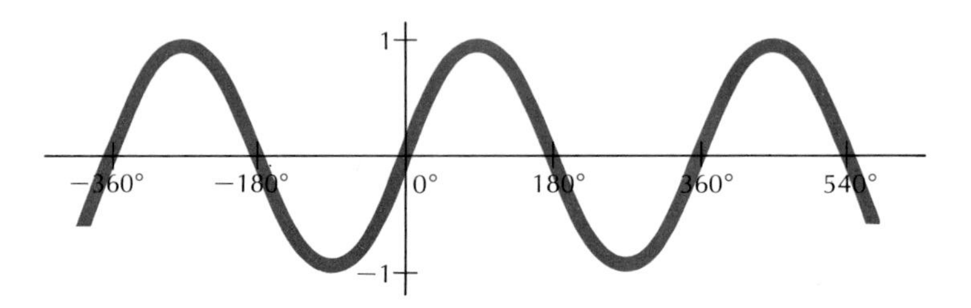

FIG. 74

From the graph of $y = \sin x$ (Fig. 74), we see that $\sin x$ is zero when $x = -540°, -360°, -180°, 0°, 180°, 360°, 540°, \ldots$. The maximum value of 1 is obtained when $x = -270°, 90°, 450°, \ldots$, and the minimum value of -1 is obtained when $x = -90°, 270°, 630°, \ldots$. We observe that each maximum and minimum repeats itself every 360°.

Now let us see what happens when we introduce a constant A. That is, how does the graph of $y = A \sin x$ compare with that of $y = \sin x$? For a given x, $\sin x$ will be the same in both cases. Thus the height y of the new curve will be A times the other. The ordinate y will still be equal to zero when $x = -180°, 0°, 180°, 360°, \ldots$, but its maximum value will be $A \cdot 1 = A$ and occurs when $x = -270°, 90°, 450°, \ldots$. The minimum value will be $A \cdot (-1) = -A$ at $-90°, 270°, \ldots$ (Fig. 75).

We learn that multiplying $\sin x$ by a constant merely stretches or contracts the curve for $\sin x$ in the vertical direction, depending on whether the constant A is greater or less than 1.

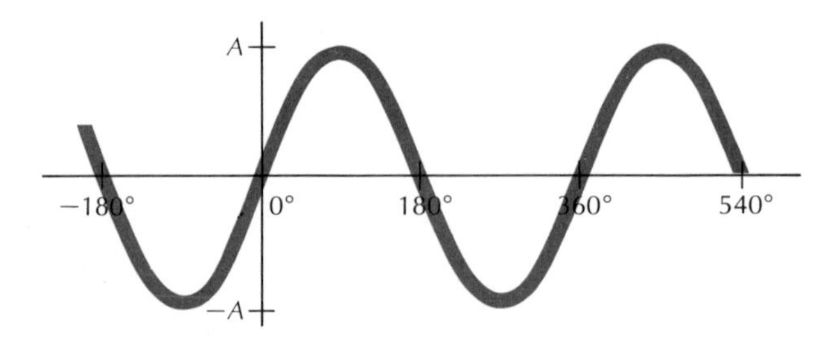

FIG. 75

What change is made in the curve $y = \sin x$ if we multiply x by a constant to get $y = \sin kx$? In order to be specific, let us discuss $y =$

sin $2x$. We observe that the maximum and minimum values will still be 1 and -1. The zeros will occur at the points where $2x = -180°$, $0°$, $180°$, $360°$, $540°$. Then

$$x = \frac{-180}{2},\ 0,\ \frac{180}{2},\ \frac{360}{2},\ \frac{540}{2}$$

or

$$x = -90°,\ 0°,\ 90°,\ 180°,\ 270°,\ \ldots$$

The maximum value of 1 occurs at points where $2x = -270°$, $90°$, $450°$, . . ., or

$$x = -135°,\ 45°,\ 225°,\ \ldots$$

Likewise, the minimum values occur at points where

$$2x = -90°,\ 270°,\ 630°,\ \ldots$$

or

$$x = -45°,\ 135°,\ 315°,\ \ldots$$

From the above discussion we can quickly sketch the curve $y = \sin 2x$ since it has the general shape of the sine curve, but the zeros, maximum, and minimum values occur at different points. We see that the curve (Fig. 76) is compressed in the horizontal direction by a factor of 2. That is, the horizontal spacing per period is $\frac{1}{2}$ that of $\sin x$. The period for $y = \sin 2x$ is 180°, which is one-half the period for $y = \sin x$.

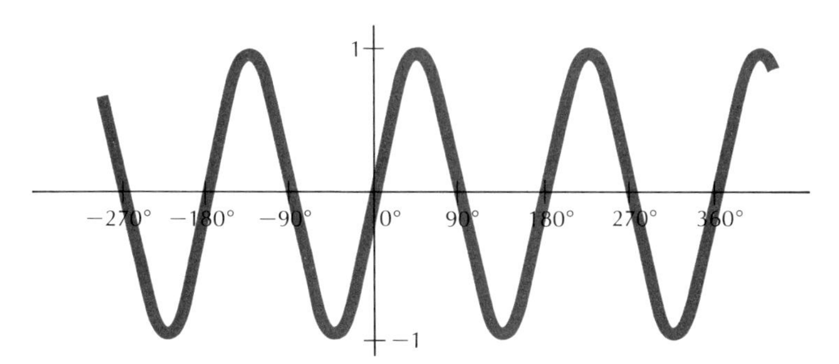

FIG. 76

We can conclude that the curve represented by the general equation $y = A \sin kx$, where A and k are constants, has the shape of the sine curve $y = \sin x$.

The effect of the constant A, which is called the *amplitude* of the wave, is to stretch ($A > 1$) or compress ($0 < A < 1$) the curve in the vertical direction. The effect of the constant k is to stretch $(0 < k < 1)$ or compress $(k > 1)$ the curve in the horizontal direction.

You need not plot the curve $y = A \sin kx$ point by point. The shape of the curve is that of $y = \sin x$, which you can readily sketch, and all that is needed is to make the proper changes made necessary by the constants A and k.

In Exercises 1–3 the student will consider the effect of introducing an A and a k into the equation $y = \cos x$ to give $y = A \cos kx$.

It should now be clear to the student *that he must be able to sketch quickly the curves for $y = \sin x$ and $y = \cos x$ for $-180° \leqq x \leqq 360°$. Then he must be able to label the points at which the curves cross the x axis and also be able to modify these curves (this is equivalent to changing the scale) when the amplitude is changed or the periodicity is altered.*

▲ EXERCISES

1. What change is made in the cosine curve (Fig. 77) if $y = \cos x$ is multiplied by a constant A, giving $y = A \cos x$?

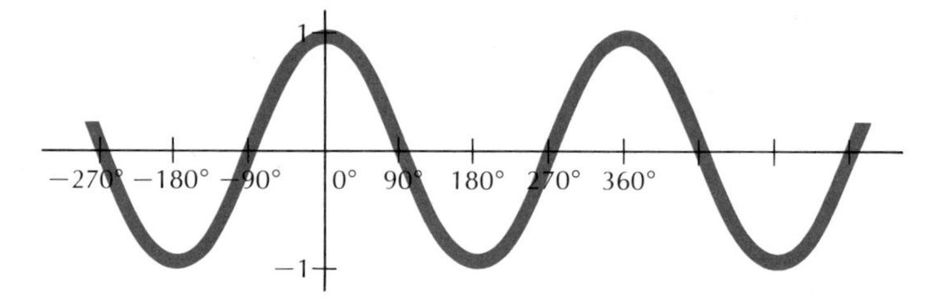

FIG. 77

2. What change is made in the cosine curve $y = \cos x$ if the x is multiplied by a constant k, giving $y = \cos kx$?
3. Make sketches for Exercises 1 and 2.
4. How many times will $y = 3 \cos 5x$ cross the x axis as x varies between 0° and 360°?
5. How many maxima (high points) will the curve in Exercise 4 have?
6. Sketch the following curves and discuss them by noting the amplitude and period:

   a. $y = \sin \dfrac{x}{2}$ b. $y = 3 \cos 2x$

   c. $y = 5 \cos \dfrac{x}{3}$ d. $y = 4 \sin 3x$

   e. $y = \cos \dfrac{x}{2}$ f. $y = -2 \sin \dfrac{x}{2}$

7. The alternating current I (in amperes) in a home is given by $I = 10 \sin \theta$

where θ is the angular frequency and takes on values $0° \leqq \theta \leqq 360°$ in $\frac{1}{60}$ second.

a. What is the maximum value of the current?
b. What is the minimum value of the current? What is its meaning?
c. What is the value of the current in $\frac{1}{240}$ second after passing through zero? $\frac{1}{120}$ second?
d. How long after passing through the zero value will the current be +5 amperes? −5 amperes?

The natives of a remote South Sea island are all members of two tribes, Abel and Babel. To a stranger they look exactly alike. But the members of the Abel tribe always tell the truth, while those of Babel blood always lie.

To this island came an explorer and met three natives.

"Of what tribe are you?" the explorer asked the first.

"Chsz cinth cstrm," replied the native.

"What did he say?" asked the explorer of the second and third natives, both of whom spoke some English.

"He say he Abel," said the second.

"He say he Babel," said the third.

To what tribes did the second and third natives belong?

CHAPTER THIRTEEN

Exponents and Logarithms

The miraculous powers of modern calculation are due to three inventions: the Arabic Notation, Decimal Fractions and Logarithms.

F. CAJORI

1 *Multiplication by Exponents*

Archimedes (287–212 B.C.), who used the complicated Greek number system which did not enable mathematicians to write large numbers, was handicapped because he was unable to multiply large numbers together. However, he did notice that some numbers could be multiplied together by using exponents. That is,

$$a \cdot a = a^1 \cdot a^1 = a^2 = a^{1+1}$$
$$a^2 \cdot a^3 = (a \cdot a) \cdot (a \cdot a \cdot a) = a^5 = a^{2+3}$$

and, in general,

$$a^m \cdot a^n = a^{m+n}$$

Thus we observe that the product of two numbers which are expressed as some powers, m and n, of the same number a, called the *base*, is given by the base a raised to the $m + n$ power. For example,

$$9 \cdot 81 = 3^2 \cdot 3^4 = 3^6 = 729$$

Also,

$$9 \cdot 81 \cdot 27 = 3^2 \cdot 3^4 \cdot 3^3 = 3^{2+4+3} = 3^9 = 19{,}683$$

Division by Exponents

Exponents may also be used in division. For example,

$$\frac{625}{25}=\frac{5^4}{5^2}=\frac{5\cdot 5\cdot 5\cdot 5}{5\cdot 5}=5^2=5^{4-2}$$

$$\frac{a^6}{a^4}=\frac{a\cdot a\cdot a\cdot a\cdot a\cdot a}{a\cdot a\cdot a\cdot a}=a^2=a^{6-4}$$

and we see that in general

$$\frac{a^m}{a^n}=a^{m-n}$$

That is, the quotient of two numbers expressed as powers of the same base is the base raised to the difference of the powers.

Hence, we are led to the general principle: when we *multiply* numbers which are powers of the same base, we *add* the exponents; and when we divide, we *subtract* exponents:

$$10^7\cdot 10^4=10^{11} \qquad \text{and} \qquad 10^7\div 10^4=10^3$$

▲ EXERCISES

1. Find the value of

   a. 3^3 b. 5^4
   c. 2^2 d. 2^7
   e. 2^4 f. 3^5
   g. 3^4 h. 4^3
   i. $(\frac{3}{2})^3$ j. $(\frac{2}{3})^2$

2. Express the following numbers as some base raised to a power n:

   a. 27 b. 125
   c. 32 d. 81
   e. 49 f. 3,125
   g. 100 h. 128
   i. 256 j. $\frac{1}{16}$

3. Find the result of the following operations:

   a. $a^2\cdot a^4\cdot a^7$ b. $3^{10}\cdot 3^2$

   c. $\dfrac{x^4\cdot x^6}{x^3}$ d. $\dfrac{10^5\cdot 10^7}{10^9}$

   e. $\dfrac{a^3\cdot a^9}{a\cdot a^4}$ f. $\dfrac{b^3\cdot b\cdot b^5}{b^2\cdot b^4}$

g. $\dfrac{10^{10}}{10^3 \cdot 10 \cdot 10^4}$

h. $\dfrac{2^3 \cdot 2^5 \cdot 2^{12}}{2^6 \cdot 2^4 \cdot 2^7}$

i. $\dfrac{2^7}{2^4 \cdot 2^3}$

j. $\dfrac{3^3 \cdot 3^4 \cdot 3^{12}}{3^5 \cdot 3^7 \cdot 3^6}$

4. Perform the following operations by using exponents:

a. $\dfrac{1{,}000 \cdot 1{,}000}{100}$

b. $\dfrac{64 \cdot 128}{16 \cdot 256}$

c. $\dfrac{27 \cdot 9}{81}$

d. $\dfrac{4 \cdot 32 \cdot 64}{256 \cdot 8}$

e. $\dfrac{16 \cdot 8 \cdot 4}{32}$

f. $\dfrac{81 \cdot 243}{27 \cdot 27}$

g. $\dfrac{25 \cdot 625}{125}$

h. $\dfrac{32 \cdot 81}{4 \cdot 9}$

i. $\dfrac{3{,}125 \cdot 243}{25 \cdot 27}$

j. $\dfrac{16 \cdot 8 \cdot 64}{32 \cdot 256}$

Problems just for fun

Mr. Commuter catches the 5 o'clock train from the city each day and is met at the station by Mrs. Commuter. One day he caught the 4 o'clock train and, on arriving at his station, started walking toward home. Mrs. Commuter started for the station at the usual time, met her husband on the way, and returned home with him. They arrived home 20 minutes before their usual time. How long did Mr. Commuter walk?

3 *Negative and Zero Exponents*

If we blindly follow our rule for division by the subtraction of exponents, we are led directly to the proper interpretation of negative and zero exponents. For example, we see that $x^9/x^9 = 1$, but following our general principle, we find that $x^9/x^9 = x^{9-9} = x^0$. Thus it seems natural to write $x^0 = 1$ and to say that any number (except zero) raised to the zero power is equal to 1. (The exception is not surprising, since we know that division by zero is not possible.) We also know that

$$\frac{a^4}{a^6} = \frac{a \cdot a \cdot a \cdot a}{a \cdot a \cdot a \cdot a \cdot a \cdot a} = \frac{1}{a^2}$$

and the general principle gives

$$\frac{a^4}{a^6} = a^{4-6} = a^{-2}$$

Thus we see that our general principle applies providing we understand that $a^{-2} = \frac{1}{a^2}$. In general, we define $a^{-n} = \frac{1}{a^n}$.

Examples

$$\frac{3^2 \cdot 3^4}{3^9} = 3^{2+4-9} = 3^{-3} = \frac{1}{3^3} = \frac{1}{27}$$

$$2^5 \cdot 2^{-11} \cdot 2^4 = 2^{5-11+4} = 2^{-2} = \frac{1}{2^2} = \frac{1}{4}$$

4 *Raising to Powers*

If we look at the operation

$$(a^2)^3 = a^2 \cdot a^2 \cdot a^2 = a^6 = a^{2\times3}$$

we see that we obtain the correct result by multiplying the two exponents 2 and 3 together. In general, we find that $(a^m)^n = a^{mn}$. We must be careful to distinguish between

$$(a^2)^3 = a^6 \qquad \text{or} \qquad (a^m)^n = a^{mn}$$

and

$$(a^2)(a^3) = a^5 \qquad \text{or} \qquad a^m \cdot a^n = a^{m+n}$$

Let us restate the three general principles of exponents involved when we combine numbers which are expressed as the *same base* raised to powers:

1. When we wish to *multiply* the numbers, we *add* the exponents.
2. When we wish to *divide* the numbers, we *subtract* the exponents.
3. When we wish to *raise* to a power, we *multiply* the two exponents.

Fractional Exponents

If we compare $\sqrt{a^4} = \sqrt{a^2 \cdot a^2} = a^2$ with $(a^4)^{1/2} = a^{4 \cdot 1/2} = a^2$ and $\sqrt{x}\,\sqrt{x} = x$ with $x^{1/2}x^{1/2} = x^{1/2+1/2} = x$, we see that in both cases the square root behaves like the fractional exponent $\frac{1}{2}$. In the same manner, we observe that the cube root behaves exactly like operations with the exponent $\frac{1}{3}$, and fourth roots like the exponent $\frac{1}{4}$. It is much easier to use our general principles and to work with fractional exponents than to work with roots. So we always change roots to fractional exponents.

■ Examples

$$\sqrt{a^8} = (a^8)^{1/2} = a^4$$

$$\sqrt[3]{64} = (2^6)^{1/3} = 2^2 = 4$$

$$\sqrt[4]{x^2} = (x^2)^{1/4} = x^{1/2} = \sqrt{x}$$

$$\sqrt{32} = (2^5)^{1/2} = 2^{5/2} = 2^2 2^{1/2} = 4\sqrt{2}$$

▲ EXERCISES

Find the results of the following operations:

1. $\dfrac{x^{11} \cdot x^4 \cdot x^3}{x^{12} \cdot x^{10}}$
2. $\dfrac{2^4 \cdot 4^2}{2^6}$
3. $\dfrac{a^5 \cdot a^2 \cdot a^3}{a^6 \cdot a^4}$
4. $\dfrac{(2^2)^5}{(2^3)^4}$
5. $(a^2)^3$
6. $\sqrt{a^{10}}$
7. $\sqrt{\sqrt{a^8}}$
8. $\sqrt[3]{16}$
9. $\dfrac{\sqrt{2^{16}}}{\sqrt{4^4}}$
10. $\sqrt{10^{-6}}$
11. $\dfrac{x^2 \cdot x^5}{x^3 \cdot x^4}$
12. $9 \cdot 3^3$
13. $\dfrac{b^4 \cdot b^7}{b^5 \cdot b^3 \cdot b^6}$
14. $(x^3)^4 \cdot x^2$
15. $\dfrac{(10^2)^3 \cdot 100}{(10^3)^3}$
16. $\sqrt[3]{x^{15}}$
17. $(x^6)^{1/3}$
18. $\sqrt{128}$
19. $\dfrac{(a^4)^{1/2}}{\sqrt{a^{-2}}}$
20. $\sqrt{10^4 \cdot 10^{-6}}$

Problems just for fun

Miss Penny Feather presides over the ribbon counter in the country store. To measure ribbons, she has a circular wooden disk whose circumference is 27 inches. On the edge of the disk are six marks so located that our dear lady can measure any number of inches from 1 to 27 by the distance around the disk between two of the marks. Can you find where the marks should be placed? There are several solutions.

6 *Powers of 10*

We wish to compute 10^x for various values of x in order to study the equation $y = 10^x$. If x is any whole number, we can easily find the value of 10^x.

$$\begin{array}{ll} x=0 & 10^0=1 \\ x=1 & 10^1=10 \\ x=2 & 10^2=100 \\ x=3 & 10^3=1{,}000 \\ x=-1 & 10^{-1}=0.1 \\ x=-2 & 10^{-2}=0.01 \\ x=-3 & 10^{-3}=0.001 \end{array}$$

We can also compute 10^x for many other values of x. For example,

$$\begin{array}{ll} x=\frac{1}{2} & 10^{1/2}=\sqrt{10}=3.16 \\ x=\frac{3}{2} & 10^{3/2}=10\sqrt{10}=31.62 \\ x=\frac{1}{3} & 10^{1/3}=\sqrt[3]{10}=2.15 \end{array}$$

Let us plot these values of $y = 10^x$ and draw a smooth curve through these points, as in Fig. 78. This curve represents $y = 10^x$, and we see that it indicates a value of 10^x for each value of x. If $x = 0.8$, we wish the value of $10^{0.8}$. Since $10^{1/2} = 3.16$ and $10^1 = 10$, we expect a value between 3.16 and 10. From the graph of $y = 10^x$, we

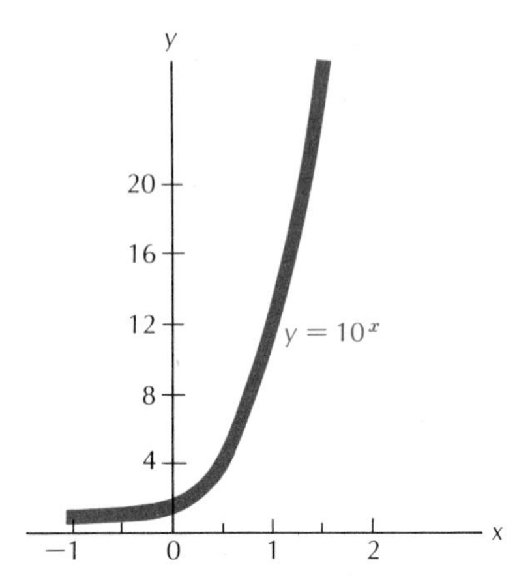

FIG. 78

find that the approximate value of $10^{0.8}$ is 6. Within the limitations of the figure, we can use the graph of $y = 10^x$ to find the value 10 raised to a positive power x. Of course, because of the size of the graph, the accuracy is limited. We can also use the graph of $y = 10^x$ to find the power x to which 10 must be raised to give a particular value of y. For example, for $y = 5 = 10^x$, how much is x? From Fig. 78 we find that x is approximately equal to 0.7.

We conclude that every positive number can be expressed as some power of 10. We can reason that, since each number can be expressed as a power of 10, we can multiply two numbers together by changing each number to the number 10 raised to the proper power and adding the exponents, that is,

$$A \cdot B = 10^m \cdot 10^n = 10^{m+n}$$

Archimedes was the first to discover this clever method of multiplying and dividing numbers by using exponents. He suggested that all numbers be expressed in powers of another number and that a table of such powers be made. Archimedes did not make such a table, and it was not until the seventeenth century that it was made. A Scotsman named John Napier desired a table similar to that suggested by Archimedes that would end the laborious work of multiplying and dividing large numbers. Briggs, in collaboration with Napier, published the first table giving x in the relation $y = 10^x$ for all values of y from 1 to 1,000, correct to 14 decimal places. Such tables are known as *logarithm tables.*

Using logarithm tables, we can multiply two numbers A and B together. We find the power a to which 10 must be raised to give A, and the power b to which 10 must be raised to give B. Then $A \cdot B =$

$10^a \cdot 10^b = 10^{a+b}$. Finally, we find from the tables the number that corresponds to 10 raised to the $a+b$ power.

The task of computing a table of logarithms, that is, the power to which 10 must be raised to give the number, appears hopeless. However, it is not so great a task as it first appears, if we observe that we may write all numbers as a number between 1 and 10 multiplied by 10 to a whole number power. For example,

$$13.27 = 1.327 \times 10^1$$
$$132.7 = 1.327 \times 100 = 1.327 \times 10^2$$
$$1{,}327 = 1.327 \times 1{,}000 = 1.327 \times 10^3$$
$$0.1327 = \frac{1.327}{10} = 1.327 \times 10^{-1}$$
$$0.01327 = \frac{1.327}{100} = \frac{1.327}{10^2} = 1.327 \times 10^{-2}$$

Then, if we knew the power of 10 which gives 1.327, we should know the power for all the other numbers above. Or if we knew the power of 10 for all numbers from 1 to 10, we should know the power for all numbers. Thus we need a table of logarithms for the numbers from 1 to 10 only. Table VII at the back of the book is such a table.

Scientific Notation

In scientific work, numbers which are very small or very large are expressed in the form used above instead of being written out in full. In this way we avoid operations with long rows of zeros. For example:

1. The estimated age of the earth is $694{,}000{,}000{,}000 = 6.94 \times 10^{11}$ years.
2. The approximate distance that light travels in 1 year is $5{,}870{,}000{,}000{,}000 = 5.87 \times 10^{12}$ miles.
3. The diameter of a certain molecule is $0.0000005 = \frac{5}{10^7} = 5 \times 10^{-7}$ centimeters.
4. The mass of an electron is 9.11×10^{-28} grams.

It is apparent that it is much easier to multiply two numbers together when they are written in the above form, which is sometimes called the *scientific notation.* If one desires to find the cross-sectional area of the molecule whose diameter is given in the third example above, one proceeds as follows:

$$\text{Area}=\frac{\pi d^2}{4}=\frac{\pi}{4}(5\times 10^{-7})^2=\frac{\pi}{4}25\times 10^{-14}=19.6\times 10^{-14}$$
$$=1.96\times 10^{-13}\text{ square centimeters}$$

It is easy to write large or small numbers in scientific notation. Move the decimal point to the right or left, obtaining a number between 1 and 10. Multiply this number by 10 with an exponent numerically equal to the number of places the decimal point has been moved. The exponent is positive if the decimal point has been moved to the left, and negative if the decimal point has been moved to the right.

▲ EXERCISES

1. Write the following numbers in scientific notation, that is, a number between 1 and 10 multiplied by 10 to a whole number power:

   a. 0.00073 b. 0.123
   c. 0.0035 d. 52,800
   e. 0.0000000068 f. 8,370,000
   g. 687,000,000,000 h. 0.0000013
   i. 3,620 j. 146,000

2. Write the following numbers in ordinary form:

   a. 5.63×10^4 b. 3.22×10^{-4}
   c. 4.6×10^{-14} d. 1.69×10^{-9}
   e. 8.32×10^{-7} f. 7.83×10^{10}
   g. 3.27×10^6 h. 2.73×10^3
   i. 1.43×10^{-3} j. 6.163×10^{22}

3. If the velocity of light is 1.86×10^5 miles per second, how far does light travel in 1 hour?
4. If the sun is 9.3×10^7 miles from the earth, how long does it take light from the sun to reach the earth?
5. If a molecule weighs 4.3×10^{-24} grams, how many molecules are there in 1 gram of the substance?

8 *Tables for $N=10^x$*

We have seen that all numbers may be expressed as 10 to some power x. We need tables only for the numbers between 1 and 10, since all numbers may be expressed as a number between 1 and 10 multiplied by 10^n, where n is a positive or negative integer. Such tables are known as logarithm tables. We shall use four-place tables, which are accurate enough for our work. If greater accuracy is needed, one can use a larger table.

If we wish to find the power x to which 10 must be raised to give the number 1,060, we first write $1{,}060 = 1.060 \times 10^3 = 10^a \times 10^3$. Then we look in the table of logarithms to find the power a to which 10 must be raised to give 1.060, using Table VII at the end of the book. We find that $a = 0.0253$. Then $1{,}060 = 10^a \times 10^3 = 10^{0.0253} \times 10^3 = 10^{3.0253}$. Likewise $425 = 4.25 \times 10^2 = 10^{0.6284} \times 10^2 = 10^{2.6284}$. The exponent 0.6284 is obtained from the table. All the numbers a obtained from the table are between 0 and 1. *We must remember to place the decimal point in front of each such number.*

Problems just for fun

Three men and their wives came to a river which they wished to cross. There was one small rowboat which could carry only two persons at a time. The husbands were jealous, so no woman could be with any man unless her husband was also present. How do they manage to cross the river? No fair building bridges or drying up the river!

▲ EXERCISES

Find the power x to which 10 must be raised to give the following numbers:

1. 268
2. 143
3. 1,070
4. 9.82
5. 7.82
6. 1.93
7. 19.3
8. 0.193
9. 193
10. 0.0193
11. 0.000193
12. 39,700
13. 7,320
14. 0.000482
15. 287
16. 0.0624
17. 84.3
18. 8.43
19. 843,000
20. 0.000843

Problems just for fun

```
  X X X
    X X
  -----
  X X X
X X 4
-------
X X X 1 7
```

In the above multiplication problem the X's are numbers to be found. Find these numbers. There are exactly two solutions.

9 *Finding the Number If Its Logarithm Is Given*

If the exponent x of 10 is known and we want to find the number N which corresponds to this number, we proceed as follows:

$$N = 10^{2.6284} = 10^2 \times 10^{0.6284}$$

Then we look in the body of the table of logarithms until we find the number 0.6284. We then look in the left-hand column to find the number, 4.25, which corresponds to $10^{0.6284}$. The number $N = 10^2 \times 4.25 = 425$. This process is just the reverse of finding the power x of 10.

If N is given by $10^{-2.3716}$ and we desire to find the corresponding N, we proceed as follows:

$$10^{-2.3716} = \frac{10^3}{10^3} \times 10^{-2.3716} = \frac{10^{0.6284}}{10^3} = 10^{-3} \times 10^{0.6284}$$

Since

$$10^{0.6284} = 4.25$$

we find that

$$10^{-2.3716} = 10^{-3} \times 10^{0.6284} = 10^{-3} \times 4.25 = 0.00425$$

Or, if $N = 10^{-0.0696}$, we write

$$10^{-0.0696} = \frac{10^1}{10^1} \times 10^{-0.0696} = 10^{-1} \times 10^{0.9304}$$

From Table VII we find that

$$10^{0.9304} = 8.52$$

So

$$N = 10^{-0.0696} = 10^{-1} \times 10^{0.9304} = 10^{-1} \times 8.52 = 0.852$$

▲ EXERCISES

Find the number which corresponds to each of the following powers of 10:

1. $10^{2.8281}$
2. $10^{1.4368}$
3. $10^{3.6652}$
4. $10^{0.5228}$
5. $10^{-3} \times 10^{0.9048}$
6. $10^{-2.8416}$
7. $10^{-1.3139}$
8. $10^{4.9175}$
9. $10^{0.8281}$
10. $10^{1.9854}$
11. $10^{-0.6378}$
12. $10^{-1.4368}$
13. $10^{-3.6728}$
14. $10^{5.7982}$

Calculations Using Logarithms

Using the idea of scientific notation, we have seen that the task of determining a table of logarithms is simplified. We need only determine the logarithms of the numbers from 1 to 10. Then we may express all numbers as some power of 10. For example

$$132 = 1.32 \times 10^2 = 10^a \times 10^2 \qquad \text{where} \qquad 10^a = 1.32$$

From the table, we find the number a to which 10 must be raised to give 1.32. Thus

$$10^{0.1206} = 1.32 \qquad 132 = 10^{0.1206} \times 10^2 = 10^{2.1206}$$

Likewise

$$173 = 10^2 \times 1.73 = 10^2 \times 10^{0.2380} = 10^{2.2380}$$

If we desire to find the product of 132 and 173, we proceed as follows:

$$132 \times 173 = 10^{2.1206} \times 10^{2.2380} = 10^{4.3586} = 10^4 \times 10^{0.3586}$$
$$= 2.28 \times 10^4 = 22{,}800$$

That $2.28 = 10^{0.3586}$ is obtained from a table. Check the extent of the accuracy by multiplying 173 by 132 in the usual manner. If we wish to obtain the quotient of 173 and 132, we proceed as follows:

$$\frac{173}{132} = \frac{10^{2.2380}}{10^{2.1206}} = 10^{2.2380-2.1206} = 10^{0.1174} = 1.31$$

Problems just for fun

In the wee small hours after a not-on-the-up-and-up poker game, the boys decided to do some plain and fancy tradin'.

Jones gave Black a no-good $10 bill and in exchange got a no-good $5 bill and a good $5 bill.

Black gave Gray a bad $20 bill and got back good $10 and $5 bills and a phony $5 bill.

Gray gave Smith a good $20 bill and got back one good and one phony $10 bill.

Smith gave Jones a good $10 bill and got back a homemade $5 bill and a good $5 bill.

Who were the winners and losers? By how much in good United States currency?

If we wish to divide 132 by 173, we proceed as follows:

$$\frac{132}{173}=\frac{10^{2.1206}}{10^{2.2380}}=10^{-1}\times\frac{10^{3.1206}}{10^{2.2380}}$$
$$=10^{-1}\times 10^{0.8826}=10^{-1}\times 7.63=0.763$$

We use a whole number power, in this case -1, so that our exponent is a positive fraction that we look up in the table.

The accuracy of our results depends upon the completeness of the table of logarithms. We can see that such tables greatly reduce the labor of multiplication and division. Between 1614 and the first quarter of the twentieth century when calculating machines were invented, logarithms were used in making all difficult calculations. They greatly shortened the calculations of astronomers and other scientists.

Multiplication and Division Using Tables of Logarithms

Let us perform the following multiplication:

$$1{,}060\times 425\times 8.76$$

We proceed as follows:

$$1{,}060=10^3\times 1.060=10^3\times 10^{0.0253}=10^{3.0253}$$
$$425=10^2\times 4.25=10^2\times 10^{0.6284}=10^{2.6284}$$
$$8.76=10^{0.9425}$$

We now add these exponents of 10, and we find that the product

$$10^{6.5962}=10^6\times 10^{0.5962}=10^6\times 3.95=3{,}950{,}000$$

Let us perform the following division:

$$\frac{1{,}060}{425}=\frac{10^{3.0253}}{10^{2.6284}}=10^{3.0253-2.6284}=10^{0.3969}=2.49$$

We try another division:

$$\frac{27.3}{1{,}390}=\frac{10^{1.4362}}{10^{3.1430}}=10^{-2}\times\frac{10^{3.4362}}{10^{3.1430}}$$
$$=10^{-2}\times 10^{0.2932}=10^{-2}\times 1.96=0.0196$$

▲ EXERCISES

1. 43×84
2. 98×27
3. $28\times 6.2\times 108$
4. $24\times 47\times 0.79$
5. $3.9\times 4.7\times 0.007$
6. $143\times 17.2\times 0.36$
7. $136\div 349$
8. $524\div 73.2$

9. $6.83 \times 43.9 \div 7.2$

10. $78.3 \times 42.6 \div 723$

11. $\dfrac{6.89 \times 193}{139 \times 8.4}$

12. $\dfrac{0.0146 \times 37.2}{0.192 \times 42.4}$

13. $\dfrac{18{,}300 \times 0.0017}{0.007 \times 2{,}930}$

14. $\dfrac{19.3 \times 0.00067}{0.038 \times 9.34}$

15. $\dfrac{1{,}110 \times 22.2}{0.007 \times 2{,}930}$

16. $\dfrac{13.5 \times 68.2}{247 \times 854}$

17. $102 \times 682 \div 1{,}470$

18. 0.0067×79

19. $1{,}793 \times 0.004$

20. 3.140×17

Problems just for fun

Three men met in a hobo jungle. The first had three loaves of bread, the second two loaves, and the third had $1 but no bread. The third offered to buy bread from the other two. After the transaction all three had the same amount of bread, and the third had spent all of his dollar. How much money did he give to each of the other hoboes?

12 *Using Tables of Logarithms To Raise a Number to a Power and To Extract the Root of a Number*

Logarithms are most useful in raising a number to a power. For example, suppose we want to find the value of 376^4. We proceed as follows:

$$\text{Let } N = 376 = 3.76 \times 10^2 = 10^{0.5752} \times 10^2 = 10^{2.5752}$$

Then

$$N^4 = (10^{2.5752})^4 = 10^{4 \times 2.5752} = 10^{10.3008}$$
$$= 10^{10} \times 10^{0.3008} = 2.0 \times 10^{10}$$

Let us use logarithms to find a root of a number. For example, let us find the cube root of 376, that is, $\sqrt[3]{376}$. We proceed as follows:

$$\text{Let } N = 376 = 3.76 \times 10^2 = 10^{0.5752} \times 10^2 = 10^{2.5752}$$

Then

$$\sqrt[3]{N} = N^{1/3} = (10^{2.5752})^{1/3} = 10^{2.5752 \div 3} = 10^{0.8584} = 7.22$$

▲ EXERCISES

Perform the following operations using logarithms:

1. $(1.23)^2$
2. $(8.43)^3$
3. $(0.93)^2$
4. $(87.4)^2$
5. $(0.017)^3$
6. $\sqrt{97.3}$
7. $\sqrt{53.8}$
8. $\sqrt{1{,}040}$
9. $38.4 \times \sqrt{42}$
10. $38.2 \times \sqrt{0.93}$
11. $(0.0832)^4$
12. $\sqrt{0.0029}$
13. $(1.83)^{3/2}$
14. $\sqrt[3]{7.93}$
15. $\sqrt{83.4} \times (34.2)^2$
16. $\sqrt{18.9} \times \sqrt[3]{298}$
17. $(3.80)^3 \times \sqrt[3]{943}$
18. $\dfrac{15.3 \times \sqrt{278}}{87}$
19. $(14.7)^3 \times \sqrt{0.005}$
20. $\dfrac{\sqrt[4]{15.3} \times (0.72)^5}{(13.2)^{3/2}}$

Four airplanes that fly at the same speed are based on an island on the equator. Each plane has tank capacity to fly halfway around the world on the equator, and any plane may transfer gasoline from its tanks to the tanks of another plane in flight. Assuming that refueling in the air or back on the island takes zero time, how can we manage to fly one plane entirely around the world and get all planes safely back to the base with no plane landing anywhere else?

Review Exercises

1. Perform the indicated operations:

a. $(\frac{2}{3})^3$ b. $(3a^3)^2(a^2)^3$
c. $8^{2/3}$ d. $(27r^3t^6)^{2/3}$
e. $(\frac{9}{16})^{3/2}$ f. $\frac{x^3\sqrt{x^8}}{\sqrt[3]{x^{12}}}$

2. Express without radicals or negative exponents:

a. $\sqrt[4]{b}$ b. $\sqrt{4a^6}$
c. $\sqrt{4a^8b^2c^4}$ d. $\sqrt[3]{27x^9y^{-6}}$
e. $\sqrt{\frac{a^6b^8}{256}}$ f. $\sqrt[4]{81x^8y^{16}}$

3. Express in scientific notation:

a. 0.000029 b. 0.00071
c. 196,000 d. 462.7
e. 0.00037 f. 393,000,000
g. 39.93 h. 392,000

4. Using logarithms, compute:

a. 347×8.42 b. 83.97×0.067
c. $\frac{8.37 \times 9.83}{5.62 \times 2.87}$ d. $\frac{8.03 \times 19.43}{1.087 \times 293.7}$
e. $\frac{\sqrt{18.7}}{18.3 \times 6.7}$ f. $\frac{7.03 \times (0.0197)^3}{(0.00783)^4}$
g. $\frac{\sqrt[3]{58.3}}{\sqrt{9.32}}$ h. $\frac{7.62 \times (0.026)^2}{(14.9)^3}$

CHAPTER FOURTEEN

Finance

In most sciences one generation tears down what another has built and what one has established another undoes. In Mathematics alone each generation builds a new story to the old structure. HERMANN HANKEL

1 *A Bit of History*

In 1626 Peter Minuit purchased Manhattan Island from the Indians for about $24. Suppose the Indians had put the money in the bank at compound interest from that date. Interest rates in new countries are higher, so we will assume a rate of 7%. This would have amounted to more than *four billion dollars* by the beginning of this century, or more than the assessed value of *all* the real estate of the borough of Manhattan.*

So, perhaps, the Indians were not cheated so much after all, and after experiencing a few traffic jams in Manhattan, we would be willing to sell the Island back to them for the amount in the savings bank. At simple interest of 7%, money will double in 14 years. But compounded annually at 7%, it will double in just over 10 years, and in one century it will grow to 868 times itself. Don't you wish your great-grandfather had put $10 in a savings bank a century ago for you?

This growth of money, which we call interest, affects all of us in many ways: we buy a car on installment payments, we put money in a savings account, we buy a house with its mortgage. In this chapter

*W. F. White, "A Scrapbook of Elementary Mathematics," pp. 47–48, The Open Court Publishing Company, La Salle, Ill., 1908.

we will look at the matter of growth of money through interest, and in the following chapter we will consider some more general problems of growth in the world around us.

Percentage in Business

The business world uses percentage in many ways. For example, a firm's or bank's balance sheet is usually given both in dollars and in per cent of total figures. Profit and loss statements show both dollars and per cent of net sales. Expressing each asset or liability of a concern as a per cent of the total assets facilitates a comparison of different firms. Merchants use percentage to indicate discounts and markups. Percentage is used to give interest rates and tax rates.

Percentage, or per cent, was defined in Sec. 7 of Chap. 2. Let us review this definition with a few problems, some from the field of finance and some from other fields.

■ **Example 1.** A grocer sells a certain brand of peas for 25 cents a can, but offers a 10% discount if they are bought in lots of a dozen or more. How much will 18 cans of this brand of peas cost?

☐ ***Solution***

Cost without discount $= \$0.25 \times 18 = \$4.50$
Amount of discount $= \frac{10}{100} \times \$4.50 = \$0.45$
Cost of peas $= \$4.50 - \$0.45 = \$4.05$

■ **Example 2.** A dealer in photographic supplies finds that he must mark up cameras $33\frac{1}{3}\%$ to cover overhead and earn a reasonable profit. In addition there is a 4% sales tax on cameras. What will be the cost to a purchaser of a camera that the dealer obtains from the manufacturer for \$60?

☐ ***Solution***

Dealer's price $= \$60$
Dealer's markup $= 33\frac{1}{3}\% \times \$60 = \$20$
Price of camera without sales tax $= \$60 + \$20 = \$80$
Sales tax on camera $= \frac{4}{100} \times 80 = \$3.20$
Purchase price of camera $= \$80 + \$3.20 = \$83.20$

■ **Example 3.** A merchant tells a salesman that he has decided to discontinue the handling of a certain magazine and that he is returning 85. magazines. If this number represents 17% of the original consignment, how many magazines did the merchant have at the start?

□ ***Solution***

$$(17\% \text{ or } \tfrac{17}{100}) \times \text{total number of magazines} = 85$$
$$\text{Number of magazines} = 85 \times \frac{100}{17} = 500$$

▲ EXERCISES

1. The cost of an article plus the cost of shipping is \$470.88. The shipping cost is 8% of the billed price. What is the cost of the article?
2. A worker's biweekly wage of \$200 was raised to \$232. What per cent increase did he receive?
3. A worker received a 15% raise in his wages. What were his wages before the raise, if his monthly wages were \$368 after the raise?
4. A factory turned out 1,890 machines instead of the planned 1,800 machines. What was the per cent of overproduction?
5. A ruler 1 yard long is allowed a 0.01-inch margin of error. What is this per cent of error?
6. Coffee beans lose 12.5% of their weight in roasting. How many pounds of coffee beans must be roasted in order to obtain 500 pounds of roasted coffee?
7. A bricklayer found that he had laid 3,532 bricks, which was 82% of the number of bricks needed for the job. How many bricks were used on the job?
8. An alloy consists of 64% copper, 31% zinc, and 5% lead. How many pounds of each metal were used in 1,000 pounds of the alloy?
9. If a 25% markdown on a hat amounted to \$3.30, determine the original and reduced prices of the hat.
10. A person's monthly salary was \$550. On Jan. 1, he received a 20% increase in salary, but on July 1, the employer was forced to cut this person's salary by 20%. What was the monthly salary after the cut?
11. A tailor bought some woolen cloth for \$1.75 per yard. He priced the cloth to allow him a profit and later reduced this selling price by one-third for a clearance sale. The tailor found that he made a 20% profit on the cost price of the cloth. What was the original selling price of the woolen cloth?
12. What percentage of profit did the tailor in Exercise 11 originally intend to make on his woolen cloth?
13. A farmer sells 400 pounds of cream, some of which is sweet and some sour. The cream tests 38% butterfat. The farmer gets 68 cents per pound of butterfat for sweet cream and 56 cents per pound of butterfat for sour cream. How many pounds of sweet cream did the farmer have if he received \$95.38 for the cream?

Problems just for fun

Smith can run around the indoor track in the Michigan Fieldhouse in 30 seconds, and Black can run around the same track in 40 seconds. If the two men start together, how many minutes will it take for Smith to be exactly one lap ahead of Black?

3 *Interest*

If every person and business could pay cash for whatever needs to be bought, there would be no such thing as borrowed money and interest on money. However, nearly all of us and all businesses find it necessary at certain times to borrow money and pay interest or rent for the use of money. We arrange a *loan* from the bank to buy an automobile or we arrange to pay for the car in *monthly installments* through a finance company. In the latter case, the finance company puts up the cash and figures the interest into the installment payments. When we buy a house, we get a *mortgage* to cover the difference between the cost of the house and our cash down payment and we pay this mortgage off over a period of 10 or 20 years with interest added. A company which wishes to build a new plant may issue *bonds* and sell them to investors. In the bond the company promises to repay the money after a certain number of years with payments of interest each year until the bond matures.

These are examples of common forms of borrowed money on which interest must be paid. It is unusual that an individual should never need to understand the fundamentals of simple or compound interest, depreciation, and time payments. We shall discuss a few of the principles basic to the growth of money by accumulation of interest and their application to savings accounts, installment buying, and amortization of mortgages.

4 *Definition of Terms*

Most of us have had some experience, either directly or indirectly, with the terminology we shall use.

Principal P is the amount of capital invested or the amount of money borrowed.

Interest I is the total amount of money paid for the use of the principal P.

Interest rate r is the per cent of the principal P that is to be paid for the use of the money P for each period of the loan. In most cases this period is 1 year, and we shall always use 1 year as the period unless another is stated. An interest rate of 4% means that 4 cents will be paid for the use of each dollar of the principal for a period of 1 year. Each state has laws which specify the highest rate of interest that may legally be charged. At present most banks pay about 5% interest on savings accounts, and the interest rate paid by the United States government on Series E bonds held to maturity is about $4\frac{1}{4}$%. Interest rates are quoted per year, and the interest is frequently calculated for each month or fraction of a month that the principal is kept.

The investor and the borrower first agree on the interest rate and the type of interest to be paid. The amount to be paid is computed on this basis. The two types of interest are *simple* and *compound*.

5 *Simple Interest*

Simple interest is equal to the product of the principal, interest rate, and time in years:

$$I = Prt$$

Simple interest is computed on the original principal only.

The amount A repaid is equal to the principal plus the interest:

$$A = P + I = P + Prt$$

■ **Example 1.** A merchant borrowed $1,500, agreeing to repay the principal with 5% simple interest at the end of 3 years. Find the amount of interest and the total sum which must be paid.

□ ***Solution.*** The amount of interest I to be paid is

$$I = Prt = (1{,}500)(\tfrac{5}{100})(3) = \$225$$

The amount the merchant must pay is

$$A = P + I = 1{,}500 + 225 = \$1{,}725$$

■ **Example 2.** Find the time required for a savings account of $1,800 to earn $360 if the simple interest rate is 6%.

□ ***Solution.*** Since

$$I = Prt$$
$$360 = 1{,}800\left(\tfrac{6}{100}\right)t$$
$$t = \tfrac{360}{108} = 3\tfrac{1}{3} \text{ years}$$

▲ EXERCISES

1. A man borrows $800 for 6 months at a simple interest rate of 6%. How much interest does he pay, and what amount is paid at the end of the 6 months?
2. What amount will $2,500 earn at 4% simple interest in 3 years? In 3 months?
3. Find the time required for $3,500 to earn $150 at 5% simple interest.
4. Find the simple interest rate that is necessary for $2,000 to earn $90 in 9 months.
5. What principal will yield $1,458 (principal and interest) in 2 years at 4% simple interest?
6. How long must $1,000 be kept at 5% simple interest to become $1,200?
7. What school tax rate must be paid by a community whose total assessed valuation is $74,800,000, if $800,000 is needed for the schools?
8. Find the assessed valuation of a city if $750,000 is to be raised by a tax at the rate of 25 mills. $\left(1 \text{ mill} = \frac{1}{10} \text{ cent} = \frac{1}{1{,}000} \text{ dollar.}\right)$
9. A man has $10,000 and invests $3,000 for 1 year at 8% interest. At what rate of interest should the remainder be invested if the income from the $10,000 is to be $660?
10. A firm invested $30,000 for 1 year at simple interest. One portion was invested at 6% and the remainder at 3%. How much money was invested at each rate of interest if the total return was the same as it would have been if all the money had been invested at 4%?
11. A man has $4,000 invested at 4.4% simple interest. How much money must he invest at 7% simple interest to make the total investment yield 6% interest each year?
12. Find the total amount a man must pay if he borrows $1,000 and agrees to pay at the end of each month $100 on the principal and simple interest at 6% per year on the principal outstanding during each month. *Hint:* The interest due with the first payment is

$$1{,}000\left(\tfrac{6}{100}\right)\tfrac{1}{12} = \$5$$

The interest due with the second payment is

$$900\left(\tfrac{6}{100}\right)\tfrac{1}{12} = \$4.50$$

13. If the lender in Exercise 12 had lent the use of the entire principal for 10 months, what simple interest rate would have yielded the same amount of interest?

14. A loan of \$2,400 is to be repaid in quarterly installments of \$400 plus simple interest at the rate of 4% per annum on the principal outstanding during each period. Find the amount of interest and the total amount paid.
15. If the lender in Exercise 14 had lent the use of the entire principal for 18 months, what simple interest rate would have yielded the same amount of interest?

Two men agree to saw a pile of logs which are 3 feet long into 1-foot lengths for \$5. How much would these men have charged to cut the same number of logs into 1-foot lengths if the logs had been 6 feet long?

6 *Compound Interest*

In many transactions where money is loaned for more than 1 year, simple interest is not used. Suppose we lend \$100 at 5% interest. At the end of the year we should expect to receive \$105, that is, our original \$100 plus \$5 interest. If the borrower wished to continue the loan for a second year and did not pay the \$5 in cash, it would be reasonable to ask him to pay interest at 5% during the second year on \$105. This is an example of compound interest. We should say that the \$100 is loaned at 5% *compounded annually*. Any case in which the interest due at the end of a stated period is added to the principal and both earn interest during the next period, and so on, is called *compound interest*. Thus in compound interest, the principal on which the interest is computed increases from period to period.

Nearly all savings accounts use compound interest, and we can best illustrate this by an example. We deposit \$1,000 in a bank at 4% interest which is to be compounded yearly, and we calculate the value of the account at the end of 10 years. At the end of 1 year, the amount is given by

$$A_1 = \$1{,}000 + \$1{,}000(\tfrac{4}{100}) = \$1{,}040$$

If we were dealing with simple interest, the amount at the end of 2 years would be

$$A_2 = \$1{,}000 + (\$1{,}000)(\tfrac{4}{100})(2) = \$1{,}080$$

But since we loaned the money at compound interest, the \$40 interest earned the first year becomes principal and also earns 4% interest during the second year. At the end of the second year the amount would be

$$A_2 = \$1{,}040 + \$1{,}040(\tfrac{4}{100}) = \$1{,}040 + \$41.60 = \$1{,}081.60$$

The principal which draws interest during the third year is \$1,081.60, and hence at the end of the third year the amount would be

$$A_3 = \$1{,}081.60 + \$1{,}081.60(\tfrac{4}{100}) = \$1{,}124.86$$

We can continue this process and find the amount in the account at the end of 10 years.

The procedure outlined above is tedious to carry out, and we shall develop a formula which will give the compound amount A at the end of n years.

Let P = the original principal and r = the interest rate, and let the compounding take place annually. Then at the end of 1 year the amount A_1 will be

$$A_1 = P + Pr = P(1 + r)$$

At the end of the second year the amount A_2 will be

$$A_2 = A_1 + A_1 r = A_1(1 + r) = P(1 + r)(1 + r) = P(1 + r)^2$$

At the end of the third year the amount A_3 will be

$$A_3 = A_2 + A_2 r = A_2(1 + r) = P(1 + r)^2(1 + r) = P(1 + r)^3$$

At the end of the fourth year the amount A_4 will be

$$A_4 = P(1 + r)^4$$

If we continue this procedure, we shall find that the amount A_n at the end of n years will be

$$A_n = P(1 + r)^n \qquad (1)$$

This formula connects four quantities, A_n, P, r, and n. If any three of them are known, the fourth may be found.

For the problem discussed previously in this section, the amount due at the end of 10 years on \$1,000 loaned at 4% interest rate compounded annually would be

$$A_{10} = \$1{,}000(1 + 0.04)^{10} = \$1{,}000(1.04)^{10}$$

In order to calculate A_{10}, we need to know the value of $(1.04)^{10}$. We shall use tables to find $(1.04)^{10}$. Table III at the back of the book gives the values of $(1 + r)^n$ for values of n from 1 to 50 years and interest rates of $1\frac{1}{2}\%$, 2%, $2\frac{1}{2}\%$, 3%, 4%, 5%, and 6%. Using this table, we find that $(1.04)^{10} = 1.4802$ and

$$A_{10} = \$1{,}000(1.4802) = \$1{,}480.20$$

Interest Compounded Semiannually

Let the original principal P be compounded semiannually (at the end of each half year) at an annual interest rate r. Since r is the annual interest rate, the interest due at the end of one-half year is $Pr(\frac{1}{2}) = Pr/2$, and the amount at the end of the first half year would be

$$A_1 = P + \frac{Pr}{2} = P\left(1 + \frac{r}{2}\right)$$

The amount at the end of the first year, the second half-year period, would be

$$A_2 = A_1 + A_1\left(\frac{r}{2}\right) = A_1\left(1 + \frac{r}{2}\right) = P\left(1 + \frac{r}{2}\right)^2$$

and the amount at the end of $1\frac{1}{2}$ years, or 3 half-year periods, would be

$$A_3 = A_2 + A_2\left(\frac{r}{2}\right) = A_2\left(1 + \frac{r}{2}\right) = P\left(1 + \frac{r}{2}\right)^3$$

In general, if the interest is compounded semiannually, the amount at the end of n years, or $2n$ half-year periods, would be

$$A_{2n} = P\left(1 + \frac{r}{2}\right)^{2n} \qquad (2)$$

■ **Example.** If \$1,000 is placed at 4% interest for 20 years and the interest is compounded semiannually, find the amount due at the end of the 20 years.

□ ***Solution.*** Using equation (2), we get

$$A_{40} = \$1{,}000\left(1 + \frac{0.04}{2}\right)^{40} = \$1{,}000(1.02)^{40}$$

From Table III we find that $(1.02)^{40} = 2.2080$ and

$$A_{40} = \$1{,}000(2.2080) = \$2{,}208.00$$

We observe that 4% compounded semiannually for 20 years is the same as a rate of 2% per half-year period for 40 periods.

8 *Interest Compounded k Times a Year*

If we have an investment P with interest at an *annual rate* of r, compounded k times a year, then one period is the kth part of a year, the interest per period is r/k, and the number of periods is kn:

$$A_1 = P\left(1+\frac{r}{k}\right)^1$$

$$A_2 = P\left(1+\frac{r}{k}\right)^2$$

$$\vdots$$

$$A_k = P\left(1+\frac{r}{k}\right)^k \qquad \text{for 1 year}$$

$$\vdots$$

$$A_{kn} = P\left(1+\frac{r}{k}\right)^{kn} \qquad \text{for } n \text{ years} \tag{3}$$

When $k=1$, or when the interest is compounded annually, equation (3) becomes equation (1). When $k=2$, or when the interest is compounded semiannually, equation (3) becomes equation (2).

▲ EXERCISES

1. The sum of $5,000 is invested at 6% interest. If the interest is compounded annually, find the value of the investment at the end of 4 years.
2. Work Exercise 1 if the interest is compounded four times a year.
3. Work Exercise 1 if the interest is compounded semiannually.
4. The sum of $200 is deposited in each of four banks, each paying 6% interest per annum. The first bank compounds interest annually, the second bank semiannually, the third bank quarterly, and the fourth pays simple interest. If no further deposits or withdrawals are made, how much is in each account at the end of 10 years?
5. A father wishes to give his son $30,000 at the age of thirty. If the son is now twelve years old, how much money must his father invest at 5% interest compounded semiannually in order to have $30,000 when his son is thirty?
6. Find the approximate time it takes for a sum of money to double itself if the sum is invested at 6% interest compounded quarterly.

7. On the tenth birthday of his daughter a man deposits $2,000 to his daughter's account in a bank that pays 5% interest compounded annually. How much money will the daughter have on her eighteenth birthday, assuming that no withdrawals or deposits are made during this time?
8. How much will $125 amount to in 15 years if the interest rate is 5% compounded semiannually?
9. A man makes deposits of $500 on March 1 for 5 years in a bank whose interest rate is 6% compounded semiannually. What will be the value of this account just after the last deposit is made?
10. Find the value of the account in Exercise 9, 5 years after the last deposit is made, if no withdrawals or further deposits are made.
11. A grandfather deposits $1,000 in a savings account at 5% compounded semiannually when his grandson is born. What will be the value of the account when the grandson reaches his twenty-first birthday?
12. If a man can lend money at 6% compounded annually, how long will it take to triple his money?
13. Prior to 1957, the Series E bond sold for $18.75 and could be redeemed 10 years later for $25. What rate of interest, compounded annually, did it pay?
14. A man has $1,000 to deposit in a savings account for 10 years. One savings association offers 6% compounded annually and another 5% compounded semiannually. Which is the better offer and by how much?
15. A father deposits $100 in a savings account for his son on each of his birthdays from age one until he reaches twenty-one. How much will he have in his account on his twenty-first birthday if the bank pays 5% compounded annually?

Problems just for fun

Three men pay $30 for a room. Later the hotel clerk discovers that he has overcharged the men and sends a $5 refund by the bellboy. The bellboy pockets $2 and delivers $3 to the men. The men have paid $27 and the bellboy has $2. What has happened to the other dollar?

9 *Present Value of Future Payments*

If a person is to receive a series of n payments, each of amount A, the first being due a year from today, he might wish to find the present value of these payments if money is worth $r\%$ compounded annually, that is, the number of dollars today which with interest would produce the payment at the future date. For example, on selling a house it is agreed that $10,000 in cash will be paid at once and that $1,000 payments will be made annually for 10 years. What is the present value of these payments if money is worth 6% compounded annually?

The present value is the sum of the cash payment plus the present value of the ten future payments. In order to find the present value of the ten future payments, we must determine the principal P that must be set aside now at 6% interest compounded annually to yield $1,000 at the time payments are due. Since

$$A=P(1+r)^n \qquad \text{or} \qquad P=\frac{A}{(1+r)^n}$$

we need to set aside (see Table IV at end of book)

$$P_1=\frac{\$1,000}{(1.06)^1}=\$943.40$$

$$P_2=\frac{\$1,000}{(1.06)^2}=\$890.00$$

$$P_3=\frac{\$1,000}{(1.06)^3}=\$839.62$$

$$\vdots$$

$$P_{10}=\frac{\$1,000}{(1.06)^{10}}=\$558.39$$

$$P_1+P_2+P_3+\cdots+P_{10}=\$7,360.09$$

Since the present value of the ten deferred payments is $7,360.09 and the down payment is $10,000, the present value of the house is $17,360.09.

Since the process outlined above involves many divisions and is tedious to carry out, tables which reduce the labor have been prepared. Table VI gives the present value of $1 per period for n periods for an interest rate r. Thus

$$\text{Present value} = \text{payment per period} \times a_{n/r}$$

where $a_{n/r}$ is the value obtained from Table VI, r is the interest rate per period, and n is the number of periods that payments are made. In most cases, as in the above problem, the period is one year.

To apply this to the previous problem, we have $n = 10$, $r = 6$. If we look in the 6% column in Table VI for the entry opposite $n = 10$, we find $a_{10/6} = 7.3601$. This is the present value of ten payments of \$1 each. Since the payments are \$1,000 each, we multiply by 1,000 to obtain the present value of \$7,360.10. The difference of 1 cent between this figure and the one obtained by the other method results from Table VI being rounded off at five figures. If it were carried to six figures, the two results would be identical.

If the payments are made oftener than once a year, for example, k times a year, the interest rate per period is r/k and the number of periods is kn. If \$100 is to be paid four times a year for 3 years and money is worth 6%, we may wish to determine the present value of these 12 payments of \$100. We use Table VI again. The interest rate is $6\% \div 4$, or $1\frac{1}{2}\%$ per period. There are 12 periods, so we look for the entry in the $1\frac{1}{2}\%$ column opposite 12 and read $a_{12/1.5} = 10.9075$. Then the present value $= \$100a_{12/1.5} = 100(10.9075) = \$1{,}090.75$.

▲ EXERCISES

1. A man buys a house and lot and pays \$5,000 down and agrees to pay \$1,500 a year for 10 years. What is the equivalent cash price if money is worth 6% compounded annually?
2. If annual sewer assessments on a piece of property amounted to \$43.35 and 8 years intervened between the first and last payments, what was the equivalent cash payment on a 6% basis at the time the first payment was made?
3. In order to purchase a secondhand car, a man pays \$360 down and \$36 a month thereafter until 12 payments, including the down payment, have been made. If the interest rate is $1\frac{1}{2}\%$ per month, what is the cash price (present value of the payments) of the car?
4. A young man eighteen years of age is to receive from an estate the sum of \$10,800 when he reaches the age of twenty-seven. Assuming that the man will live to receive his inheritance, what is the present value of his interest in the estate if money is worth 5% compounded annually?
5. A man buys a piece of property at \$10,000 cash or \$2,000 down and \$1,000 a year for 11 years for a total of \$13,000. Assuming that money is worth 6% compounded annually, which is the best buy and by how much?
6. An appliance sells for \$255 cash or \$30 down and \$10 a month for 30 months. What is the rate of interest compounded monthly on the time payments?
7. A company issues bonds of \$100,000 to be redeemed at the end of 5 years. How much money must it deposit in a savings account at the end of each year at 4% compounded annually in order to have an account to pay off the bonds?

Problems just for fun

On a shopping trip, Mary Jane spent one-half the money that was in her pocketbook. When she got home she had just as many cents as she had had dollars and half as many dollars as she had had cents when she left home. How much money did Mary Jane have at the start of the shopping trip?

10 *Amortization of a Debt*

It is a common practice to pay off an interest-bearing debt by means of a series of payments which are usually of equal amounts. Each payment must be larger than the interest on the original debt, and the amount of payment in excess of the current interest for any period is used to decrease the principal, which thus becomes smaller and is finally reduced to zero. A debt paid off in this manner is said to be *amortized.*

■ **Example 1.** A debt of $1,000 is to be paid, principal and interest, in five equal annual payments. What is the amount of each payment if the rate of interest is 6%? Write up a schedule showing the yearly condition of this debt.

☐ ***Solution.*** The present value of the debt is \$1,000, and since

Present value = yearly payment times $a_{n/r}$

we have

\$1,000 = yearly payment times $a_{5/6}$ (Table VI)

$$\text{Yearly payment} = \frac{\$1{,}000}{4.2124} = \$237.39$$

Table 33 shows the operation of the payments.

TABLE 33

Year	*Principal outstanding at beginning of year*	*Interest at 6%*	*Payment*	*Principal repaid*
1	\$1,000.00	\$60.00	\$237.39	\$177.39
2	822.61	49.36	237.39	188.03
3	634.58	38.07	237.39	199.32
4	435.26	26.12	237.39	211.27
5	223.99	13.44	237.39	223.95

The difference of 4 cents is due to rounding off the annual payment and interest to the nearest cent.

■ Example 2. A debt of \$1,000 (without interest) is to be paid at the end of 5 years. We plan for it by paying five equal yearly payments into a savings account which pays 6% compounded annually. What is the amount of each payment?

☐ ***Solution.*** The future value of the debt is \$1,000 and we have

\$1,000 = yearly payment times $a_{n/r}$ (Table V)
= yearly payment times $a_{5/6}$

$$\text{Yearly payment} = \frac{\$1{,}000}{5.6371} = \$177.40$$

Table 34 shows the operation of the payments.

TABLE 34

End of year	*Payment*	*Interest at 6%*	*Total at end of 5 years*
1	$177.40	$46.57	$223.97
2	177.40	33.88	211.28
3	177.40	21.93	199.33
4	177.40	10.64	188.04
5	177.40	None	177.40
			$1,000.02

Let us examine carefully the difference between the last two examples. In Example 1 we are paying interest to the lender on the principal still owed at the beginning of the year. In the second example we are putting money in a bank which will pay to our account the interest earned each year and this money will, in turn, earn interest.

▲ EXERCISES

1. A man pays off a $10,000 mortgage, interest and principal, by equal payments made at the end of each year for 10 years. What is the annual payment if interest is paid at 6% per annum compounded annually? Make a schedule showing the part of each payment which goes for interest and principal, respectively.
2. An electric range costs $345.90. The down payment is $50, and the balance is to be paid in 15 equal monthly payments. If the interest rate on the unpaid balance is 18% per annum, determine the monthly payment. Make a schedule showing the part of each payment which goes for interest and principal, respectively.
3. A mortgage of $16,000 is paid off, interest and principal, by equal payments, made at the end of each year, for 20 years. What is the annual payment if interest is paid at 6% per annum on the unpaid balance?
4. A standard-quality washing machine may be purchased for $237.36 cash or a down payment of $38 and the remainder to be paid in 10 monthly installments. Find the amount of each installment if the interest rate is 24% per annum on the unpaid balance.
5. A washing machine may be purchased for $265.55 cash or a down payment of $45 and the remainder to be paid in 12 monthly installments of $21.50. Find the interest rate that is being paid on the unpaid balance at any time.
6. What is the present value of a $1,000 Series E bond maturing in 7 years if we figure its interest at 4% compounded annually?
7. An electric refrigerator may be purchased for $275 cash or a down payment of $27.50 and the remainder to be paid in 15 monthly install-

ments of \$19.25. Find the interest rate that is being paid on the unpaid balance at any time.

8. Determine the total amount of interest paid for the refrigerator in Exercise 7.
9. A contractor is engaged to build a plant addition for which he is to be paid \$250,000 at the completion of construction 2 years from now. If money earns 6% compounded quarterly, how much must the company set aside each quarter to have the cash available to pay the contractor?
10. A company takes \$500,000 from its surplus to construct an addition to the plant. How much cash must be put back into surplus each year if the total amount is to be paid back in 10 years and if the interest rate is 4% compounded annually?
11. The amount charged against one piece of property for erecting a sewage-disposal system is \$2,296.30. The owner may pay this amount in cash or extend the payments over 15 years at 5% interest compounded annually. If the owner uses the installment plan, what are his annual payments? How much additional money will he pay for the privilege of spreading the payments over 15 years?
12. A company issues \$100,000 worth of 6% bonds payable in 10 years. How much cash must it plan to set aside at the end of each year in order to pay the interest on the bonds each year and to build up a reserve to redeem the bonds at maturity if it can earn 5% compounded annually on its reserve?

When asked the time, an old gentleman replied, "If you add one-quarter of the time from noon until now to one-half the time from now until noon tomorrow, you will get the exact time." What was the time?

CHAPTER FIFTEEN

Growth

Mathematics is a science continually expanding; and its growth, unlike some political and industrial events, is attended by universal acclamation. H. S. WHITE

1 *How Quantities Grow*

In the preceding chapter we have observed how money placed at interest increases with both simple and compound interest. This is one example of the change in a quantity as time passes which we call *growth.* In this chapter we wish to study the general process of growth —the growth of a tree, plant or person; growth of the population of a country; etc. Sometimes the quantity may grow smaller as time passes, as the volume of water in a lake during a dry season or as the radioactive emanation from a radioactive substance. This is *decay* or *negative growth.*

Growth takes place in many ways. Sometimes the changes are extremely complicated and depend upon many factors. In such cases no one is able to describe fully and explain the growth. Sometimes a quantity grows in a manner that is comparatively simple, and we can understand and interpret the process. It is this latter type of growth that we wish to study.

We all have some ideas—perhaps hazy—about growth. For example, the more money we have invested at compound interest, the greater the amount that accumulates; the larger the population of a state, the greater the expected number of births; the better the climatic conditions, the faster the growth of a plant or a tree; etc.

But we know that in order to apply mathematics to a problem, we must state and interpret the problem more carefully.

We have previously studied the variation or growth processes which can be expressed by the relations

$$y = kx \qquad \text{and} \qquad y = kx^2$$

Two examples of growth expressed by the first relation are (*a*) the price of a meat roast increases with the weight in pounds and (*b*) the paycheck of a person who has a fixed hourly rate of pay increases with the number of hours worked. Two examples of growth expressed by the second relation are (*a*) the surface area of a balloon increases with the square of the diameter and (*b*) the distance a stone falls from rest increases with the square of the time of fall.

Interest Compounded Continuously

We have studied the way a sum of money grows when invested at simple interest and at compound interest. We found that when a principal P was invested at an interest rate r compounded k times a year, the amount A due at the end of n years was given by the formula

$$A = P\left(1 + \frac{r}{k}\right)^{nk} \tag{1}$$

Now let us see what change is made in the amount A if the money is compounded exceedingly often, that is, if the value of k becomes very large. When k becomes very large, for example a trillion, the compounding must be done every trillionth of a year, or about every 0.00003 second. That is, if k becomes large without limit, then we say that the money is compounded continuously.

Let us see how a dollar invested at 100% interest grows during 1 year if the number of times k that interest is compounded is increased. The amount A due on \$1 at the end of 1 year is given by

$$A = \left(1 + \frac{1}{k}\right)^{k} \tag{2}$$

Table 35 gives the value of A for different values of k.

TABLE 35

k	1	10	100	1,000	10,000	100,000
A	2.000	2.594	2.704	2.717	2.718	2.718

We notice that k increases very rapidly, but the amount A, while it increases, does not increase very rapidly and appears to be approaching a limiting value of about 2.718. In fact, mathematicians have proved that as k increases indefinitely, the value of A correct to eight decimal places is equal to 2.71828183.

Since this limiting value of $\left(1+\frac{1}{k}\right)^k$ as k increases indefinitely is encountered in many branches of mathematics, it is denoted by the letter e. Thus

$$e = 2.71828183 = \text{limiting value of } \left(1+\frac{1}{k}\right)^k$$

as k becomes very large. For our purposes, we will take $e = 2.718$. Thus the amount due on \$1 at the end of 1 year if the interest is compounded continuously at 100% is \$2.72. As depositors, we wish that our savings bank would pay interest compounded continuously.

3 *Effect of Compounding Continuously*

Equation (1) gives the amount due at the end of n years when the interest is compounded k times a year.

$$A = P\left(1+\frac{r}{k}\right)^{nk} \tag{1}$$

In the preceding section we found that $A = 2.718$ when k increases indefinitely and $P = 1$, $r = 100\%$, $n = 1$. Now, what will be the effect on this equation if k is increased indefinitely but if we do not give special values to P, n, and r?

If we let $r/k = 1/z$, then $k = rz$, and equation (1) becomes

$$A = P\left(1+\frac{1}{z}\right)^{nrz} = P\left[\left(1+\frac{1}{z}\right)^z\right]^{nr} \tag{3}$$

If k increases indefinitely, z will also increase indefinitely. We have just seen above that the limit of $\left(1+\frac{1}{k}\right)^k$ as k becomes very large is e. Therefore, when k, and hence z, become exceedingly large, equation (3) becomes

$$A = Pe^{nr} \tag{4}$$

This equation (4) gives the amount A of any principal P after n years if the interest is *compounded continuously* at an annual rate r.

That is, the amount A is equal to the principal P invested, multiplied by the number $e=2.718$ raised to the nr power.

For example, if $r=10\%$, $n=5$ years, and $P=\$100$, the amount a savings bank would owe us is

$$\begin{aligned} A &= 100e^{5(1/10)} = 100e^{1/2} \\ &= 100 \times 1.649 = \$164.90 \end{aligned}$$

The Exponential Functions e^{kx} and e^{-kx}

When any quantity y varies with another quantity x in such a way that the rate of change in y is constantly proportional to the value of y, it is said to vary in a manner similar to the continuous-interest law. The functional relation between y and x is given by

$$y = Ae^{kx} \tag{5}$$

where A and k are constants.

This relation (5) is important and is known as the *law of continuous growth.* It is commonly found in problems of finance, growth of timber, growth of bacteria, etc. If the exponent is negative, the relation

$$y = Ae^{-kx} \tag{6}$$

is called the *law of continuous decay.*

This law of decay is used to determine the altitude when the atmospheric pressure is given, the healing of the surface area of a wound, the decomposition of radium, transmission of light through various media, damped vibrations of a pendulum, loss of heat or electric energy, etc.

The two functions given by equations (5) and (6) are called *exponential functions.* Whenever the nature of the functional dependence between two variables is known to be exponential, the exact dependence is given by one of these equations.

The Graph of the Functions $y=e^x$ and $y=e^{-x}$

The values of e^x and e^{-x} are defined for all values of x and have been determined for many values of x. These values may be found in tables. Let us plot each of these functions (Fig. 79).

The two curves in Fig. 79 are known as the *standard exponential curves.* We observe that each curve crosses the y axis at the point $y=1$. The curve for $y=e^x$ rises very rapidly as x increases and ap-

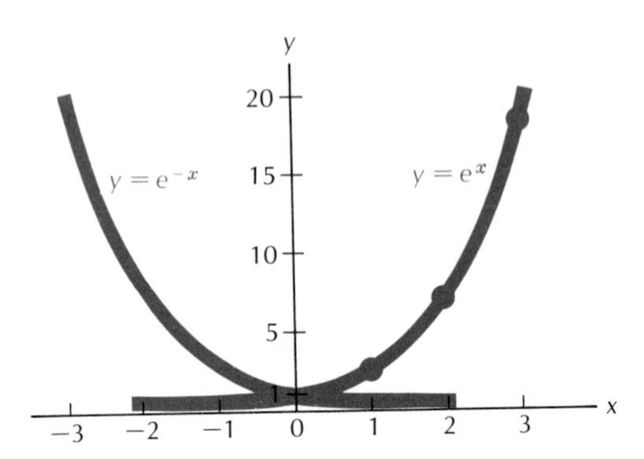

FIG. 79

proaches the x axis as x becomes large and negative. The curve for $y = e^{-x}$ rises very rapidly as x becomes large and negative, and approaches the x axis as x becomes large and positive. Neither curve crosses the x axis.

▲ EXERCISES

1. Using the same coordinate axis and scale, plot the curves for the following functions:

   a. $y=x+1$ b. $y=x^2+1$
   c. $y=x^3+1$ d. $y=e^x$

2. Compare the four curves in Exercise 1 to notice which exhibits the greatest growth.
3. Using the same coordinate axes and scale, plot the curves for the two functions $y=2^x$ and $y=x^2$.
4. Which of the two functions in Exercise 3 increases the most when x changes from 2 to 3?
5. The two curves in Exercise 3 will intersect at three points. Find the coordinates of the three points.
6. By means of Table VIII at the end of the book, find the value of the following:

   a. $e^{1.2}$ b. $e^{7.5}$
   c. $e^{2.5}$ d. $e^{-0.4}$
   e. $e^{-2.5}$ f. $e^{-1.1}$
   g. $e^{-2.1}$ h. $e^{4.5}$

7. Using Table VIII at the end of the book, find the value of x for which the following equations are true:

   a. $e^x=1.82$ b. $e^x=0.18$
   c. $e^x=8.2$ d. $e^x=0.082$
   e. $e^x=1{,}090$ f. $e^x=0.006$
   g. $e^x=0.20$ h. $e^x=0.37$

Problems just for fun

At exactly 12 o'clock, 2 bacteria are placed in a growing medium. One minute later there are 4 bacteria, in another minute they have increased to 8, in another to 16, etc. At exactly 1 o'clock, the growing mass measures 1 gallon. At what time will there be 1 quart of bacteria?

6 *Growth in Nature*

In nature many quantities grow or increase in a manner similar to the way a sum of money grows when the interest is compounded very often or continuously. Mathematicians have proved that whenever a physical quantity P has a rate of growth that is proportional to the amount present at any time, the amount at any time is given by an equation similar to equation (5). That is, if a quantity P grows at a rate r per unit of time to a value Q after t units of time, then Q is given by the relation

$$Q = Pe^{rt} \tag{7}$$

In many instances there is a continuous decrease in the quantity instead of an increase. When a quantity P, which decreases continuously at a constant rate r per unit of time, has decayed to a value Q after t units of time, then Q is given by the relation

$$Q = Pe^{-rt} \tag{8}$$

■ **Example 1.** The number of bacteria in a culture of yeast increases continuously at a rate per minute which is always 12% of the number of bacteria then present. If the original number of bacteria is 1,000, how many bacteria are in the culture at the end of 10 minutes?

□ ***Solution.*** The number Q of bacteria present at any time is given by the relation

$$Q = 1{,}000e^{0.12t}$$

When $t = 10$, $Q = 1{,}000e^{1.2} = 1{,}000(3.32) = 3{,}320$. The value $e^{1.2} = 3.32$ is obtained from Table VIII at the end of the book.

■ **Example 2.** In a treated culture, the number of bacteria present was 1,000 to start with, and 5 minutes later the number of bacteria was 670. If the bacteria decreased continuously at a rate $r\%$ per minute, find the rate r and the time it takes the number of bacteria to decrease to 100.

□ ***Solution.*** The number Q of bacteria present at any time is given by the relation

$$Q = 1{,}000e^{-rt} \tag{9}$$

When $t = 5$, $Q = 670$. Hence

$$670 = 1{,}000e^{-5r} \qquad \text{and} \qquad e^{-5r} = 0.67$$

From Table VIII, we find that $e^{-x} = 0.67$ when $x = 0.4$, and

$$r = \frac{0.4}{5} = 0.08 = 8\%$$

Equation (9) becomes

$$Q = 1{,}000e^{-0.08t}$$

When $Q = 100$, we have

$$100 = 1{,}000e^{-0.08t} \qquad \text{and} \qquad e^{-0.08t} = 0.1$$

From Table VIII, we find that $e^{-x} = 0.1$ when $x = 2.3$. Hence $0.08t = 2.3$ and $t = 28.75$ minutes, which is the time for the number of bacteria to decrease to 100.

■ **Example 3.** A snowball rolls down a steep mountainside. The snowball weighs 5 pounds originally, weighs 9 pounds after 400 feet, and increases its weight continuously at a rate of 80% every 400 feet. What will be the weight of the snow mass after $\frac{1}{2}$ mile?

□ ***Solution.*** This is a problem in growth, but the units involved are distance rolled rather than time. The equation

$$Q = Pe^{rd}$$

applies, where $P = 5$ pounds and $Q = 9$ pounds when $d = 400$ feet. Then

$$9 = 5e^{400r}$$
$$e^{400r} = 1.8$$

From Table VIII, $400r = 0.6$ and $r = 0.0015$. Then our equation of growth is

$$Q = 5e^{0.0015d}$$

and we wish to find Q when $d=\frac{1}{2}$ mile $=2{,}640$ feet. This we do as follows:

$$Q=5e^{0.0015\times 2,640}=5e^{3.96}=5\times 50=250 \text{ pounds}$$

where we have estimated $e^{3.96}=50$ from $e^{4.0}=54.6$.

▲ EXERCISES

1. The number of bacteria in a certain culture increases at a rate per hour equal to 60% of the number. If the original number of bacteria was 200, find the number after 2 hours and after 10 hours.
2. The growth of a certain type of tree is such that its cross-sectional area increases continuously at an annual rate of 10%. Find the cross-sectional area of the tree at the end of 12 years, if the original cross-sectional area was 4 square inches.
3. Find the amount that would be produced by $1,000 if it could be invested at 6% interest compounded continuously for 10 years.
4. Compare the answer for Exercise 3 with the amount received from $1,000 invested at 6% compounded semiannually for 10 years.
5. A building depreciated continuously at a constant rate of 6% per year. If the original value was $50,000, what is the formula for its value after t years? What was the value of the building after 20 years?
6. Sunlight transmitted down into deep water loses its intensity continuously and has approximately 50% of its surface intensity at a depth of 100 feet. Find the depth at which the intensity of the light will be 20% of its intensity at the surface.
7. The pressure P as measured by the height of a mercury column in inches is given by the equation

   $$P=29.92e^{-h/26,200}$$

   where h is the elevation above sea level expressed in feet. Find the pressure in inches of mercury at the following altitudes:

   a. Sea level, $h=0$ feet b. $h=2{,}620$ feet

   c. $h=29{,}000$ feet d. $h=10$ miles

8. If a metal ball is cooled by a moving stream of air, the difference in temperature between the ball and the air decreases continuously at a rate of k degrees per second, where k depends on the radiation qualities of the metal. If $k=0.002$, how long will it take for the difference in temperature between the ball and the air to drop from 38 to 23°C?
9. Radium disintegrates continuously with a half-life of 1,600 years; that is, after 1,600 years, one-half of the amount remains and after another 1,600 years, one-quarter of the original amount remains, etc. Find the amount of radium remaining in 1 gram after 5,000 years.
10. Radioactive sodium disintegrates continuously with a half-life of 15

hours. If 10 milligrams of radioactive sodium is administered to a patient, how much of the sodium remains at the end of 1 hour? After 5 hours? After 30 hours?

11. A sheet hanging in the wind on a certain day loses its moisture continuously at the rate of 50% per hour. How much moisture remains in the sheet at the end of the fourth hour?
12. The charge on a condenser decreases continuously by 10% of the charge on the condenser each second. If the original charge is 250 units, how long will it take for the charge to decrease to 125 units?
13. A radioactive substance disintegrates continuously at the rate of 5% per hundred years. About how long will it take for one-half of a lump to disintegrate?
14. If the number of insects in a colony increases continuously at 5% per day, and if the colony now has 1,000 insects, how many insects will it contain 1 week from now?
15. Some bacteria which double themselves every $\frac{1}{2}$ hour are placed in a pint jar. After 10 hours, the jar is half full. How long will it be before the jar is filled?

Arrange the digits 1 through 9 in the square to make a perfect square in which the sums of each row and each column are the same number.

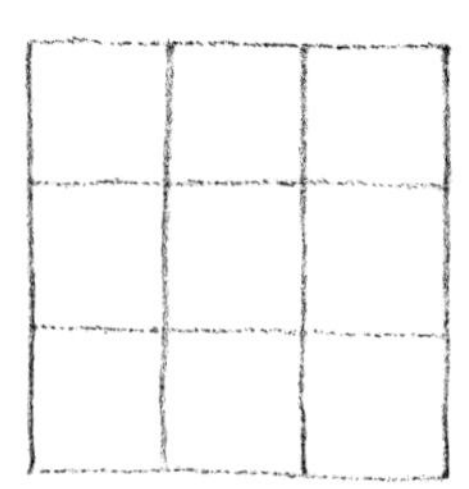

CHAPTER SIXTEEN

Probability

It is remarkable that a science [probability], which began with the consideration of games of chance, should have become the most important object of human knowledge.

LAPLACE

1 *Probability, Or Chance*

"Do you think we will win our football game Saturday?" "Do you think it will rain?" "Professor, what are my chances of making an A in this course?" "Will I be alive ten years from today?" "If we produce this new model, how many will we sell?" "Holding the ace and queen of trumps, should I finesse or play the ace hoping to drop the singleton king?" These are some of the questions that illustrate the popular notion of probability or chance. Unless we have a crystal ball in which to read the future, we cannot give any reliable answers to the first four questions. However, every expert bridge player knows the odds on the fifth question depending on the number of trumps in the opponents' hands. And based on market research, the manufacturer may obtain an estimate of the number he may sell.

We are aware that figuring the odds of something happening or not is important in bridge, poker, craps, and other games of chance. However, this branch of mathematics, which is called *probability,* has also proven to be very important in many areas of business and science, for example, in statistical studies of physical, biological, and social phenomena, the theory of errors of measurements, etc. Important applications of the theory of probability are found in fire and life insurance, the Mendelian law in genetics, the making of a good pension or retirement program, manufacturing processes, the military science of bombing and artillery fire, etc.

The insurance business, one of the largest businesses in the world, is based upon the ability to establish the likelihood of the occurrence of certain events. On the basis of past records, the life insurance company establishes the average life expectancy of a person and bases its premiums upon the probability of his reaching a certain age.

In order to develop a good pension or retirement fund, we need to know the probability that we will live 5, 10, or 15 years after retirement. After the amount of money expected to be needed after retirement is determined, we simply calculate the amount needed to be set aside each month at a given interest rate to yield the desired sum at retirement.

2 *Historical Development of Probability*

Historically, the theory of probability had its origin in games of chance or gambling. Apart from a few casual remarks on the subject by Galileo (1564–1642), we find the origin of the science of probability in the personal correspondence between two great men, Blaise Pascal (1623–1662) and Pierre de Fermat (1601–1665). Chevalier de Méré, a French nobleman and a man of great experience in gambling, asked Pascal to explain an apparent contradiction between his theoretical reasoning and his observations gathered from the gambling tables. The problem was: Suppose a player throws two dice several times and loses if he throws double six on any one throw. Of course the odds favor the player on a few tosses and favor the house if a large number of throws is required. At how many throws do the odds change in favor of the house? Pascal solved the problem by showing that the odds favor the player if 24 throws are required but favor the house slightly if 25 throws are required. Fermat became interested in this problem and in those of a similar nature. In their private correspondence, these two men laid the foundation of the science of probability.

3 *Basic Ideas of Chance*

Differing from many problems in mathematics where we can state exactly what will happen, we are concerned in probability with the likelihood that a certain event will occur, but we cannot be certain what will happen. If we have 100 red balls and just 1 blue ball in a bag, we can say it is almost certain that a random draw will give a red ball. Yet we know it could happen that we get the blue ball, although it is not likely. What, then, do we mean by probability? To begin, let us look at some simple coin-tossing experiments.

■ **Example 1.** If we toss a coin in the air, what is the chance, or probability, that it will fall "heads"?

We know that a tossed coin will fall in one of two ways, that is, with "heads" or "tails" showing. We are assuming that the coin is an honest one and that there is no reason to expect that it is more likely to fall one way than the other. We write the total list of possible events of a toss as H, T and we are concerned with one of these two ways. Thus we say that the chance, or probability, of a head is $\frac{1}{2}$. By this we mean that if a coin were tossed many times, we should expect the number of heads tossed divided by the total number of tosses to be about $\frac{1}{2}$.

■ **Example 2.** If we toss two coins in the air, what is the chance that both will fall heads?

Here there are four possible outcomes (HH, TT, HT, TH), by which we mean that both may fall heads, both tails, the first coin heads and the second tails, or the first tails and the second heads. Of these, only the first gives our desired result, or is a *success*. Since the total list of four possible outcomes contains one *success* and three *failures*, we say that the chance is $\frac{1}{4}$.

■ **Example 3.** If we toss two coins, what is the chance that they will fall one head and one tail?

In the preceding discussion we see that two of the four possible outcomes are successes and we say that the chance is $\frac{2}{4}=\frac{1}{2}$.

■ **Example 4.** If we have tossed a coin 30 times and 20 of these have produced heads, what is the chance that the next toss will be heads?

Here is one of the common errors in thinking about probability. We expect that the result of a long series of tosses will produce about 50% heads and we reason that, when the number of heads exceeds tails, a tail is more likely on the next toss. Not so. Our coin has no way of *knowing* what has gone on before. On *any* toss there are two possible outcomes (H, T) and the probability of heads is still $\frac{1}{2}$.

"When you toss a coin to decide who is going to pay the check, let your companion do the calling. 'Heads' is called seven times out of ten. The simple law of averages gives the man who listens a tremendous advantage." (Henry Hoyns, quoted by Bennett Cerf in *The Saturday Review of Literature*.)

Is Mr. Hoyns correct?

Problems just for fun

In a hand of bridge, which is the more likely to happen: (*a*) You and your partner have all the hearts, or (*b*) you and your partner have no hearts?

4 *Definition of Probability*

If, on any one trial, an event can happen in a different ways and can fail to happen in b different ways, and if all the $a+b$ ways are equally likely, the ratio

$$p=\frac{a}{a+b}$$

of the favorable to the total number of ways is called the probability p that the event will occur on a particular trial. The ratio

$$q=\frac{b}{a+b}$$

of the ways the event can fail to happen to the total number of ways the event may happen is called the probability q that the event will fail to occur on the trial. That is, the probability of an event equals the number of ways it can occur divided by the total number of ways it can occur and fail to occur.

Notice that

$$p+q=\frac{a}{a+b}+\frac{b}{a+b}=\frac{a+b}{a+b}=1$$

If an event is certain to happen, its probability is 1, for in this case there are no ways in which it can fail and $p=a/a=1$. If an event is certain to fail, $a=0$ and $p=0$, but $q=1$. In all other cases, a is greater than zero and less than $a+b$, so that the probability p is a positive fraction whose value is never less than 0 nor greater than 1.

■ **Example 1.** An ordinary die is thrown. What is the probability that an even number, that is, 2, 4, or 6, will turn up?

The total number of ways in which the die may fall is 6 and we assume the die is not loaded so that each way is equally likely. The possible outcomes are 1, 2, 3, 4, 5, 6, and 2, 4, 6 are successes. Then the probability that an even number will come up is $\frac{3}{6}=\frac{1}{2}$.

■ **Example 2.** A number from 1 to 10 is chosen at random. What is the probability that it is divisible by 4?

There are ten numbers (1,2,3,4,5,6,7,8,9,10) and only two (4,8) are divisible by 4. Hence the probability of getting a number divisible by 4 on one draw is $\frac{2}{10}=\frac{1}{5}$.

In order to apply the definition of probability to an experiment, we need to list or count the total number of equally likely outcomes of the experiment and then to count the number of these outcomes that are successes, that is, ways in which the desired event occurs.

We must be certain that these are *equally likely*. If we toss two coins, there are three possible outcomes—two heads, one head, no heads. We might conclude that the probability of two heads is $\frac{1}{3}$. However, these three possibilities are not equally likely because there are four equally likely falls—HH, HT, TH, TT. Thus the true probability of two heads is $\frac{1}{4}$.

▲ EXERCISES

1. If we name a date at random, what is the chance that this date falls on Sunday?
2. If 1 card is drawn from an ordinary deck of 52 bridge cards, what is the chance of drawing a heart?
3. A number from 1 to 10 is selected at random. What is the probability that it is divisible by 2? By 3? By 5?
4. What is the probability
   a. Of throwing a 6 when rolling a single die?
   b. Of naming at random a date that falls on Tuesday?
5. If three coins are tossed, what is the probability of getting
   a. Three heads?
   b. Three tails?
   c. Two heads and one tail?
6. If a number between 1 and 12, inclusive, is selected at random, what is the probability that it is not divisible by any number except itself and 1?
7. If a number between 1 and 30, inclusive, is selected at random, what is the probability that it is

a. Divisible by 2?
b. Divisible by 3?
c. Divisible by 5?
d. Not divisible by any number except itself and 1?

8. a. List all possible two-letter combinations from the word "COED."
   b. What is the probability that the two-letter combination contains an O?
   c. What is the probability that the two-letter combination contains neither an O nor a C?

9. A coin and a die are tossed into the air.

   a. In how many different ways can they fall?
   b. Assuming both coin and die to be ideal, what is the probability that the coin will show heads and that the die will show an even number?

10. A card is drawn from a 52-card deck.

   a. What is the probability that it is a ten-spot?
   b. What is the probability that it is the 10 of spades?
   c. If the first card drawn is a ten-spot and it is held out by the drawer, what is the probability of drawing another ten-spot on the next draw?
   d. What is the probability of drawing a face card (king, queen, or jack) in a single draw?

Without taking your pencil from the paper, draw 4 straight lines which will pass through all 10 of the dots.

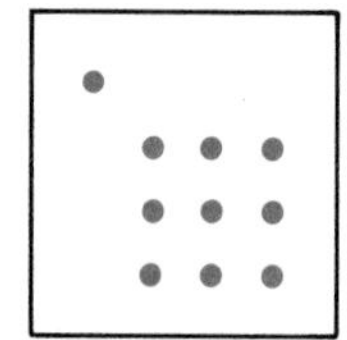

More Problems on Probability

In order to determine the probability that a certain event will happen as the result of an experiment, we often need to draw a diagram or picture to show all possible outcomes. The following problems will serve as illustrations.

■ **Example 1.** If two dice are thrown, what is the probability of both showing the same number?

☐ ***Solution.*** Let us draw a diagram to show all the possible outcomes:

1,1	1,2	1,3	1,4	1,5	1,6
2,1	2,2	2,3	2,4	2,5	(2,6)
3,1	3,2	3,3	3,4	(3,5)	3,6
4,1	4,2	4,3	(4,4)	4,5	4,6
5,1	5,2	(5,3)	5,4	5,5	5,6
6,1	(6,2)	6,3	6,4	6,5	6,6

In each of these 36 possibilities, the first number represents the spot on the first die and similarly for the second. Of these 36, exactly 6 show the same numbers. These are the ones underlined. Then the probability is $\frac{6}{36} = \frac{1}{6}$.

■ **Example 2.** If two dice are thrown, what is the probability that the sum of the spots showing will be 8?

☐ ***Solution.*** If we refer to the diagram of Example 1, we see that exactly five possible outcomes are successes, namely, those which are in parentheses. Then the probability is $\frac{5}{36}$.

■ **Example 3.** If two balls are drawn at random from a bag containing five red balls and two black balls, what is the chance that (*a*) two red balls will be drawn? (*b*) That two black balls will be drawn? (*c*) That one red and one black ball will be drawn?

☐ ***Solution.*** Let us designate the red balls by R_1, R_2, R_3, R_4, and R_5 and the two black balls by B_1 and B_2. The total number of outcomes is 21 as shown by:

R_1,R_2					
R_1,R_3	R_2,R_3				
R_1,R_4	R_2,R_4	R_3,R_4			
R_1,R_5	R_2,R_5	R_3,R_5	R_4,R_5		
R_1,B_1	R_2,B_1	R_3,B_1	R_4,B_1	R_5,B_1	
R_1,B_2	R_2,B_2	R_3,B_2	R_4,B_2	R_5,B_2	B_1,B_2

The successes under (*a*) are found in the top four rows and the chances of selecting two red balls out of the seven balls is $\frac{10}{21}$. The successes of (*c*) are found in the bottom two rows except the right entry, and the chance of drawing one red and one black ball is $\frac{10}{21}$. Only one success is found for (*b*) and the chance of drawing two black balls is $\frac{1}{21}$.

We observe that $\frac{10}{21} + \frac{10}{21} + \frac{1}{21} = 1$. This checks with the observation that every draw must result in two red, two black, or one red and one black ball.

■ **Example 4.** Three different pairs of gloves are placed in a box. If two gloves are drawn at random from the box, what is the chance of drawing a matched pair? Drawing a glove for the right and left hand? Drawing two gloves for the right hand? Drawing two gloves for the left hand? for the same hand?

□ ***Solution.*** Let us use R_1,L_1, R_2,L_2, and R_3,L_3 to represent the three pairs of gloves. The possible ways of drawing two gloves are:

R_1,R_2				
R_1,R_3	R_2,R_3			
R_1,L_1	R_2,L_1	R_3,L_1		
R_1,L_2	R_2,L_2	R_3,L_2	L_1,L_2	
R_1,L_3	R_2,L_3	R_3,L_3	L_1,L_3	L_2,L_3

There are 15 ways of selecting two gloves and only 3 ways of obtaining a matched pair. Hence the probability of obtaining a matched pair is $\frac{3}{15} = \frac{1}{5}$.

There are 9 ways of selecting a glove for the right and left hand. Hence the chance of drawing a glove for each hand is $\frac{9}{15} = \frac{3}{5}$.

There are 3 ways of drawing two gloves for the right hand and the probability is $\frac{3}{15} = \frac{1}{5}$. Similarly for the left hand the chance is $\frac{1}{5}$.

There are 6 possibilities for two gloves for the same hand and the chance is $\frac{6}{15} = \frac{2}{5}$.

▲ EXERCISES

1. If a number is drawn at random from the integers 1 to 20, inclusive, what is the probability that it will

   a. End in 5?
   b. Be even?
   c. Be divisible by 3?
   d. Be divisible by 4?

2. What is the probability that a single throw of a die will result in a number under 4? Under 5?
3. What is the probability of throwing a pair of even numbers with two dice?
4. If two dice are thrown, what is the chance of throwing the dice so that faces add to

   a. 2
   b. 3
   c. 4
   d. 5
   e. 6
   f. 7
   g. 8
   h. 9
   i. 10
   j. 11

5. If two dice are thrown, what is the probability of throwing

   a. Two 1's?
   b. Two 3's?

6. A word is chosen at random from the words "girl, book, school, class, week, year." What is the probability that the word chosen will

   a. Contain only one vowel?
   b. Contain two vowels?
   c. Be a four-letter word?
   d. Be a five-letter word?
   e. End in the letter "l"?

7. If two balls are drawn at random from a bag containing three white and three black balls, what is the probability that

   a. Both will be black?
   b. One will be black and one white?

8. If three balls are drawn from a bag containing four black and three red balls, what is the probability that

   a. All will be black?
   b. Two will be red and one black?
   c. All will be red?

9. A bag contains three red balls with the numbers 1, 2, and 3 painted on them and two white balls with the numbers 1 and 2 on them. If two balls are drawn at random, what is the probability that

   a. Both balls will be red?
   b. The two balls with the figure 2 will be drawn?
   c. The sum of the two numbers will be 4?
   d. The sum of the two numbers will be 5?

10. Four aces, four kings, and four queens are shuffled and three cards are drawn. What is the probability that

    a. All will be aces?
    b. Two will be aces and one a king?
    c. There will be one ace, one king, and one queen?

Problems just for fun

Three natives made a raid on a neighboring village and stole some bananas. It became dark before they reached their own village, so they decided to sleep and divide the loot the following morning. The thieves posted a pet monkey nearby as lookout and went to sleep. The first thief woke up and, not trusting the others, divided the large pile into three equal piles and had one banana left over. He fed the extra banana to the monkey, hid his pile. and left the remainder. A short time later the second thief woke up, divided the bananas that remained into three equal piles, and found that there was an extra banana, which he fed to the monkey. This thief hid his pile and went back to sleep. The third thief awoke and divided the remaining pile into three equal piles, with one banana left over, which he gave to the monkey. He hid his share and went back to sleep. In the morning the three thieves divided the small remaining pile into three equal piles with one banana left over, which they fed to the monkey. How many bananas did the thieves steal?

6 *The Problem of Choice*

To find the probability that a certain event will happen, we need to determine all the possible ways that the event can occur and fail to occur. If the number of ways is small, we can list all the possibilities. But as the number of possibilities increases, it takes too long and is sometimes almost impossible to name all the ways. For example, to how many patrons can a telephone company give telephone numbers having four digits if the first number cannot be zero? How many code "words" can we form by using any six different letters of the alphabet? How many car licenses can a state have if the license consists of one letter followed by five numbers?

We shall devote the next few sections to learning how to count the possible arrangements for such problems. We shall first illustrate the general methods in some simple cases where we can check the new method by enumeration of all possibilities.

■ **Example 1.** If we wish to ascend a certain mountain, we may go by car, by foot, or by a cograil car from a town at the foot of the mountain to a point A part way up the mountain, but to reach the top of the mountain from there, we must either walk or ride the cograil car. In how many ways is it possible to reach the top of the mountain?

□ ***Solution.*** First let us make a simple sketch showing the ways it is possible to reach point A and the top of the mountain.

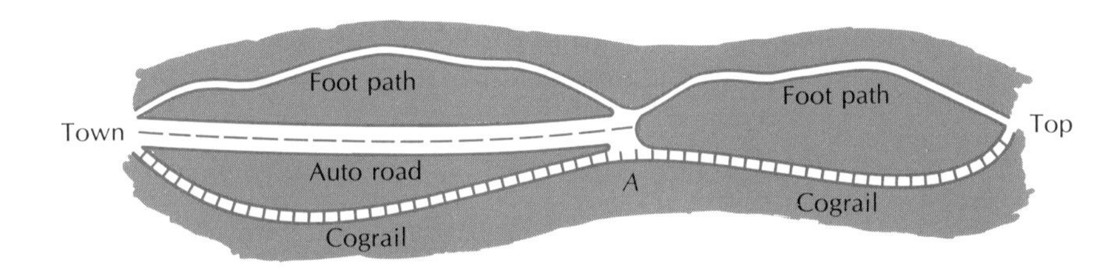

FIG. 80

By observing the drawing (Fig. 80), we can write down all the possible ways of ascending the mountain.

1. Foot path—foot path
2. Foot path—cograil
3. Auto road—foot path
4. Auto road—cograil
5. Cograil—foot path
6. Cograil—cograil

We find that there are six, and only six, ways of ascending this mountain. Notice that the answer of six ways can be obtained by multiplying 3, the number of ways of going from the town to point A, by 2, the number of ways of going from point A to the top of the mountain.

We might obtain this solution in another way. For each of the three ways of going from the town to point A, there are two ways of going from A to the top of the mountain. Hence the number of ways to climb the mountain is $2+2+2=6$.

■ **Example 2.** Our friend the milkman dresses in the dark so as not to awaken his wife. If he has four pairs of shoes, in how many ways is it possible for him to leave the house with a left and a right shoe from different pairs?

□ ***Solution.*** He may select any one of four left shoes. With the left shoe which he may select, he may choose any one of three right shoes from the remaining pairs of shoes. Hence the number of ways of selecting a left and a right shoe not from the same pair is

$$3+3+3+3=4\times 3=12$$

We can check this answer by actually enumerating all possible choices. Let the pairs of shoes be designated by

$$L_1,R_1 \qquad L_2,R_2 \qquad L_3,R_3 \qquad L_4,R_4$$

where L_1 means the left shoe and R_1 means the right shoe belonging to the first pair of shoes and similarly for the other pairs. The solution above states that with L_1 the milkman can select R_2, R_3, or R_4. Proceeding in this way, we can write down the 12 possible selections

$$\begin{array}{cccc} L_1,R_2 & L_2,R_1 & L_3,R_1 & L_4,R_1 \\ L_1,R_3 & L_2,R_3 & L_3,R_2 & L_4,R_2 \\ L_1,R_4 & L_2,R_4 & L_3,R_4 & L_4,R_3 \end{array}$$

■ **Example 3.** Three flags are to be placed in a vertical row on a mast as a signal. If there are seven different flags, how many signals are possible?

□ ***Solution.*** We can select the first, or top, flag in seven different ways. Having selected the top flag, there remain six ways of selecting the second flag. Having selected the top two flags, the bottom flag can be selected in five ways from the remaining flags. We see that for each way of selecting the top flag there are six ways of selecting the second flag, and hence there are $7\times 6=42$ ways of selecting the first two flags. Likewise, for every way of selecting the first two flags there are five ways of selecting the bottom flag. Hence there are $7\times 6\times 5=210$ ways of selecting the three flags, and 210 signals can be arranged. It would be a real task to enumerate the 210 arrangements.

7 *Fundamental Principle of Choice*

In order to help us solve problems similar to the examples above, we now state a fundamental principle.

Fundamental principle of choice. *If one selection can be made in any one of n ways, and if, after one of the n ways has been selected, a second selection can be made in any one of m ways, then the two selections can be* *made together, in that order, in $n \times m$ ways. If, in addition, a third selection can be made in k independent ways, the three selections can be made together in $n \times m \times k$ ways.*

Proof. Corresponding to the first one of the n ways of making the first selection, there are m ways of making the second selection. For the second one of the ways of making the first selection,there are m ways of making the second selection, and so on—for each of the n ways of making the first choice, there are m ways of making the second choice. There are then

$$\underbrace{m+m+m+m+\cdots+m}_{n \text{ terms}}=n \times m$$

ways that the two selections can be made together in that order.

Sweet Sue types four letters for her boss and addresses four envelopes. As it is late in the afternoon and she is late for a date with Boyfriend, she hurriedly inserts the four letters into the envelopes at random. What is the probability that exactly three of the letters will go into the correct envelopes?

▲ EXERCISES

1. If a nickel, a dime, and a quarter are tossed into the air together, in how many ways may they fall?
2. Between Milwaukee, Wis., and Chicago there are two steam railroads, one electric railroad, a boat line, a bus line, and an airplane line. In how many different ways can one make a round trip between the two cities?
3. From a group of nine students a president, a vice-president, and a sec-

retary are to be elected. How many sets of officers are possible if no student may hold two offices?

4. If a die is tossed, in how many ways can it fall?
5. Five people enter a bus at the same time. If there are three vacant seats, in how many ways may these people be seated?
6. How many signals can be made using two different flags if each flag can be held in any one of five different positions? Each different position of the two flags (except together) represents a signal.
7. If one coin and one die are thrown, in how many ways can they fall?
8. How many mixtures can be made using one of three spices and one of four flavoring extracts?
9. Ten men compete in a race in which the first five places win prizes. If there are no ties, in how many ways may the prizes be awarded?
10. How many straight lines are determined by 5 distinct points on a given circle? By 10 points? By 20 points? If the solution is not obvious, try drawing a diagram and counting.

A man and his wife, who weigh 150 pounds apiece, have two sons, each of whom weighs 75 pounds. They own a boat which is capable of carrying only 150 pounds. If this family wishes to cross a river, how can they do it?

8 *Ordered Counting or Permutations*

Let us now determine the number of ways that 2 letters can be selected from the 4 letters *a, b, c,* and *d*. We know that there are 4 ways of selecting the first letter, and after this choice there are 3 ways of making the second selection. Hence the number of ways of selecting 2 letters from 4 letters is

$$4 \times 3 = 12 \text{ ways}$$

We can easily list all the possible arrangements.

ab	*ba*	*ca*	*da*
ac	*bc*	*cb*	*db*
ad	*bd*	*cd*	*dc*

Each ordered arrangement, such as *ab*, is called a *permutation*. Hence there are 12 different permutations possible from 4 letters taken 2 at a time.

The number of permutations of n objects taken r at a time is denoted by the symbol $P_{n,r}$. Thus, in the example above, the permutations possible from 4 letters taken 2 at a time are denoted by $P_{4,2}$, which we know is equal to 4×3. Hence

$$P_{4,2} = 4 \times 3 = 12$$

Let us reconsider the examples in Sec. 6. From Example 3, there are 7 different flags to be selected 3 at a time. Hence the number of permutations possible is

$$P_{7,3} = 7 \times 6 \times 5 = 210$$

These examples suggest that the number of permutations of n objects taken r at a time is the product of r factors whose first term is n, second term is $n - 1$, third term is $n - 2$, etc., until there are r factors. That is,

$$P_{n,r} = n(n-1)(n-2)(n-3) \cdots (n-r+1) \tag{1}$$

To prove this formula for $P_{n,r}$, we can reason as follows: Since the number of ways of selecting the first object is n, the number of ways that the second object may be selected is 1 less than n, or $n - 1$ ways. The third object may then be selected in $n - 2$ ways. We continue in this way, and the rth object can then be selected in n minus the previous $r - 1$ ways, that is, in $n - (r - 1) = n - r + 1$ ways. The fundamental principle of choice, in Sec. 7, tells us that the total number of ways of selecting r objects at a time from n objects is the product of the ways of selecting each object. That is,

$$P_{n,r} = n(n-1)(n-2)(n-3) \cdots (n-r+1)$$

■ **Example 1.** Twelve boys compete in a race in which the first five places win prizes. If there are no ties, in how many ways may the prizes be awarded?

□ ***Solution.*** We have 12 objects, and we want to find the number of arrangements, or permutations, that are possible if the objects are taken 5 at a time. We want the value of $P_{12,5}$. Using equation (1), we find that

$$P_{12,5} = 12 \times 11 \times 10 \times 9 \times 8 = 95{,}040 \text{ ways}$$

■ **Example 2.** How many possible ways are there to arrange 10 books which just fit on a shelf 18 inches long?

□ ***Solution.*** We have 10 objects, and these 10 objects are to be taken 10 at a time. Hence the number of permutations is $P_{10,10}$. From equation (1) we find that

$$P_{10,10} = 10 \times 9 \times 8 \times 7 \times 6 \times 5 \times 4 \times 3 \times 2 \times 1 = 3{,}628{,}800$$

Problems just for fun

If three points are chosen at random on a sphere, what is the probability that all three will lie in the same hemisphere?

9 *Factorial Notation*

A convenient way of expressing the products of n consecutive numbers, such as

$$n(n-1)(n-2) \cdots 5 \cdot 4 \cdot 3 \cdot 2 \cdot 1$$

is the factorial notation $n!$, and we call this product *factorial n.* That is, we use $n!$ to represent the product

$$n! = n(n-1)(n-2) \cdots 5 \cdot 4 \cdot 3 \cdot 2 \cdot 1 \tag{2}$$

We observe that

$$\begin{aligned}
1! &= 1 \\
2! &= 2 \cdot 1 = 2 \\
3! &= 3 \cdot 2 \cdot 1 = 6 \\
4! &= 4 \cdot 3 \cdot 2 \cdot 1 = 24 \\
5! &= 5 \cdot 4 \cdot 3 \cdot 2 \cdot 1 = 120 \\
&\vdots \\
10! &= 10 \cdot 9 \cdot 8 \cdot 7 \cdot 6 \cdot 5 \cdot 4 \cdot 3 \cdot 2 \cdot 1 = 3{,}628{,}800
\end{aligned}$$

It is easy to see that

$$(n+1)! = (n+1)n! \tag{3}$$

for

$$(n+1)! = (n+1)(n)(n-1)(n-2) \cdots 4 \cdot 3 \cdot 2 \cdot 1 = (n+1)n!$$

Since $(n+1)! = (n+1)n!$, we can use this equation to define $0!$ If we let $n=0$, we obtain

$$(0+1)! = (1)0!$$

or

$$1 = 0!$$

Thus equations (2) and (3) define the factorial of all non-negative integers 0, 1, 2, 3, ...

We can use this factorial notation to write $P_{n,r}$ in a different form. We may write

$$P_{n,r} = n(n-1)(n-2)\cdots(n-r+1) \times \frac{(n-r)(n-r-1)\cdots 5\cdot 4\cdot 3\cdot 2\cdot 1}{(n-r)(n-r-1)\cdots 5\cdot 4\cdot 3\cdot 2\cdot 1}$$

because the last quotient is 1 and does not alter the value of the expression for $P_{n,r}$. We notice that the numerator is $n!$ and that the denominator is $(n-r)!$ We may write

$$P_{n,r} = n(n-1)(n-2)\cdots(n-r+1) = \frac{n!}{(n-r)!}$$

The number of permutations of n objects taken r at a time is usually expressed by the equation

$$P_{n,r} = \frac{n!}{(n-r)!} \tag{4}$$

The solution to Example 1 of Sec. 8 is

$$P_{12,5} = \frac{12!}{7!} = 95{,}040$$

The solution to Example 2 of Sec. 8 is

$$P_{10,10} = \frac{10!}{0!} = 10!$$

since $0! = 1$.

The values of factorial m have been tabulated, but they are easily worked out for small numbers, so we shall not use tables.

▲ EXERCISES

1 Evaluate

a. $P_{10,6}$ b. $P_{8,4}$
c. $P_{6,6}$ d. $P_{3,3}$

2. Simplify

a. $\dfrac{P_{5,3}}{P_{3,3}}$ b. $\dfrac{P_{n,r}}{P_{r,r}}$

3. Write out all permutations of the letters a, b, and c taken two at a time.
4. Write out all permutations of the letters a, b, and c taken three at a time.
5. A shelf will hold only 9 of 13 books which are all the same size. In how many ways may the shelf be filled?
6. A man has a combination lock with 40 numbers on it. He forgets the combination but remembers that it requires 4 different numbers to open it. Would he be wise to try all the combinations in order to find the one which opens the lock? If each try takes 1 minute and he is unlucky enough to hit the right combination on the last try, how long will it take?
7. How many numbers of four digits can be made using the digits 1, 4, 7, and 9 if no numeral is used twice in a number? How many numbers of three digits from the same four numerals?
8. An automobile manufacturer has nine different colors available. He uses different colors for the fenders, body, and wheels of each car. How many color combinations can he produce?
9. A boy has three suits, four shirts, five ties, and two hats. In how many ways may this lad dress by changing suits, shirts, ties, and hats?
10. In a football game, how many different signals may a quarterback call using the numbers 4, 5, 6, 7, and 8 if he uses four numbers at a time?
11. How many batting orders are possible for a baseball team (9 players)?
12. If 15 men are on the basketball squad, in how many ways can a team be chosen? (Team has center, right guard, left guard, right forward, left forward.)
13. Messages are sent from one ship to another by shining three colored lights in a row. Each light may be red, green, or white, and the same color may be repeated. Calculate the number of different arrangements.
14. A classroom contains 10 rows of 8 seats each. The seats are assigned at random to a class. What is the probability that the first individual assigned will be given a seat in the first row?
15. In how many different ways can the eight members of a university crew be seated in their shell?

The Mind Reader

"Choose five letters out of the alphabet," said Mind Reader, "and I will name one of your five letters. Of course the odds are heavily against me, but I will bet you $10 even that I can name one correctly if you will give me five chances." Is this a good bet?

10 *Restricted Arrangements*

Whenever an arrangement is to be made which is subject to some restriction, the restricted group must be considered first. An example will illustrate this.

■ **Example.** A bus has 20 seats on each side. In how many ways can 40 people be seated if 12 of them insist upon sitting on the shady side?

□ ***Solution.*** We must first assign 12 of the 20 shady seats to the 12 restricted people in some order, and the number of ways this can be done is $P_{20,12}$. The other 28 people may sit any place in the remaining 28 seats, and the number of ways they may be seated is $P_{28,28}$. For each seating arrangement of the 12 restricted persons on the shady side, there are $P_{28,28}$ seating arrangements for the other 28 people. Hence, the number of ways for seating all 40 people is

$$P_{20,12} \times P_{28,28} = \frac{20!}{8!} \times 28!$$

Warning. In working any problem, it is necessary to analyze it and not to try merely to fit it into a special type or formula.

▲ EXERCISES

1. In how many ways can a baseball team be selected from 15 players if 3 of the men can only pitch and the rest can play any of the other positions?
2. In how many ways can a baseball team be selected from 15 players if 3 men can only pitch and 2 can only catch, but the rest of the players can play any of the other positions?
3. How many five-place numbers can be formed from the digits 1, 2, 3, 4, and 5 if 3 is always to occupy the middle place?
4. In how many ways can eight books be arranged on a shelf if two of the books must be kept side by side?
5. In how many ways can five men and four women be seated in a row if a man is always seated at each end?
6. In how many ways can 11 football players be arranged if 3 of the men can play in a 7-man line only and 2 of the players can play only in the backfield?
7. A jeweler wishes to display the 12 birthstones. How many ways can he arrange them in a row if he wishes to keep the diamond and the ruby together in the center?
8. There are five different kinds of pie and three different kinds of cake listed on the menu.

a. How many ways can you eat three desserts?
b. How many ways can you eat three desserts only one of which is a piece of cake?

9. How many even numbers of three places can be formed with the digits 1, 2, 3, 4, and 5 if repetition of digits is not allowed in the same number?
10. In how many orders may a janitor clean the blackboards of 12 rooms if the four blackboards on the first floor must be cleaned some time during the course of cleaning the first eight boards?

Problems just for fun

The game of Tac Tics is played with 16 counters arranged in a square as shown in Fig. 81. Two players play alternately by removing 1 to 4 counters. All the counters removed in a single turn must come from one column or one horizontal row and must be adjacent counters with no gaps in between. The player taking the last counter loses. In Figs. 82 and 83, two games are in progress and it is your move. On each board find a move which guarantees that you will win.

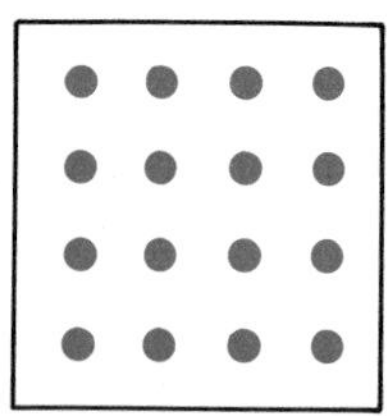
FIG. 81

FIG. 82

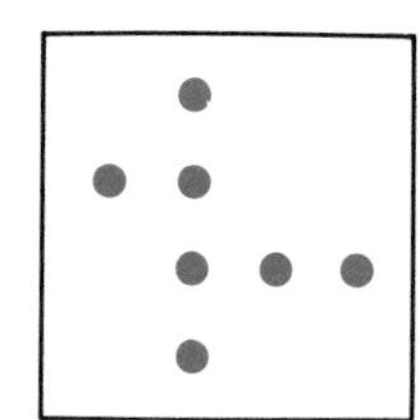
FIG. 83

11 *Combinations or Unordered Groups*

In Sec. 8, we found the possible permutations of the letters *a*, *b*, *c*, and *d* taken two at a time to be:

ab	*ba*	*bc*	*cb*	*cd*	*dc*
ac	*ca*	*bd*	*db*		
ad	*da*				

We considered the permutation *ab* as being distinct and different from the permutation *ba*. But considered as a group or combination, there is no difference between *ab* and *ba*. Likewise, when considered as a combination, there is no difference between *ac* and *ca*, *ad* and *da*, *bc* and *cb*, *bd* and *db*, or *cd* and *dc*. Hence, if we disregard order, there are only six possible combinations of the letters *a*, *b*, *c*, and *d* taken two at a time, namely:

ab *bc* *cd*
ac *bd*
ad

We often need to know the different groups or combinations that can be chosen from n objects taken r at a time. *A combination is regarded as different if a single individual of the group is changed, but the different orders or permutations that can be made within the group do not change the combination.*

■ **Example 1.** How many committees consisting of a chairman, a secretary, and a committeeman may be chosen from a group of six people?

□ ***Solution.*** Since all the positions are different, order must count, and hence this is a problem in permutations. The number of ways of selecting this committee is

$$P_{6,3}=\frac{6!}{3!}=120$$

■ **Example 2.** In how many ways can a committee of 3 members be selected from 6 people?

□ ***Solution.*** We have just found that the number of permutations of 6 objects taken 3 at a time is $P_{6,3}=120$. But our solution does not depend entirely upon the number of possible arrangements, because a person is on the committee regardless of whether he is the first, middle, or last man. That is, order of arrangement within the groups of three has no meaning. Since the arrangement within each group of three is 3! or 6 ways, the number of permutations of the 6 people selected 3 at a time is six times too large. Hence the number of combinations or possible ways of selecting a committee of 3 people from 6 people is $\frac{120}{6}=20$.

A Way of Calculating the Number of Combinations

We denote the number of combinations of r objects that can be selected from n objects by

$C_{n,r}$

We know that the number of ordered permutations possible using n objects r at a time is

$$P_{n,r}=\frac{n!}{(n-r)!} \tag{5}$$

We have also seen that within a group or combination the order is not important and is not considered. That is, *abcd* is the same group as *bdca* or *dcab*. Since the number of possible arrangements within each combination or group of r objects is $r!$, the number of permutations $P_{n,r}$ is $r!$ times the number of combinations. That is,

$$P_{n,r}=r!C_{n,r}$$

And we may write

$$C_{n,r}=\frac{P_{n,r}}{r!}=\frac{n!}{r!(n-r)!}$$

■ **Example 1.** Find the number of straight lines that may be drawn through 10 points no 3 of which lie on the same line.

□ ***Solution.*** Since a line is determined by 2 points, and since the order of the choice of the points is immaterial, the problem is clearly one of finding the number of combinations of 10 objects taken 2 at a time. Thus we need to find the value of

$$C_{10,2}=\frac{P_{10,2}}{2!}=\frac{10!}{8!2!}=\frac{1\cdot2\cdot3\cdot4\cdot5\cdot6\cdot7\cdot8\cdot9\cdot10}{1\cdot2\cdot3\cdot4\cdot5\cdot6\cdot7\cdot8\cdot1\cdot2}=45$$

■ **Example 2.** In how many ways can a hand of 13 cards be drawn from the usual pack of 52 cards so as to contain precisely 5 spades?

□ ***Solution.*** Any 5 of the 13 spades might be drawn, and this can be done in $C_{13,5}$ ways. But the other 8 cards may be any 8 of the 39 clubs, diamonds, or hearts. These 8 cards can be drawn in $C_{39,8}$ ways. Each set of 5 spades can go with any set of the 8 other cards. Hence the total number of hands possible is

$$C_{13,5}\times C_{39,8}=\frac{13!}{8!5!}\times\frac{39!}{31!8!}=79{,}181{,}063{,}676$$

—an enormous number!!

▲ **EXERCISES**

1. Calculate
   a. $C_{25,4}$
   b. $C_{10,3}$

c. $C_{7,3}$
d. $C_{8,8}$
e. $C_{5,5}$

2. What is the essential difference between permutations and combinations?
3. How many straight lines are determined by five points on a circle? Check by actually counting the number.
4. A club has 12 members. How many committees of 4 members may be chosen?
5. On an examination, a student is asked to answer any 10 questions out of 12 questions. In how many ways may the student choose the questions?
6. If you were to choose any three pictures from a collection of seven, all different, how many choices would you have?
7. A box contains nine differently colored balls. How many sets of three can be taken from the box?
8. If a student must answer eight questions true or false, how many different ways may he answer the eight questions?
9. How many different recitation schedules could a student make if he registered for three out of five elective subjects which meet at different times?
10. Find the number of different poker hands of 5 cards that one can draw from a deck of 52 cards.
11. How many poker hands are there containing two kings?
12. How many poker hands are there containing only spades?
13. How many poker hands are there containing only face cards (kings, queens, and jacks)?
14. How many poker hands are there containing four queens?
15. How many different bridge hands can be dealt? Express your answer and estimate its magnitude.
16. If one draws 5 balls at random from a bag containing 10 red and 5 white balls, in how many ways may one get 3 red and 2 white balls?
17. If one draws 8 cards from a pile containing 11 spades and 6 hearts, in how many ways may he get 3 spades and 5 hearts?
18. A committee to consider labor-management problems is to be made up of 3 outsiders, 3 employers, and 3 workers and is to be selected from 8 employers, 15 workers, and 10 outsiders. In how many ways may this committee be made up?
19. How many batting orders for a baseball team are possible if the pitcher bats last and the three outfielders bat at the head of the batting order?
20. A committee of 5 is to be chosen from 10 Democrats and 6 Republicans. In how many ways can the committee be selected if it is to have a majority of Democrats?
21. The company president walks for exercise and his office is six blocks east and four blocks north of his home (Fig. 84). In how many different ways may he walk to his office by a shortest path, i.e., always walking north or east?

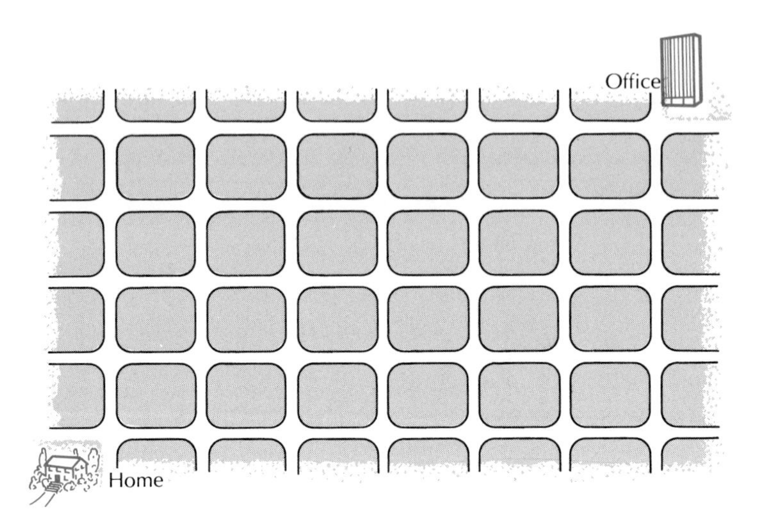

FIG. 84

Problems just for fun

Here we see three cards face down. We know that

1. There's at least one queen just to the right of a queen
2. There's at least one queen just to the left of an ace
3. There's at least one spade just to the right of a heart
4. There's at least one spade just to the left of a spade

Can you name the three cards?

13 *More Problems on Probability*

We have learned a method for determining the number of arrangements that can be made from a group of objects without enumerating these arrangements. We are now ready to solve problems in probability when the number of ways that an event can happen or fail to happen is large.

■ **Example 1.** If 5 balls are drawn at random from a bag containing 7 red and 6 black balls, what is the chance of drawing precisely 3 red balls?

□ ***Solution.*** Any 3 of the 7 red balls can be drawn, and the number of different sets of 3 red balls that can be drawn from the 7 red balls is $C_{7,3}$. The other 2 (out of the 5 drawn) must be black, and these black balls can be drawn from the 6 black balls in $C_{6,2}$ ways. The total number of ways we can draw 5 balls, 3 of which are red and 2 of which are black, is $C_{7,3} \times C_{6,2}$, which gives us the total number of ways our drawing can be successful. The total possible number of ways of drawing 5 balls from the 13 balls is $C_{13,5}$. Using the definition of probability, we find that the probability of drawing to produce 3 red balls is

$$p=\frac{C_{7,3}\times C_{6,2}}{C_{13,5}}=\frac{\dfrac{7!}{3!4!}\times\dfrac{6!}{2!4!}}{\dfrac{13!}{5!8!}}=\frac{175}{429}=\frac{2}{5}\text{ approximately}$$

There are approximately 2 chances in 5 to pick the 5 balls so that 3 are red.

■ **Example 2.** Three cards are drawn from a suit of thirteen cards. *(a)* What is the chance that neither king nor queen is drawn? *(b)* What is the chance that a king or queen is drawn, one or both?

□ ***Solution.*** *(a)* The total number of ways of drawing 3 cards from a suit of 13 cards is

$$C_{13,3}$$

There are 11 cards in the suit other than the king or queen. Hence, there are $C_{11,3}$ ways of drawing 3 cards which include neither the king nor the queen. Therefore the chance of drawing neither the king nor the queen is

$$p=\frac{C_{11,3}}{C_{13,3}}=\frac{15}{26}$$

(*b*) This event can occur when the event described in (*a*) fails to occur. Hence the required probability is

$$1-\frac{15}{26}=\frac{11}{26}$$

▲ EXERCISES

1. What is the probability of drawing two hearts from an ordinary deck of playing cards?
2. What is the chance of selecting a man when choosing one person by lot from a group of seven men and five women?
3. Five persons are to be selected for a committee out of a group of 15 men and 10 women.

   a. What is the probability that 4 men and 1 woman will be selected?
   b. What is the probability that 3 men and 2 women will be selected?

4. If two dice are thrown, what is the probability that

   a. The sum will be seven?
   b. The sum will be eleven?
   c. The sum will be less than seven?
   d. Both will be odd?

5. If two balls are drawn at random from a bag containing one red and two white balls, what is the probability that

   a. The red ball will be drawn?
   b. The red ball will not be drawn?
   c. At least one white ball will be drawn?

6. A bag contains a red ball, a white ball, and a blue ball. If one ball is drawn at random, noted, and replaced, and then a second ball is drawn at random, what is the probability that

   a. Both will be the red ball?
   b. Both will be the same ball?
   c. One will be red and one blue?

7. If a person draws 3 cards from a bridge deck of 52 cards, what is the chance that

   a. All will be red?
   b. All will be hearts?
   c. Two will be red and one black?
   d. Two will be hearts and one a spade?

8. A deck of cards is shuffled and four cards are taken from it.

   a. What is the probability that all of them will be face cards?

b. What is the probability that all of them will be black cards which are not face cards?

9. Write an expression for the probability that a 13-card hand from an ordinary 52-card deck will contain exactly 4 aces. Give your answer in terms of symbols only.
10. A box contains 10 beads with numbers from 1 to 10 on them.

    a. In how many ways can 2 beads be drawn from the box?
    b. What is the probability that the sum of the numbers on the 2 beads is 6?
    c. What is the probability that 2 even numbers are drawn?

11. Two boys and three girls take seats at random in a row of five seats. Find the probability that boys and girls are sitting alternately in the seating arrangement.
12. A box contains 11 cards. Six of the cards are plain, but the other five are marked with an X. If one draws 4 cards from the box, what is the probability of drawing

    a. 2 plain cards and 2 marked with an X?
    b. All plain cards?
    c. All marked with an X?

13. A deck of 52 cards is numbered 1 to 52. If we draw 5 cards one at a time from the shuffled deck, what is the probability that the numbers on the 5 cards will be in increasing order?
14. A gambler says, "The chance of rolling a six with one die is $\frac{1}{6}$. So the chances of rolling at least one six with three dice is three times as much or even money. I will bet even money you will not roll a six." What are the true odds?

Compound Probabilities

Most complex problems in probability are solved by determining the probabilities of their simpler parts. We can see the ideas easily by considering two questions when dice are cast. We ask, "What is the probability that in two throws a 5 will show both times?" There are six possible cases on the first throw, and one shows a 5, so the probability of a 5 on the first throw is $\frac{1}{6}$. Similarly the probability of a 5 on the second throw is $\frac{1}{6}$. How may we combine these two simple probabilities to answer our question? As we demand that both happen, the probability is $\frac{1}{6} \times \frac{1}{6} = \frac{1}{36}$. Each of the six possible throws on the first die may be followed by any one of the six on the second throw, so there are a total of 36 equally probable throws. Only the throw of 5, followed by a throw of 5, is a success for our problem. Thus $\frac{1}{36}$ is the correct probability. We may also see this by writing out the cases and see that only the underlined 5,5 is a success:

Problems just for fun

Four schoolboys living in Switzerland start home from school on their skis. They leave the school by the four doors marked *A, B; C,* and *D,* and boy *A* goes to home *A,* boy *B* to home *B,* etc. How do they ski home without having two tracks cross and without leaving the shaded area shown?

1,1	1,2	1,3	1,4	1,5	1,6
2,1	2,2	2,3	2,4	2,5	2,6
3,1	3,2	3,3	3,4	3,5	3,6
4,1	4,2	4,3	4,4	4,5	4,6
5,1	5,2	5,3	5,4	5,5	5,6
6,1	6,2	6,3	6,4	6,5	6,6

We may ask, "What is the probability of an even number on one throw?" This may occur with a 2, 4, or 6. The probability of each one is $\frac{1}{6}$. Any one gives an even number, or a success. Then the probability of an even number on one throw is $\frac{1}{6}+\frac{1}{6}+\frac{1}{6}=\frac{1}{2}$. We may verify this by counting the cases.

These two types of compound probabilities are stated as theorems:

► **Theorem I.** The probability that both of two events will happen is the product of their separate probabilities.

► **Theorem II.** The probability that one or both of two events will happen is the sum of the two separate probabilities minus the probability that both will happen.

The following examples will help to explain these ideas.

■ **Example 1.** Two equally matched tennis players, Smith and Jones, play three sets. What is the probability that Smith will win all three sets?

□ ***Solution.*** The probability that Smith will win any given set is $\frac{1}{2}$, and the events are independent, for one set is in no way affected by the winning or losing of another set. Then the probability that Smith will win all three sets is

$$\frac{1}{2}\times\frac{1}{2}\times\frac{1}{2}=\frac{1}{8}$$

■ **Example 2.** What is the probability of throwing at least one 5 on two throws of a die?

□ ***Solution.*** On each throw the probability of a 5 is $\frac{1}{6}$. But the correct answer is not $\frac{1}{6}+\frac{1}{6}=\frac{1}{3}$, for if we count the cases where a 5 occurs in the 36 possibilities above, we find that there are 11 of them and the correct probability is $\frac{11}{36}$. The trouble is that a 5 may occur on both throws and the chance of this happening is $\frac{1}{36}$. By Theorem II, we have

$$\frac{1}{6}+\frac{1}{6}-\frac{1}{36}=\frac{11}{36}$$

■ **Example 3.** In drawing one card from a bridge deck, what is the probability that it will be a king or a spade (or both)?

□ ***Solution.*** The probability that it will be a spade is $\frac{13}{52}=\frac{1}{4}$. The probability that it will be a king is $\frac{4}{52}=\frac{1}{13}$. But it may be both, that is, it may be the king of spades and the probability of this is $\frac{1}{52}$. Thus by Theorem II, the correct probability is

$$\frac{1}{4}+\frac{1}{13}-\frac{1}{52}=\frac{16}{52}=\frac{4}{13}$$

We may check this by observing that the deck of 52 cards contains 13 spades plus 3 other kings, so the chance of success is $\frac{16}{52}$.

■ **Example 4.** In drawing one card from a bridge deck, what is the probability that it will be a face card or a spade (or both)?

□ ***Solution.*** The probability of a spade is $\frac{1}{4}$ and the probability of a face card is $\frac{12}{52}=\frac{3}{13}$. But three face cards are also spades, so by Theorem II, we have

$$\frac{1}{4}+\frac{3}{13}-\frac{3}{52}=\frac{22}{52}=\frac{11}{26}$$

■ **Example 5.** What is the probability of drawing two red balls in succession from a bag containing 6 black and 5 red balls?

□ ***Solution.*** The probability of drawing one red ball is $\frac{5}{11}$. We may then propose as our solution

$$\frac{5}{11}\times\frac{5}{11}=\frac{25}{121}$$

This is correct if whatever ball was drawn first is replaced in the bag before the second draw. However, if the ball is not replaced, our second draw will be from a bag containing 6 black and 4 red balls and the probability of a red draw will be $\frac{4}{10}$. Thus the probability of drawing two red balls in succession when the first ball drawn is not replaced is

$$\frac{5}{11}\times\frac{4}{10}=\frac{2}{11}$$

■ **Example 6.** From the bag in Example 5, what is the probability of drawing a red and a black ball in two draws?

☐ ***Solution.*** This requires a somewhat different analysis, for the first draw is a success whether it is black or red and the chances on the second draw depend on what happened on the first draw. If the first draw was red, and the chance of this is $\frac{5}{11}$, we have a bag of 6 black and 4 red balls and hope to draw a black. The chance of this is $\frac{6}{10}$. If the first draw was black, and the chance of this is $\frac{6}{11}$, we have a bag of 5 black and 5 red balls and hope to draw a red. The chance is $\frac{1}{2}$. Then the probability of getting 1 red and 1 black by the first method is $\dfrac{5}{11}\times\dfrac{6}{10}=\dfrac{3}{11}$ and by the second method is $\dfrac{6}{11}\times\dfrac{1}{2}=\dfrac{3}{11}$, or the probability by one method or the other is

$$\frac{3}{11}+\frac{3}{11}=\frac{6}{11}$$

We may check this by writing out the 55 cases and counting to see that 30 have 1 red and 1 black ball.

▲ EXERCISES

1. What is the probability that a coin tossed three times will fall heads all three times?
2. On a Saturday in May, the Wisconsin and Northwestern baseball and track teams are competing against each other. The probability that the Wisconsin baseball team will win its game is $\frac{2}{3}$. The chance that the Wisconsin track team will win its meet is $\frac{1}{4}$. What is the probability that both Wisconsin teams will win?
3. One box contains five white and three black balls, and a second box contains seven white and five black balls. At random a person selects one box and draws one ball. What is the probability that it will be drawn from the first box and be white?
4. Now suppose all the balls are dumped into one box and one ball is drawn at random. What is the probability that it is white? Compare with Exercise 3.
5. In Exercise 4, what is the probability of drawing two white balls in succession if the first ball drawn is not replaced?
6. What is the probability of throwing an odd number with a die if we have two tries?
7. If two balls are drawn without replacement from a box containing four black and two red balls, what is the probability that one will be black and the other red?
8. If two balls are drawn without replacement from a box containing four black and three red balls, what is the probability that one will be red and one will be black?

9. On the throw of two dice, what is the probability of a 7 or an 11?
10. On the throw of two dice, what is the probability of a double or a total of 6?
11. On the throw of two dice, what is the probability of a total that is a 7 or an even number?
12. Two cards are drawn from a bridge deck and the first card is replaced before the second card is drawn.

    a. What is the probability that both will be spades?
    b. What is the probability that at least one of the cards will be a spade?

13. A box contains five white and four black balls. Four balls are drawn at random. What is the probability that

    a. All four are white?
    b. At least one is white?
    c. Two are white and two are black?

14. Box A contains 10 white balls, box B contains 7 black and 3 white balls, box C contains 5 black and 5 white balls. A box is chosen at random and one ball is drawn from it. The ball is white. What is the probability that it came from box A? Box B? Box C?
15. A group consists of seven Democrats of whom three are women, and nine Republicans of whom four are women. A treasurer is chosen by lot. What is the probability that the treasurer will be a Republican? If the treasurer proves to be a woman, what is the probability that she will be a Republican?
16. John says, "I have two children and one is a boy." Henry says, "I have two children and the older is a boy." What is the probability in each family that both children are boys?

Problems just for fun

A grocer sells a pound of sugar to two customers, but his balance scale has arms which are not quite the same length. The first time he puts the weight in one pan and the sugar in the other; the second time he reverses the procedure. Does he gain or lose?

Mathematical Expectation

An important use of the theory of probability is the determination of the mathematical value of a person's expectation. If a person wagers that a certain event will happen, *the mathematical value of his expectation* is defined as the product of the probability that an event occurs and the value of the prize the person receives if he wins.

If a person pays an amount to take part in a game of chance, and if this amount is more than the value of his expectation, he has made a bad deal. If this person continues to gamble and to pay more than his mathematical expectation, he will end up in the red on Skid Row.

■ **Example.** A person pays 20 cents to draw a card from a well-shuffled deck of cards. He is to receive \$2 if he draws an ace. Is he paying too much for this chance?

□ ***Solution.*** The probability of drawing an ace is $\frac{4}{52}$ and his expectation is $\frac{4}{52} \times 2 = \frac{2}{13}$ of a dollar, or approximately 16 cents. This person pays too much for the chance to play.

▲ EXERCISES

1. A person is to receive \$8 if he tosses two coins and they both come up heads. What is the value of his expectation?
2. A person pays \$1.25 for the privilege of drawing three balls from a box known to contain five white and four black balls and is to receive \$10 if he draws three white balls. Does he pay too much for this privilege?
3. In a lottery, 2,000 tickets are sold for 10 cents each. A prize worth \$150 is to be paid to the person holding the lucky number. Do these lottery tickets cost too much?
4. A fancy bedspread valued at \$50 is to be raffled off at a church bazaar. If 250 chances at 25 cents per chance are sold, is this price too high?
5. A person is to receive 10 cents if he throws a total of 8 using two dice. What should he pay to play?
6. A player in a dice game throws two dice and will win \$4 if the total number of spots showing on the two dice is 10 or more. What can he afford to pay for a chance to play?
7. If an ordinary die is thrown 20 times, what is the expected number of times that a 4 will turn up?
8. In a dice game in which a simple die is rolled, a player wins \$1 if he rolls a 4 or a 5. How much should he pay to play?
9. If the player in Exercise 8 receives nothing when he rolls a 2, 3, or 6, one dollar for rolling a 4 or 5, and two dollars for rolling a 1, how much should this player pay to play the game?

Problems just for fun

A carnival spieler shows his audience three cards. One card is red on both sides, one blue on both sides, and one red on one side and blue on the other. He has one of his audience shuffle the cards and place one on the table in such a way that no one knows what color is on the bottom side. The top side is red. The spieler says, "Obviously this is not the blue-blue card. Then it is either the red-red card or the red-blue card. I will bet even money that it is the red-red card." Is this a fair bet?

10. A man pays one dollar and draws two cards from a bridge deck. He wins two dollars if either is a spade. Is this a good bet?
11. A man bets even money that he will throw a sum of 8 with two dice before he throws a sum of 7. Is this a good bet? (In craps language, he bets he will make his point of 8 before he craps out.)
12. A man is to receive $10 if he draws three face cards from a bridge deck. What is the value of his expectation?
13. Toss 3 coins. If you throw 3 heads or 3 tails, you collect $15. Anything else, you pay $5. Is this a fair game?
14. Pick the kings, queens, and jacks from a deck of cards. Shuffle well and draw two cards face down. The dealer says, "I am a ladies man and I will bet even money that one of these cards is a queen." Is this a fair bet?

Review Exercises

1. A family of five owns a four-passenger car. How many seating plans are there for filling the car?
2. a. Find the numerical value of the symbols $P_{12,3}$ and $C_{7,4}$.
   b. Give a verbal meaning to each of the two numbers you obtained.
3. How many numbers of seven digits each can be made from the digits 0, 1, 2, 4, 6, 7, 9
   a. If the 0 is always in the middle and no digit may be used more than once?
   b. If the 0 is always in the middle, the 1 is always at the beginning, and no digit may be used more than once?
4. The Greek alphabet contains 24 letters. How many different fraternity names can be obtained by using the letters three at a time
   a. If repetitions are allowed?
   b. If repetitions are not allowed?
5. A club with five members decides to choose each week a different set of three members to play three-handed hearts (the other two members kibitz). How many weeks will they go before repeating?
6. A boat is to be manned by eight men. If two men can row only on the right side and one man can row only on the left side, in how many ways can the crew be arranged?
7. In how many ways can four prizes be awarded to four boys if
   a. Each boy can receive only one prize?

b. There is no restriction to the number of prizes any boy may receive?

8. A four-floor department store has both a stairway and an elevator between each floor. An escalator operates between the first and second floors only. How many ways can a customer ascend from the first to the fourth floor?
9. In making out a 10-minute quiz, an instructor chooses 3 problems at random from a list of 10 problems which are numbered from 1 to 10. What is the probability that one of the selected problems will be problem number 6?
10. From a pack of 52 cards, a card is drawn at random and not replaced. A second card is then drawn. Find the probability that the first card is a spade and the second card is a heart.
11. From a bag containing seven black balls and two white balls, two balls are drawn. Find the probability that

    a. Both balls are black.
    b. Both balls are white.
    c. One ball is white and one is black.

12. In the 1958 World Series, Milwaukee won the first two games from New York. For each of the remaining games, assume that one team is just as likely to win as the other. What is the probability that New York can win the Series? (The first team winning four games is the winner.)
13. A shipment of 1,000 shirts contained 150 defective shirts with cuffs of the wrong material. The retailer receiving them decides to sell them with no returns at a reduced price rather than sort them out. The perfect shirts are worth $4.50 while the defective ones are worth nothing. What is the fair price?
14. Two decks of cards are shuffled separately and a card is taken from each. What is the probability that at least one of the cards is a heart?
15. A carnival spieler shows his crowd a box containing eight black and two white balls and offers to pay $10 if you draw the two white ones. He says, "One-fifth of the balls are white so the usual price of a try is $2. However, to get the game started I will let the first five players play for $1 each." Is he likely to lose? What is the fair value of a chance?
16. If a person draws 5 cards from a deck of 52 cards

    a. How many different hands can be chosen?
    b. What is the probability that all 5 cards will be black?
    c. What is the probability that all 5 cards will be spades?

d. What is the probability that 3 of the cards will be black and 2 of the cards will be red?

A cube 8 inches on each edge is to be cut into 27 cubes 2 inches on each side. After each cut the pieces may be piled in any way before making the next cut. What is the smallest number of cuts possible?

CHAPTER SEVENTEEN

Introduction to Statistics

Battalions of figures are like battalions of men, not always as strong as is supposed.

M. SAGE

1 *Empirical Probability*

In Chap. 16, we found the probability of an event by determining the exact number of ways the particular event could happen or fail to happen. But many problems in real situations are so complicated that it is impossible to reason or deduce the expected results by the methods used in the last chapter. We also need to determine the probability of an event in problems for which we cannot enumerate the number of ways the event could happen or fail to happen. For this type of problem the probability must be determined by an actual examination of particular cases.

For example, a firm which is manufacturing men's shirts would like to know the probability that a potential customer wears a collar size of 16, in order to know how many shirts of this collar size to make. The way this probability can be determined is to find experimentally the collar size of the shirts worn by a large number of men who are selected at random from potential consumers so as not to favor a special class. If b men out of N men in a group wear a 16 collar size, the probability that another man selected at random from this group would wear a 16 collar is b/N.

Probabilities which are determined by an actual examination of the particular cases are called *empirical*.

The insurance companies base many of their calculations upon empirical probability. The data are contained in mortality tables which are based upon case records. This table applies to large groups of people and furnishes no surety to an individual person.

Table 36 is based upon 1,000,000 people observed from their first birthday until death. From the table, we see that 971,804 live to the age of ten. Then the empirical probability that a person age one will live to age ten is

$$\frac{971{,}804}{1{,}000{,}000} = 0.972 = 97.2\%$$

TABLE 36

At age	*Number surviving*	*At age*	*Number surviving*
1	1,000,000	60	677,771
10	971,804	70	454,548
20	951,483	75	315,982
25	939,197	80	181,765
30	924,609	85	78,221
35	906,554	90	21,577
40	883,342	95	3,011
50	810,900	99	125

The empirical probability that a person age one will live to be twenty-five is

$$\frac{939{,}197}{1{,}000{,}000} = 0.939 = 93.9\%$$

The empirical probability that a person age twenty-five will live to be thirty-five is

$$\frac{906{,}554}{939{,}197} = 0.965 = 96.5\%$$

▲ EXERCISES

1. What is the probability that a person age forty will not live to age sixty?
2. What is the probability that a person age eighty will live another 10 years?
3. Of 7,820 men attending a certain university, 5,432 were under 6 feet in height. What is the probability that a man chosen at random from this group will be 6 feet or taller?

4. In a certain community there were 8,932 children under five years of age. If 6,280 were girls, what is the probability that a child chosen at random from this group will be a boy? A girl?
5. A baseball player is currently "batting 300."

   a. What is the probability that he will get a hit the next time at bat?
   b. What is the probability that he will fail to make a hit the next time at bat?
   c. What is the probability that he will make a hit each of the next two times at bat?
   d. What is the probability that he will get at least two hits out of his next three times at bat?

6. A baseball player has made 30 hits in 120 times at bat.

   a. How many hits is he expected to make in the next 10 times at bat?
   b. What are the chances that he will make 3 hits in the next 4 times at bat?

7. Out of a group of 100,000 persons alive at age twenty, approximately 97,000 will be alive at age 30.

   a. What is the probability that a person of this group will live to be thirty?
   b. What is the probability that a person age twenty of this group will not live to be thirty?
   c. If we choose two persons age twenty of this group, what is the probability that both will live to be thirty?

8. An instructor's record shows that he turned in failing grades for 15 of 120 students. What is the probability that a student in this instructor's class will pass? Will fail?
9. In a certain community there are 4,806 people forty years of age. According to Table 36, what is the probable number of people who will live to reach fifty, sixty, seventy, and ninety, if the people in this community are typical of the group in the table?
10. A university graduating class of 1,200 students has an average age of 23. How many may we expect to be alive to attend their 25th reunion? Their 50th reunion? (In Table 36 compute the figure for age 23 by assuming that the same number of persons die each year from ages 20 to 25.)
11. A city has 100,000 people at age 40 in 1970. How many deaths may it expect from this group in the next 10 years?
12. Assuming that a child lives to be 10 years old, what is the probability that he will live to age 70?
13. What chance does a person age 70 have to live to age 95?
14. Of 10,000 children age 10, how many will survive to age 60?
15. Of 1,000 persons alive at age 70, how many will survive to age 75? 80? 85? 90?

Problems just for fun

Find the numbers which are represented by asterisks in the following problem in division:

```
          * * 8 * *
     ____________________
 * * | * * * * * * * *
       * * *
       -----
           * *
           * *
           ---
             * * *
             * * *
             -----
```

The Field of Statistics

The determination of empirical probability must be made by an examination of a number of cases which have been selected at random from a large group and which are considered representative of the larger group. It is assumed that an empirical probability will approach a true and unknown probability as the number of observed cases increases.

We find that we must work with a collection of data which may contain many items, and one purpose of statistics is to simplify large collections of numerical data. The general problems of accumulating, summarizing, and interpreting such data are included in the *field of statistics.* The bounds of mathematical statistics are not sharply defined, but we shall regard the field as including all the mathematics applied to the analysis of quantitative data obtained from observation.

This branch of mathematics is interesting because of the numerous applications of statistics. We are bombarded daily with statistical information on the radio and TV and in newspapers and magazines. Statistics may be encountered in articles on economics, education, politics, social, biological, and physical science, business, etc. In fact the methods of statistics are now applied to almost every phase of our existence.

3 *The Two Basic Problems of Mathematical Statistics*

A statistician who has been given a set of measurements obtained by observation is usually asked *(a)* to summarize, analyze, and interpret this data and *(b)* to draw conclusions about the whole *population* or *universe* from which the set of measurements was obtained.

In case *(a)*, the data may or may not include the whole population, but in case *(b)*, the available data are usually a small fraction of the population. In either case, the observed data must have been collected accurately if the statistical worker is to be able to give a valid report. In the second case, it is essential that the relatively few observations that the statistician uses be as nearly as possible a *random sample* of the universe under consideration. We say that an object is selected randomly from a population whenever each object in the population has an equal chance of being selected. A set of such selections is called a *random sample.* It is not an easy task to ensure that a sample be truly a random sample, although various means have been devised to accomplish this.

Many tools and techniques have been developed for summarizing, analyzing, and interpreting data obtained from observations. There are also a large number of techniques for drawing conclusions about a population on the basis of the information given by a random sample. We shall study a few of the more common techniques and fundamental ideas of both phases of statistics.

The data with which the statistician works must be accurately and legitimately obtained. The statistician must be extremely careful in all computations that he makes. Actually, this computation is only arithmetical in nature, but it is often so involved and lengthy that mechanical computers and automatic calculating machines are used.

We must keep in mind that even if the data are faultlessly collected and the computations are errorless, the conclusions and interpretations drawn from these data may be incorrect and distorted. Sometimes this happens because the statistical worker is ill-informed, and sometimes because he presents a misleading report of his work. The latter case gives rise to the common statement that "a statistician can make the figures show whatever he pleases." One objective of our study will be to observe how one can be led to fallacious conclusions concerning the measurements under study.

One study of graduates of Princeton University showed an average of 1.8 children per graduate. A similar study of graduates of Wellesley College showed 1.3 children per graduate. One careless observer concluded from these figures that men had more children than women!

Types of Problems

Let us look at some examples of the two broad problems of mathematical statistics named in Sec. 3 and the questions we might be called upon to answer.

■ **Example 1.** In the 1967 season, the SMU basketball players attempted the number of free throws shown in Table 37.

TABLE 37

Phillips	176	Beasley	60
Holman	138	Jones	27
Begert	92	Rainer	21
Voight	67	Sibley	3
Higgenbotham	62	Morris	2

Question. What can be said about this squad *as a group* concerning the number of free throws attempted?

■ **Example 2.** Thirty students in a mathematics class made the test grades shown in Table 38.

TABLE 38

78	63	92	48	80	74
96	85	71	90	87	82
58	67	58	70	49	72
34	75	74	76	86	67
77	83	62	52	60	57

Question. How can these data be summarized, and what can be said about the class as a whole?

■ **Example 3.** A random sample of 200 voters in Tippecanoe County showed that 55% of them were Democratic.

Question. Does this fact disprove the statement that "Tippecanoe County is predominantly Republican"?

■ **Example 4.** The average grade point index of a random sample of 50 fraternity members was 3.84 and was 3.96 for a random sample of 40 independent students.

Question. Do these figures show that the independent students do better in their academic work than the organized students?

5 *Average—the Measure of Central Tendency*

When quiz papers are returned, the students always ask what is the average of the class. When considering a new field of occupation, persons inquire as to the average beginning salary, the average salary after a few years' experience, etc. Each person is seeking to find a number on which he can make comparisons or make decisions. If the individual's quiz grade is higher than the class average, he may decide he is doing a good job. Or, if the average salary in a certain occupation is to his liking, he may decide to take up that profession. We need to find out what the term "average" means and how to calculate it.

Usually the first step in the analysis of a set of observations is to find a single number which is more or less representative of all the observations. This single number is called an *average*, or *measure of central tendency*. There are a number of different kinds of average, but we shall consider just three of them.

Let us look at the data in Table 37. If we add the 10 observations given in the table and divide the result by 10, we obtain 64.8, which is the *arithmetic mean* average, or simply the *mean* of the 10 figures. Likewise, we find that the *mean* grade for the grades given in Table 38 is 70.8.

It is conventional to assign some letters, say x, to the data. We indicate the first figure by x_1, the second by x_2, etc. Each observation is designated by x_i, where $i = 1, 2, 3, 4, \ldots$. Then we use $\overline{x}$ to represent the arithmetic mean of all the data represented by the letter x_i. In using the arithmetic mean, we have each time to write the sum of the terms x_i. This becomes tiresome, so we shorten the task by using the symbol Σ to represent the sum of terms following, or

$$\sum_{i=1}^{N} x_i = x_1 + x_2 + x_3 + x_4 + \cdots + x_N$$

The symbol Σ is the Greek letter for S and thus suggests "sum." We read the symbol above as "sum of the x_i's from $i = 1$ to N."

Example 1. If $x_i = i$, then

$$\begin{aligned}\sum_{i=1}^{5} x_i &= x_1 + x_2 + x_3 + x_4 + x_5 \\ &= 1 + 2 + 3 + 4 + 5 = 15\end{aligned}$$

Example 2. If $x_i = i^2$

$$\sum_{i=1}^{4} x_i = x_1 + x_2 + x_3 + x_4$$
$$= 1 + 4 + 9 + 16 = 30$$

Then the *arithmetic mean*, or *mean* $\overline{x}$, is given by

$$\overline{x} = \frac{1}{N} \sum_{i=1}^{N} x_i = \frac{x_1 + x_2 + x_3 + \cdots + x_N}{N} \tag{1}$$

For the data in Table 37

$$\overline{x} = \frac{1}{10} \sum_{i=1}^{10} x_i = \frac{648}{10} = 64.8 \text{ free throw attempts}$$

The arithmetic mean is the most commonly used average. Where the spread of the data is fairly uniform, this is quite satisfactory and is the best possible single number. The popularity of the arithmetic mean is an indication that the numbers of most sets of data are grouped about the mean in a regular or normal manner. We shall discuss this in considerable detail in Chap. 19. If we look at the numbers in Table 37, we see that three of the numbers are very close to the mean 64.8. Two are far off on the high side and two are far off on the low side. Hence 64.8 is a satisfactory single number to represent this group.

Now suppose that we drop Sibley and Morris from consideration in Table 37 since they obviously played only a few minutes and the remaining eight were the actual playing squad. Then the arithmetic mean becomes 80.4. Does this number represent the squad as a whole? Not very well. No player is close to this figure and five of the eight are considerably below. The trouble is that the two high figures unduly influence the arithmetic mean in this case.

Whenever we wish to avoid the undue influence of extreme values in the set of numbers, we use the *median average*, or simply the *median*, to represent the set. To find the median of a set of numbers, we arrange the numbers in order from the smallest to the largest, and the median is the middle number in the array, so that half of the numbers are larger than the median and half are smaller. When there are an even number in the array, by convention, we take the median as being halfway between the two middle values. For the eight numbers left in Table 37 we have 21, 27, 60, 62, 67, 92, 138, and 176; and the median is 64.5. Observe that this number is more representative of the numbers as a group.

In a small industrial village in England the average (arithmetic mean) annual family income is \$8,824. We might conclude from this that this is a very prosperous community. However, the truth is that 50 workmen make \$5,000 each and the owner of the factory has an income of \$100,000. Here the mean is a very poor average and the median of \$5,000 gives a much truer picture of the village.

Sometimes we use a third measure of central tendency, which is called the *mode*. The mode of a set of observations is simply that value which occurs most frequently. Sometimes a set of observations does not have a mode, and sometimes it has more than one mode. For the data in Table 37 there is no mode. The mode is more useful when there are a large number of observations in the set, perhaps thousands. Often the mode is more useful than either the mean or the median. For example, a collar manufacturer may find it more useful to know that more men wear a 15 collar size than any other collar size than to know that the mean collar size is 15.2865.

When we are given a single number representing the "average" of a number of observations, we must insist on knowing the type of "average"; that is, does this number represent the arithmetic mean, the median, or the mode?

▲ EXERCISES

1. Find the value of $\sum_{i=1}^{N} x_i$

   a. When $x_i = i$ and $N = 8$
   b. When $x_i = i^3$ and $N = 4$
   c. When $x_i = i - 1$ and $N = 7$
   d. When $x_i = i^2 - i$ and $N = 5$
   e. When $x_i = 4$ and $N = 5$
   f. When $x_i = 8 - i$ and $N = 4$
   g. When $x_i = i^2 - 4$ and $N = 6$
   h. When $x_i = 16 - i^2$ and $N = 5$

2. Find the value of the arithmetic mean

$$\bar{x} = \frac{1}{N} \sum_{i=1}^{N} x_i$$

   a. When $x_i = i$ and $N = 10$
   b. When $x_i = i^2$ and $N = 5$
   c. When $x_i = i + 10$ and $N = 8$
   d. When $x_i = 4$ and $N = 10$
   e. When $x_i = i^2 - 4$ and $N = 6$
   f. When $x_i = 16 - i^2$ and $N = 5$

3. Find the median of each of the sets x_i of Exercise 1.
4. Find the median of each of the sets x_i of Exercise 2.
5. Find the median and mean of the set of grades given in Table 38 (page 243).
6. Which average of Exercise 5 best represents the set of grades? Why?
7. In a recent "Letter to the Editor" in a newspaper, a telephone-company worker complained that although the company asserted that the "average" pay of its employees was \$90 per week, he got only \$72 per week and was the highest paid man in his department. What are the possibilities for accounting for this apparent discrepancy?
8. Take a group of 10 coins and toss the group 25 times. After each toss

count the number of heads. Prepare a table showing the number of times you obtained 0, 1, 2, 3, . . ., 9, and 10 heads. Find the mode for your data.

9. Count the number of words in each line of page 242 of this book. Arrange your data to show the number of lines having 1, 2, 3, . . . words. Find the mean, mode, and median for your data. Which average best describes the data?*

10. A student has received grades of 96, 87, and 47 on three tests. What grade must he get on the next test in order to have an average (mean) grade of 80?

11. On five tests a student has an average grade of 72. What grade did he get on the next test if his average (mean) dropped to 70?

12. In a co-op house there are 20 boys. The average weight of the 6 seniors is 165 pounds. The average weight of the 3 juniors is 170 pounds. The average weight of the 7 sophomores is 155 pounds. The average weight of the 4 freshmen is 147 pounds. What is the average weight of the 20 boys living in this co-op house?

13. An instructor increased the scores on each test paper on a certain test by 10 points. What did this do to the class average or mean?

14. If the instructor in Exercise 13 had increased each test score by 10%, what would this have done to the class average or mean?

An army officer decided to arrange his men to form a solid square. He found that he had 27 men left over. When the officer increased the number of men on a side by 1, he found that 30 additional men were needed to complete the square. How many men did the officer have?

*Notice that the instructions for gathering the data are not precise. To make them precise we must agree on the meaning of a "word." Is a numeral to be counted as a word? Is a word divided between two lines to be counted as part of the first line or the second? Should a hyphenated word be counted as one word or two? One of the important requirements of good statistics is that we be precise on the instructions for gathering data.

Variability of Data—Standard Deviation

Usually it is not enough to give only the average value of a set of observations.

Let us consider another basketball squad of 10 men that had attempted the number of free throws shown in Table 39.

TABLE 39

A	75	*F*	77
B	63	*G*	67
C	62	*H*	59
D	49	*I*	60
E	67	*J*	69

The mean for this set is 64.8, which is the same as the mean for the data in Table 37 for the SMU squad. But when we compare these two basketball squads, we find a striking difference between them when we notice how scattered, or variable, the two sets of data are. The range of free-throw attempts for the SMU squad is 2 to 176 and for the other squad 49 to 77. This is the simplest way of showing the variability of the data. Whenever we use an average to summarize a set of data, we should also give the extent to which the values in the set of data vary among themselves. In fact, variability is the essence of statistics. If there were no variability, there would be nothing to study about the data, because the knowledge of the one item would then tell the entire story.

The range of a set of observations is defined as the difference between the largest and smallest observations. Thus we readily see that the SMU squad with a range of 174 is much more variable than the second squad whose range is only 28.

Often, however, the range fails to give a true picture of the variability of a set of observations. It is highly influenced by extremes of the data and makes no use of any of the observations between the extremes. Consider the two sets of observations *A* and *B* in Table 40.

Here we see that both sets of observations have the same mean ($x = 30$) and the same range ($R = 40$), and yet surely we should agree

TABLE 40

Set *A*	10	20	30	40	50
Set *B*	10	29	30	31	50

that set A contains more variability than set B. How then can we determine this? What is needed is some quantity which will incorporate the fact that the values 20 and 40 in set A are farther from the mean than the values 29 and 31 in set B. This is usually done by computing the *standard deviation*. If x_i represents one observation, and the mean x, calculated by equation (1) of Sec. 5, is the mean of a set of N observations, then the deviation of the observation x_i from the mean is $x_i - \bar{x}$. In order to find the average of all the deviations, we cannot add them and divide by N, because we should always get zero for an answer no matter how variable the data might be. Explain why this is so.

Since some of the deviations will be plus and some minus, we shall square each of the deviations and obtain the average of the squares of the deviation. This number is called the *mean square average*, or *variance* of the data. The measure of variability that is most often used is the square root of the variance; this is called the *standard deviation* of the data and is denoted by the symbol σ_x.

$$\sigma_x = \text{standard deviation of the } x\text{'s} = \sqrt{\frac{1}{N} \sum_{i=1}^{N} (x_i - \bar{x})^2} \qquad (2)$$

In general, the average along with the standard deviation will summarize the information contained in a small set of observations. In a later section we shall discuss further the use of the standard deviation.

For the data in Table 37, the mean is 64.8. Calculating the variance and standard deviation, we get the results shown in Table 41.

TABLE 41

Player	x_i	$x_i - \bar{x}$	$(x_i - \bar{x})^2$
Phillips	176	111.2	12,365.44
Holman	138	73.2	5,358.24
Begert	92	27.2	739.84
Voight	67	2.2	4.84
Higgenbotham	62	−2.8	7.84
Beasley	60	−4.8	23.04
Jones	27	−37.8	1,428.84
Rainer	21	−43.8	1,918.44
Sibley	3	−61.8	3,819.24
Morris	2	−62.8	3,943.84

$$\sum_{i=1}^{10} (x_i - x)^2 = 29{,}609.60$$

$$\text{Variance} = \sigma_x^2 = 2{,}960.96$$

$$\text{Standard deviation} = \sigma_x = \sqrt{2{,}960.96} = 54.4$$

The column $x_i - x$ should always add to zero. This gives a good check on our numerical work. Also we can simplify our numerical work by rounding off 64.8 to 65. The errors resulting from this rounding off tend to cancel each other so that σ_x is affected very little. If we do this rounding off for the basketball squad of Table 41, the sum of the column $(x_i - \overline{x})^2$ is 29,610 and we get the same $\sigma_x = 54.4$.

The value of σ_x gives us a measure of the spread or dispersion of the data. The larger the value of σ_x, the more scattered are the figures of the data. Closely grouped numbers give a small value of σ_x. In Chap. 19 we shall see how the statistician uses the standard deviation σ_x in drawing useful conclusions.

In Exercise 8 of the next set of exercises, you will develop another formula for computing the standard deviation σ_x. This is known as the *short form* and is

$$\sigma_x = \sqrt{\frac{1}{N} \sum_{i=1}^{N} x_i^2 - \overline{x}^2} \qquad (3)$$

The short form gives the same value for σ_x as does equation (2) and saves a lot of time when the number N of figures is large, because we do not need to compute the differences $(x_i - \overline{x})$. This is especially true if we have a machine to compute our squares.

Employing formula (3) and the data of Table 41, we get the results shown in Table 42.

TABLE 42

Player	x_i	x_i^2
Phillips	176	30,976
Holman	138	19,044
Begert	92	8,464
Voight	67	4,489
Higgenbotham	62	3,844
Beasley	60	3,600
Jones	27	729
Rainer	21	441
Sibley	3	9
Morris	2	4
	648	71,600

$$\overline{x} = \frac{1}{N} \sum_{i=1}^{10} x_i = \frac{1}{10}(648) = 64.8$$

$$\sigma_x^2 = \frac{1}{N} \sum_{i=1}^{10} x_i^2 - \overline{x}^2 = \frac{1}{10}(71{,}600) - (64.8)^2$$

$$\sigma_x^2 = 7{,}160 - 4{,}199.04 = 2{,}960.96$$

$$\sigma_x = \sqrt{2960.96} = 54.4$$

▲ EXERCISES

1. When will $\sum_{i=1}^{N} (x_i - \bar{x})^2$ be zero?

2. Compute the mean, median, and standard deviation for the set 82, 72, 78, 96 of four quiz grades in mathematics.
3. Nine holes on a golf course have the lengths in yards:

440	120	410
380	510	170
430	240	540

Find the mean, median, mode, and standard deviation.

4. Find the standard deviation for the data in Table 39.
5. What conclusions do you draw from the standard deviations of the data in Tables 37 and 39?
6. Compute the standard deviation for the data in Table 38.
7. The following sample of nine breakfast checks, in cents, was taken from a cafeteria line:

35	32	30
42	26	20
40	15	12

Compute the mean and standard deviation.

8. Show that

$$\frac{1}{N}\sum_{i=1}^{N} (x_i - \bar{x})^2 = \frac{1}{N}\sum_{i=1}^{N} (x_i^2 - 2x_i\bar{x} + \bar{x}^2)$$

$$= \frac{1}{N}\sum_{i=1}^{N} x_i^2 - 2\bar{x}\frac{1}{N}\sum_{i=1}^{N} x_i + \bar{x}^2$$

$$= \frac{1}{N}\sum_{i=1}^{N} x_i^2 - \bar{x}^2$$

Then

$$\sigma_x = \sqrt{\frac{1}{N}\sum_{i=1}^{N} x_i^2 - \bar{x}^2}$$

Problems just for fun

A commuter train leaves New York fairly well filled. At the first stop half of the passengers get off and 12 new passengers board the train. At the second stop half the passengers leave the train and 8 people climb aboard. At the third stop again half leave the train and 10 board the train. At the fourth stop half leave the train and 4 board the train. At the fifth stop half leave as before but only 2 new passengers appear. At the sixth and last stop 14 passengers leave the train. How many passengers were on the commuter train when it left New York?

9. Use the short form for σ_x developed in Exercise 8 to compute the standard deviation of the data given in Exercise 7. Which formula do you prefer?
10. Use the short form for σ_x to compute the standard deviation for the sets A and B in Table 40.
11. Calculate the standard deviation for the data obtained in Exercise 8, page 246.
12. If each coin used is just as likely to fall heads as tails, it can be shown that the mean number of heads of a large number of tosses of 10 coins will be 5 and that the standard deviation will be $\frac{1}{2}\sqrt{10}$. Compare your mean and standard deviation in Exercise 11 with these theoretical results.
13. An experimental plot of corn produced the following heights in inches:

48	57	64	73	72	53	62	61
74	78	72	71	81	71	73	71

Find the mean, median, and mode. Which of these averages do you consider best represents the experimental plot?
14. Find the standard deviation of the heights in the experimental plot.
15. Suppose the experimental plot had the same mean but a standard deviation of 4 inches. What conclusions would you draw about the variability of the heights?

CHAPTER EIGHTEEN

Frequencies and Distributions

It is a truth very certain that, when it is not in our power to determine what is true, we ought to follow what is most probable.

RENE DESCARTES

1 *Types of Statistical Data*

The data with which we are concerned in mathematical statistics are obtained by making *(a)* measurements or *(b)* counts.

Measurements. Scientists and engineers need to measure accurately the magnitude of the physical quantities with which they work, for example, the weight and physical dimensions of objects, the strength of an electric current, the distance between two points, the charge on the electron, the mass of the atom, etc. Many of us would be inclined to make a single measurement and to accept this value as being the correct value. For careful scientific work, however, a number of independent measurements are made under the same conditions, and an average of these measurements is then taken as the correct value. Any attempt to measure the magnitude of a quantity is subject to error. The error is the difference between the unknown real or correct value and the measured value. But we have no way of knowing what the correct value is, and no one measurement is exactly correct except by chance. The chance errors in measurement are such that some of the measurements are too large and some of them are too small. If we have many measurements, the positive and negative errors will tend to balance each other. In contrast to chance errors, we have errors which persist and are always in the same direction. For example, if a grocer's scales are out of adjustment and read 1 pound when 15 ounces is placed on the pan, the error in any number of weighings will always be in the same direction. If a clock runs too fast we will not get a correct measurement of time, since the errors will all be in the same direction. Such errors are called persistent error, constant error, or biased error. Biased error can sometimes be discovered and eliminated, but random error can never be eliminated, although its effect may be reduced by taking a large number of observations or measurements.

Counts. Other data may be obtained by counting, and these data seem to be different from those obtained by measurement. If we say that there are 25 students in a mathematics class, the figure 25 is supposed to be absolutely exact and not an approximation. But in practice, this distinction is more apparent than real, for there are many probable causes of error in a count, and when the numbers involved are large, the counts are only approximations. If a city decides to determine its population by an actual count, the count will give only an approximation of the population, because the population will be different at different times of the day, and some people will not be counted at all while others may be counted twice.

The statistical data which have been collected are classified as either *(a)* discrete or *(b)* continuous.

Distributions of data in which only certain values are possible are called *discrete*. For example, if we were listing the number of rooms in the houses in a certain district, we would get 5-room houses or 4-room houses, but there would be no house with 4.6 rooms. If the data listed the number of bets made at the $2, $5, $10, and $20 windows at a horse race, no bet of $7.50 could be listed.

In contrast to discrete distributions, we may have *continuous distributions*, in which any intermediate value may occur. For example, heights of women do not necessarily fall at 61, 62, or 63 inches or at any other particular value. It is possible for a woman to be 62.67 inches in height or any other possible value between the tallest and the shortest heights.

With discrete data there is a bunching of values at particular points because no intermediate values can occur. Sometimes we get a similar bunching for continuous data. For example, the U.S. Bureau of the Census has found that when asked to give his age, a person usually gives whole numbers of years and that there is a tendency to give ages in multiples of 5 and 10. That is, a person says he is 50 years of age even though he may be 51, 52, 53, or 54 years old. Or even 70!

Problems just for fun

When Barnacle Bill and his pals leave ship to go ashore, it is low tide and they climb down 7 rungs of the ship's ladder to reach the small boat. When they come back about 6 hours later, it is high tide. If the water level has risen 9 feet and the rungs of the ladder are 18 inches apart, how many rungs do Bill and his pals have to climb to get aboard?

Accuracy of Measurements and Calculations

If we are required to summarize a set of observations, we should estimate the probable accuracy of the data, whether they have come from measurements or from counts. It may well happen that the observations themselves are stated with spurious accuracy. For example, suppose we are given data which claim to represent the waist measurements of various individuals. We should not be surprised if the numbers were, say, 29 inches, 30.5 inches, 32.75 inches, etc. But if the numbers were given as 29.7128 inches, 30.4157 inches, and 33.0182 inches, etc., we should very likely question whether it was possible to obtain such apparently accurate measurements. Or if the data claim that a certain city had yearly populations of 43,516, 45,782, 49,103, etc., we might doubt whether these counts could have been taken this accurately.

Assuming that the observed data do possess their claimed accuracy, the person who summarizes the data may give his results with spurious accuracy. For example, if we are told that family incomes for 158 families are $1,800, $2,500, $2,800, $3,600, $2,900, $4,200, etc., the arithmetic mean of these incomes might turn out to be $4,151.898. It is apparent from the original data that these incomes have been given to the nearest $100 and that a reported average of $4,151.898 gives a false impression of the accuracy of the observations. In this case, we could prevent such a situation by giving the average as $4,150, or even $4,200. Although there are no hard and fast rules for rounding off numbers, we should realize that the results computed from data are limited by the accuracy of the individual items included in the data.

▲ EXERCISES

Discuss the possibilities for spurious accuracy in the following statements concerning observed data:

1. The population of the United States in 1940 was 132,322,408.
2. The average height of 63 students is 66.312581 inches.
3. The indebtedness of the United States government on Mar. 29, 1950, was $255,745,196,000.
4. A newspaper report declares that "the exact center of United States population in 1940 was on top of a fence post in a southern Indiana county."
5. In 1808 Alexander Wilson said that he saw a great flight of 2,230,272,000 pigeons pass over Kentucky in 4 hours.

3 *Arrangement and Presentation of Data—Frequency Distribution*

A long list of numerical data, such as the grade averages of 200 individual students, is more confusing than helpful because of the sheer mass of figures. We cannot see the forest for the trees. A natural way to avoid this mass of figures is to group them into a *frequency table* or *frequency distribution*. For example, in the case of grade averages, we might choose the ranges of numbers that give the classes A, A−, B+, B−, C+, C−, D+, D− and count the number of student averages in each range. We could see much better the true distribution of grades from the number of students in each of these eight ranges than from the list of 200 individual grades.

It is apparent that if we group the data into too few classes, that is, make the class width too large, we gain simplicity but lose detail and information. There is a danger of using too few or too many classes. When we use a few classes, the number of cases in each class is enough so that the erratic variations tend to disappear and the pattern of the distribution becomes plainer. Of course, we can go too far and choose too few classes so that no pattern at all is evident. If the class widths are too small, we do not have any distribution at all, but we have only "lined up" the numbers in an array.

A statistician wants the number of classes to be large enough so that all the items in a class may reasonably be treated as equal without too much error and so that the general pattern of the distribution is not obscured.

The first step is to choose *classes* or *class intervals* which cover the entire range of the data and are of equal width. For each interval we have *class limits* so that there are no gaps or overlaps between the intervals. For example, 1.0 to 2.0, 2.0 to 3.0, 3.0 to 4.0, 4.0 to 5.0 would be satisfactory only if none of the data were 2.0 or 3.0. If these do occur, we must change our limits to clarify where such members are placed. This might be done by making our limits 1.0 to 1.9, 2.0 to 2.9, etc., if our data are given to one decimal place.

The next step is to tally the numbers into the classes just as we would tally ballots. For example, in a typical A = 4, B = 3, C = 2, D = 1, F = 0 grade system, our 200 grade averages would range from 0.0 to 4.0. We might choose our class limits as 0 − 0.4, 0.5 − 0.9, etc. Our frequency table is given in Table 43, and we can see how much clearer the distribution of grades is shown by this grouping than by the 200 individual averages.

Once we have tallied the numbers into a frequency table, we discard the original data and work entirely from the table to compute averages and standard deviations. To do this, we call the mid-point of each interval the *class mark* and assume that all numbers in the interval have this value. Naturally this introduces some error, but if the values cluster close to the center or are nearly evenly distributed through the interval, this error is minor. Sometimes we can reduce this error by judicious choice of the class limits. For example, if the

TABLE 43

Class limits	*Frequency tally*	*Frequency*
0–0.4	////	4
0.5–0.9	~~////~~ ~~////~~ ///	13
1.0–1.4	~~////~~ ~~////~~ ~~////~~ ~~////~~ ~~////~~ //	27
1.5–1.9	~~////~~ ~~////~~ ~~////~~ ~~////~~ ~~////~~ ~~////~~ ~~////~~ ~~////~~ ~~////~~	45
2.0–2.4	~~////~~ ~~////~~ ~~////~~ ~~////~~ ~~////~~ ~~////~~ ~~////~~ ~~////~~ ~~////~~ ~~////~~ /	51
2.5–2.9	~~////~~ ~~////~~ ~~////~~ ~~////~~ ~~////~~ ~~////~~ ~~////~~ /	36
3.0–3.4	~~////~~ ~~////~~ ~~////~~ ////	19
3.5–4.0	~~////~~	5
	Total frequency	200

numbers in our data tend to cluster around 10, 20, 30, etc., we should choose class limits so that the class marks as mid-points are these numbers. In the example given in Table 43, the class marks are 0.2, 0.7, 1.2, 1.7, etc.

Relative and Cumulative Frequency Distributions

It is sometimes useful to know the relative and cumulative frequencies of the distribution. The *relative frequency* corresponding to a particular class mark is its frequency divided by the total number of figures in the data. The relative frequency of a test grade average between 1.5 and 1.9 in Table 43 is $\frac{45}{200} = 0.22$. The relative frequency is often expressed in per cent. The *cumulative frequency* of a particular value is the sum of all measurements less than or equal to the given value. The cumulative frequency corresponding to a grade average between 1.5 and 1.9 in Table 43 is $4 + 13 + 27 + 45 = 89$. We may add the relative frequencies to get the *cumulative relative frequency*, which is the cumulative frequency of a particular value divided by the total number of counts. The cumulative relative frequency of the range 1.5 to 1.9 is $\frac{89}{200} = 0.44$. Cumulative relative frequencies must be used when one is comparing two cumulative frequency curves which are based on different numbers of items.

TABLE 44

Class limits	*Class mark*	*Frequency*	*Relative frequency*
0–0.4	0.2	4	0.02
0.5–0.9	0.7	13	0.06
1.0–1.4	1.2	27	0.14
1.5–1.9	1.7	45	0.22
2.0–2.4	2.2	51	0.26
2.5–2.9	2.7	36	0.18
3.0–3.4	3.2	19	0.09
3.5–4.0	3.7	5	0.03
		200	1.00

We shall add these new types of frequencies to the data in Table 43 to get Tables 44 and 45.

TABLE 45

Upper limit	*Class mark*	*Cumulative frequency*	*Relative cumulative frequency*
0.4	0.2	4	0.02
0.9	0.7	17	0.08
1.4	1.2	44	0.22
1.9	1.7	89	0.44
2.4	2.2	140	0.70
2.9	2.7	176	0.88
3.4	3.2	195	0.97
4.0	3.7	200	1.00

Graphical Representation of Frequency Distributions

The frequency tables are often presented in graphical form, because a graph gives a better visual picture of the distribution than a table. Charts are used to show the fluctuations in the data. Statisticians use the histogram or the frequency polygon to graph a frequency table. The *histogram* is a bar graph with the bars centered over the class marks and with heights equal to the class frequencies. The height of each bar is proportional to the number of observations falling in this class. A *frequency polygon* is a broken-line graph connecting the points whose coordinates are class marks and class frequencies.

The frequency polygon is always drawn down to the axis at the class mark of the first empty class. The relative frequency is shown by an appropriate vertical scale on the right side of the figure. Figure 85 is a histogram with a frequency polygon superimposed as a color line for the data in Table 43.

The histogram is often used for popular presentation of a frequency distribution because it is more striking and causes the individual classes to stand out more clearly. If we look at Table 43 and Fig. 85, we see that the frequency tally itself gives a rough form of the

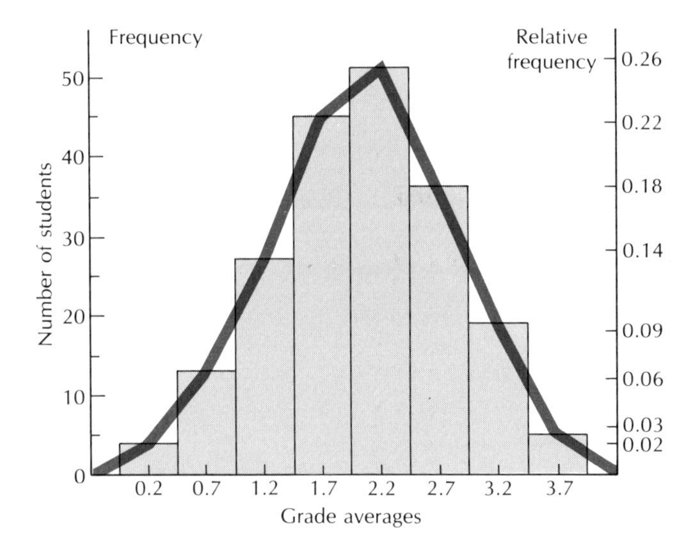

FIG. 85

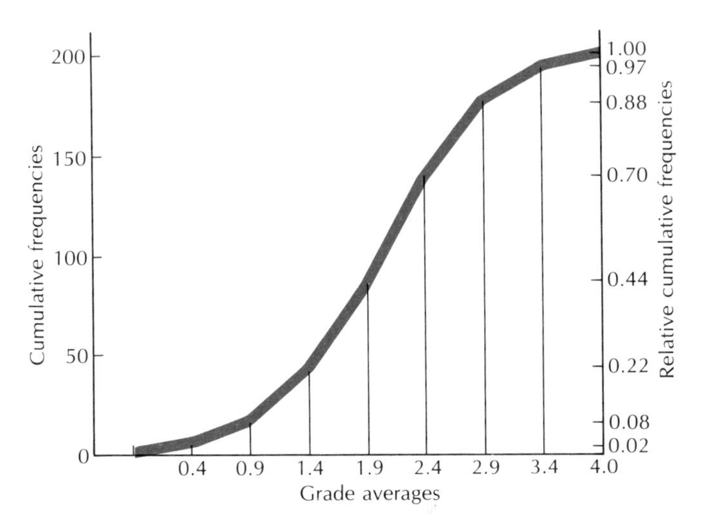

FIG. 86

histogram. If we need to compare two different frequency distributions, the frequency polygons are more effective than the histograms.

The data in Table 45 are plotted by constructing ordinates at the upper limit of the class boundary. The height of the ordinate is equal to the cumulative frequency. The tops of these ordinates are joined by straight lines. The relative cumulative frequency is shown by an appropriate vertical scale on the right side of the figure. The graph of the cumulative frequency of a distribution is called an *ogive*. Figure 86 is the *ogive* for the data in Table 45.

▲ EXERCISES

1. A class of students tossed eight coins 240 times and obtained the results shown in Table 46. Take the class limits to be −0.5, 0.5, 1.5, 2.5, . . ., 7.5, 8.5. Draw the relative frequency histogram and relative cumulative frequency ogive for this distribution.

TABLE 46

No. of heads in a toss	*No. of times occurring*
0	2
1	11
2	32
3	45
4	65
5	53
6	26
7	5
8	1

2. Table 47 gives the weight in pounds of 100 men at an induction center. Make a frequency table for this data.

TABLE 47

124	132	142	151	153	149	154	171	160	167
158	155	195	172	154	161	168	137	146	159
157	185	174	162	169	138	143	152	174	146
163	167	150	175	159	183	155	176	164	168
134	145	140	158	152	177	165	164	155	175
178	156	176	166	163	140	155	148	176	162
133	152	148	187	187	145	155	173	166	160
173	144	179	156	176	154	189	157	146	154
153	135	154	157	164	176	161	181	139	150
162	153	147	159	171	148	165	153	145	163

3. Make a histogram and frequency polygon for the data in Table 47.
4. Plot the ogive for the cumulative frequency of the data in Table 47.
5. From the ogive of Exercise 4, estimate the percentage of men weighing more than 160 pounds. Less than 150 pounds.
6. Table 48 gives the dressed weight in pounds of 30 chickens received at a meat counter. Make a frequency table for this data.

TABLE 48

6.6	1.7	5.1	1.2	4.2	2.3
5.0	2.5	1.3	1.6	2.4	1.9
4.8	2.4	1.5	1.8	1.8	2.1
2.8	3.0	1.8	2.1	2.0	1.5
3.2	3.5	2.6	2.7	1.6	2.9

7. Draw a histogram and frequency polygon for the data of Table 48.
8. Draw the ogive for the cumulative frequency of the data of Table 48.
9. From the ogive of Exercise 8, estimate what percentage of the chickens weigh more than 3 pounds. More than 2 pounds.
10. One hundred entering freshmen boys at a university had the weights given in Table 49. Make a frequency table for this date.

TABLE 49

169	165	110	153	187	160	172	231	190	203
183	155	168	195	172	178	169	133	178	170
180	215	156	133	185	139	170	205	153	180
241	165	165	210	210	174	145	150	145	185
195	150	185	185	216	162	128	152	142	139
199	223	173	141	155	240	120	134	165	140
156	149	181	117	189	145	172	162	136	168
171	143	165	145	126	220	145	146	213	205
155	147	200	185	170	159	160	190	180	185
183	150	169	172	185	200	140	143	194	178

11. Draw a histogram and frequency polygon for Table 49.
12. Draw the ogive for the cumulative frequency of Table 49.
13. One hundred entering freshmen girls at a university had the weights given in Table 50. Make a frequency table for this data.

TABLE 50

115	201	104	114	123	128	115	139	127	118
120	105	120	116	120	113	121	104	110	169
120	130	142	129	110	114	135	153	125	116
110	106	130	133	118	122	115	120	105	110
174	124	119	122	123	121	120	120	125	144

TABLE 50 Cont.

120	120	116	149	117	115	107	135	137	120
126	115	160	125	130	125	110	182	96	95
160	130	108	143	108	128	146	132	115	132
139	120	122	121	127	124	135	136	126	115
120	115	120	154	131	107	115	123	109	111

14. Draw a histogram and frequency polygon for Table 50.
15. Draw the ogive for the cumulative frequency of Table 50.
16. One hundred entering freshmen had the following College Board Mathematics Scores (rounded off to the nearest 10, that is, 537 becomes 54). Make a frequency table for this data.

TABLE 51

47	55	62	73	57	58	46	64	51	51
55	46	66	41	55	52	67	66	51	59
65	60	67	53	47	45	58	75	56	61
54	46	54	58	57	57	56	41	62	48
45	53	55	50	57	36	64	55	35	60
47	58	72	50	48	57	59	48	59	45
41	52	66	56	63	67	75	33	52	62
52	58	70	54	52	47	56	49	56	51
49	50	66	47	50	67	44	54	53	45
59	58	66	53	39	45	44	47	55	51

17. Draw a histogram and frequency polygon for Table 51.
18. Draw the ogive for the cumulative frequency of Table 51.
19. One hundred entering freshmen at a university had the following College Board Verbal Scores (rounded off to the nearest 10, that is, 583 is rounded off to 58). Make a frequency table for this data.

TABLE 52

64	56	66	48	46	61	46	46	46	53
62	48	74	38	49	53	69	64	46	63
56	44	66	56	46	44	60	66	48	46
58	50	44	56	58	40	50	53	62	58
40	60	54	37	61	45	63	46	44	42
46	57	59	52	44	51	56	53	62	48
51	41	51	44	70	55	70	30	48	50
51	60	56	42	73	57	56	42	46	68
53	46	40	37	58	44	41	66	57	60
59	53	64	62	40	49	39	38	58	41

20. Draw a histogram and frequency polygon for Table 52.
21. Draw the ogive for the cumulative frequency of Table 52.

Prove that the following enrollment data are inconsistent:

1,000 total enrollment in college
525 freshmen
312 male students
470 married students
42 male freshmen
147 married freshmen
86 married male students
25 married male freshmen

6 *Averages for Grouped Data*

When working with unclassified data, we found it necessary to compute both an average for the data and a measure of the variability of the data. We should also learn how to compute an average for a frequency distribution and a measure of the extent of its variability.

When the raw data have been classified in a frequency table, we discard the raw data and no longer know the value of a single item but only the number of items that fall in various classes. We must make some assumptions in order to find the various averages and the standard deviation of the grouped data. We assume that all the items in a class interval have the value of the class mark. Of course this introduces some error, but if the number of individual items is distributed almost uniformly within the class interval, the class mark will come close to representing the average of that class. For example, we assume that the 45 students in Table 44 that had grade averages between 1.5 and 1.9 all had averages of 1.7, which is the class mark for this interval.

How shall we find an average—mean, mode, or median—for a

set of observations which have been grouped into a frequency distribution? The *mode* is defined as that class mark which has the greatest frequency. The mode can be found by noting which class mark is under the highest point on the frequency polygon. For example, from the frequency polygon in Fig. 85 (page 261), we find that 2.2 is the mode of the data in Table 43. The frequency distribution need not have a mode, or it may have several.

The *median* is defined as that value of the variable which has a relative cumulative frequency of 0.50 or a cumulative frequency of $N/2$.

We can find the median from the *ogive* by locating the point 0.50 on the relative cumulative frequency axis, moving parallel to the horizontal axis until we reach a point Q on the ogive, and then moving down to the axis to locate the point x which is the value of the median. The median for the data graphed in Fig. 86 is about 2.1, and the method of obtaining this value is illustrated in Fig. 87.

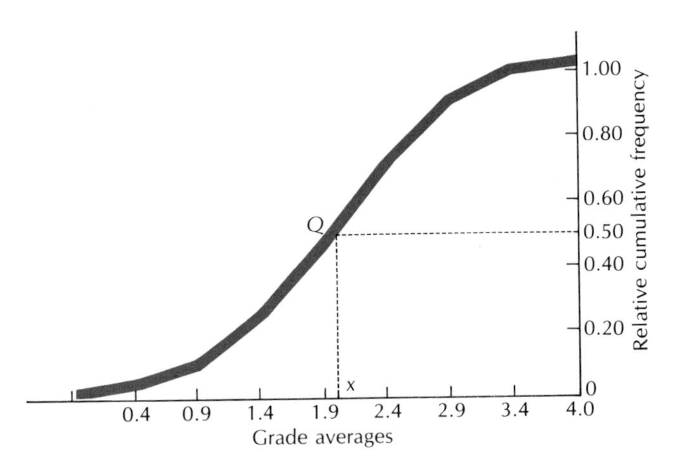

FIG. 87

In order to find the mean from a frequency table, we have assumed that each class mark x_i ($i = 1, 2, 3, \ldots, k$, where k is the number of classes) represents all the measurements which fall in its class. In order to get the sum of all the observations in a class, we multiply the class mark x_i by the frequency f_i of the class. The mean $\overline{x}$ is the sum of these products $x_i f_i$ ($i = 1, 2, 3, \ldots, k$) divided by the total number of observations N; that is,

$$\overline{x} = \frac{1}{N} \sum_{i=1}^{k} x_i f_i \tag{1}$$

where k is the number of classes, and N is given by

$$N = \sum_{i=1}^{k} f_i \tag{2}$$

For the data in Table 44,

$$N = \sum_{i=1}^{8} f_i = 4 + 13 + 27 + 45 + 51 + 36 + 19 + 5 = 200$$

$$\sum_{i=1}^{8} x_i f_i = (0.2)(4) + (0.7)(13) + (1.2)(27) + (1.7)(45)$$
$$+ (2.2)(51) + (2.7)(36) + (3.2)(19) + (3.7)(5) = 407.5$$

and

$$\overline{x} = \frac{1}{N} \sum_{i=1}^{8} x_i f_i = \frac{1}{200}(407.5) = 2.04$$

▲ EXERCISES

1. If we computed the mean of the 200 individual grade averages tallied in Table 43, we would expect to get a number different from 2.04. Why? By how much might we find it differing?
2. Find the mode, median, and mean for the data in Table 46, as classified in the frequency table of Exercise 1 of page 262.
3. Find the mode, median, and mean for the data in Table 47, as classified in Exercise 2 on page 262.
4. Find the mode, median, and mean for the data in Table 48, as classified in the frequency table of Exercise 6 on page 263.
5. Find the mode, median, and mean of the data in Table 49, as classified 101–110, 111–120, 121–130, etc., of Exercise 10 on page 263.
6. Find the mode, median, and mean of the data in Table 50, as classified 91–100, 101–110, 111–120, etc., of Exercise 13 on pages 263–264.
7. Find the mode, median, and mean of the data in Table 51, as classified 30–35, 36–40, 41–45, etc., of Exercise 16 on page 264.
8. Find the mode, median, and mean of the data in Table 52, as classified 30–35, 36–40, 41–45, etc., of Exercise 19 on page 264.

Variability of Frequency Distributions

We shall use the standard deviation to measure the spread, or variability, in a frequency distribution. We must remember that the x_i's are class marks and that each class mark represents f_i figures. The square of the standard deviation is given by

$$\sigma_x^2 = \frac{(x_i - \overline{x})^2 f_1 + (x_2 - \overline{x})^2 f_2 + \cdots}{f_1 + f_2 + f_3 + \cdots} = \frac{\sum_{i=1}^{k} (x_i - \overline{x})^2 f_i}{\sum_{i=1}^{k} f_i}$$

The standard deviation σ_x is easily found by extracting the square root of σ_x^2.

Let us calculate σ_x for the data in Table 43. We have found that $\overline{x} = 2.04$ for these data.

Problems just for fun

A certain tribe uses finger reckoning to multiply numbers between 5 and 10. They extend the number of fingers equal to the excess of one number over 5 and do the same with the other number on the other hand. The sum of the extended fingers gives the first figure of the product, and the product of the unextended fingers gives the second figure. For example, if we wish to multiply 8 times 6, we extend three fingers on one hand and one on the other. These are the excess over 5 of 8 and 6. Then $3 + 1 = 4$ is the sum of the extended fingers, and $2 \times 4 = 8$ is the product of the unextended fingers. We get 4 and 8, or 48, the product of 8 times 6. Can you explain why this method is correct?

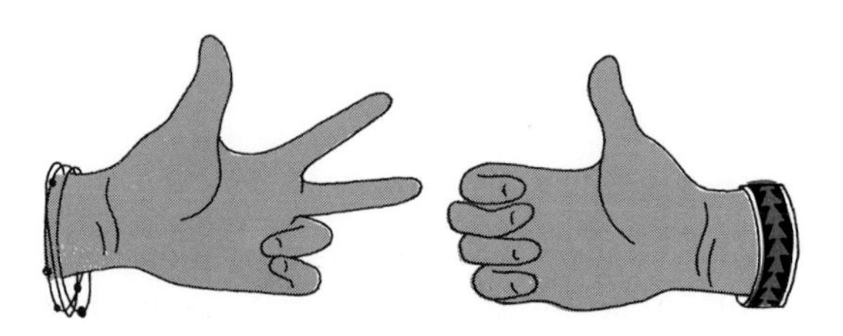

TABLE 53

Class marks, x_i	*Frequency,* f_i	$(x_i-\bar{x})$	$(x_i-\bar{x})^2$	$(x_i-\bar{x})^2 f_i$
0.2	4	−1.84	3.3856	13.5424
0.7	13	−1.34	1.7956	23.3428
1.2	27	−0.84	0.7056	19.0512
1.7	45	−0.34	0.1156	5.2020
2.2	51	0.16	0.0256	1.3056
2.7	36	0.66	0.4356	15.6816
3.2	19	1.16	1.3456	25.5666
3.7	5	1.66	2.7556	13.7780

$$N=\sum_{i=1}^{8} f_i=200 \quad \text{and} \quad \sum_{i=1}^{8}(x_i-\bar{x})^2 f_i=117.4702$$

$$\sigma_x^{\,2}=\frac{117.4702}{200}=0.58735$$

$$\sigma_x=\sqrt{0.58735}=0.77$$

▲ EXERCISES

1. Compute σ_x for the frequency distribution of the data in Table 46.
2. Compute σ_x for the frequency distribution of the data in Table 47.
3. Compute σ_x for the frequency distribution of the data in Table 48.
4. Compute σ_x for the frequency distribution of the data in Table 49.
5. Compute σ_x for the frequency distribution of the data in Table 50.
6. Compute σ_x for the frequency distribution of the data in Table 51.
7. Compute σ_x for the frequency distribution of the data in Table 52.
8. John measured the length of a field 10 times by pacing. His 10 measurements have a mean of 320 feet, with a standard deviation of 8 feet. George did the same thing and also got a mean of 320 feet, but the standard deviation of his 10 measurements was 2 feet. Comment.

Problems just for fun

Which will cost more, a dozen dozen balls at half a dozen dimes a dozen or half a dozen dozen balls at a dozen dimes a half dozen?

Review Exercises

1. For 20 different automobiles the gasoline mileage was measured and the frequency distribution in Table 54 was constructed. Compute the cumulative frequencies and draw the ogive. Compute the median, mean, and standard deviation.

TABLE 54

Class mark, miles per gallon	*Frequency*
10	3
15	8
20	7
25	2

2. In 1937, a survey of 100 hardware stores showed the gross margin of profit given in Table 55.

   a. What are the class marks for this frequency distribution?

TABLE 55

Per cent gross profit	*Number of stores*
0–3.9	3
4.0–7.9	10
8.0–11.9	16
12.0–15.9	30
16.0–19.9	20
20.0–23.9	16
24.0–27.9	5

   b. Draw the histogram and find the mode.
   c. Draw the ogive and find the median.
   d. Compute the mean.
   e. Compute the standard deviation.

3. After the home-coming football game, the line in the college cafeteria showed the age distribution given in Table 56. For this distribution, do the tasks specified in the previous exercise.

TABLE 56

Ages in years	*Number of persons*
0–8	3
9–17	18
18–26	44
27–35	23
36–44	17
45–53	10
54–62	8
63–71	5

4. The distribution of lunch checks at a restaurant is given in Table 57. For this distribution, do the tasks specified in Exercise 2.

TABLE 57

Luncheon checks, cents	*Number of checks*
50–74	12
75–99	34
100–124	47
125–149	53
150–174	39
175–199	22
200–224	10
225–249	3

Problems just for fun

A dealer had 6 barrels containing 15, 16, 18, 19, 20, and 31 gallons. Five of these barrels contained beer and one wine. He sold the barrels of beer to two customers, one receiving twice as many gallons as the other. Which barrel contained wine?

CHAPTER NINETEEN

The Normal Curve and Sampling

The most important questions of life are, for the most part, really only problems of probability. LAPLACE

1 *Common Shapes of Frequency Curves*

A frequency polygon can assume almost any shape, but statisticians have found from experience that the frequency polygons of most distributions fall into one of a small number of types. The most common of the shapes is found to be humpbacked or bell-shaped, with the larger frequencies occurring near the center and the smaller frequencies occurring at the extreme values of the distributions. Often this mound-shaped distribution is symmetrical about its maximum frequency; that is, the right-hand portion of the frequency polygon is a mirror image of the left-hand portion (see Fig. 88). But sometimes

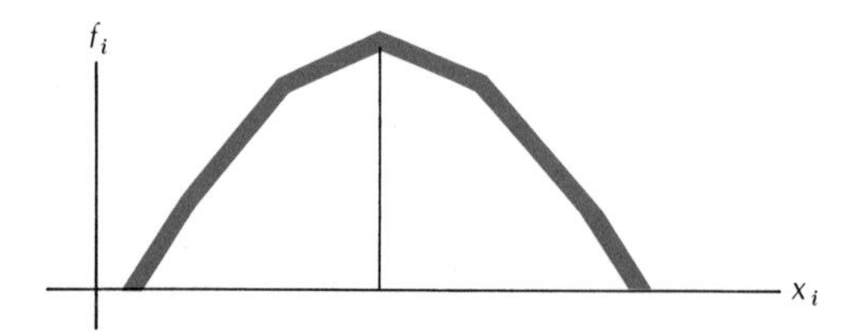

FIG. 88

the high point on the frequency polygon is not halfway between the extreme values, and the polygon is said to be *skewed.* In this case, the frequencies decrease more rapidly on one side of the value of the maximum frequency than on the other side. The two frequency polygons in Fig. 89 are skewed.

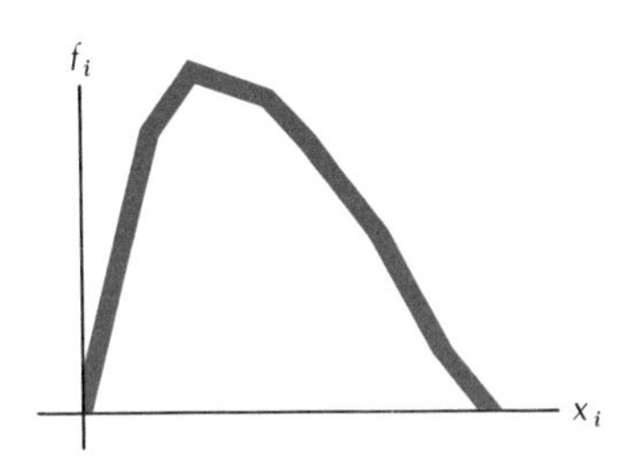

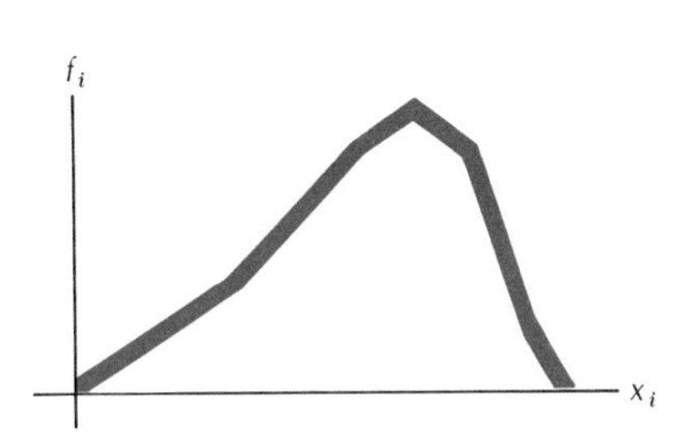

FIG. 89

If we have a frequency distribution for a large number of measurements or counts arranged so that there are many classes, each of small class width, the bases of the rectangles in the histogram will be small and the rectangles will rise in such tiny steps that the frequency polygon is almost a smooth curve (Fig. 90).

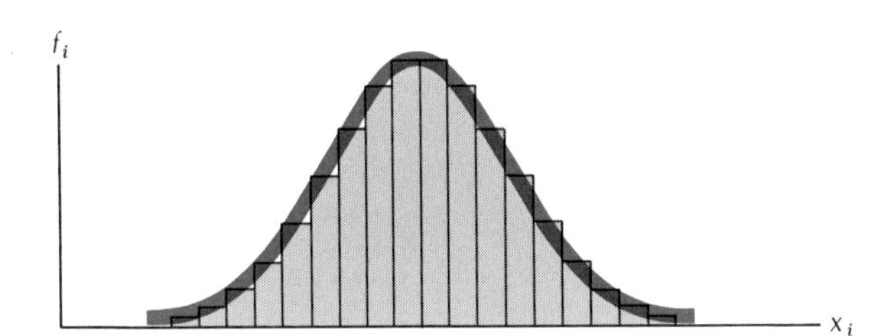

FIG. 90

If we increase the number of classes in the frequency distribution, the frequency polygon and the smooth curve will become very close together. We can now assume that the area under the smooth curve is the same as the total area of the histogram.

In all the work that follows, we shall draw frequency polygons as smooth curves.

If a mound-shaped distribution is of the symmetrical type, the mean, mode, and median fall together. If the distribution is skewed, the mode lies under the highest point of the frequency polygon, and the median and the mean lie in that order toward the longer branch of the polygon. The positions of the mean, median, and mode are shown in Fig. 91. The mean always falls farthest from the mode because it is affected the most by the extreme values.

The differences between the values for the mode, median, and

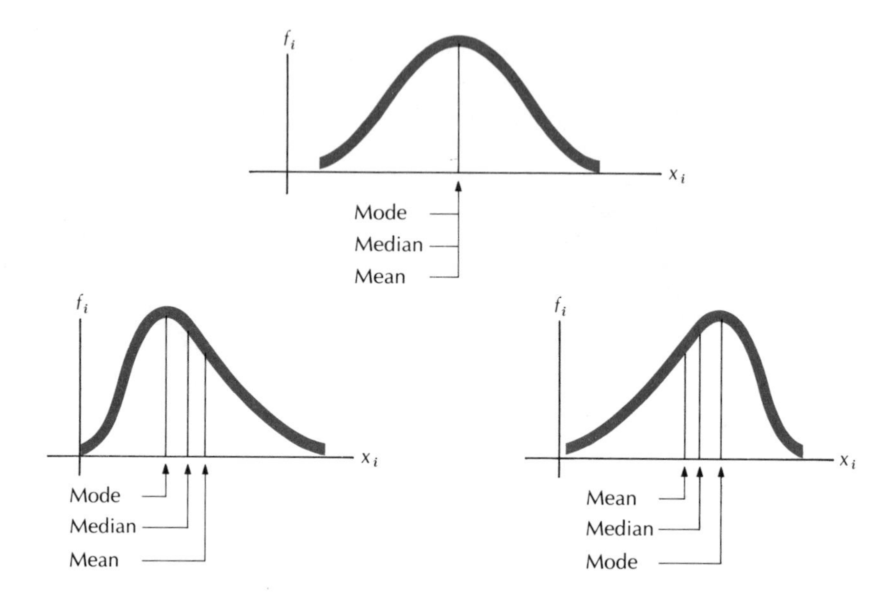

FIG. 91

mean are an indication of the symmetry of the distribution. When there is a marked skewness—that is, when the mode, median, and mean differ greatly in value—the mean is not a good representative of the data, and the median or mode will give the better indication of the central tendency of the distribution.

Problems just for fun

Joe, an inveterate gambler, said to his friend Jim, "I'll bet you half the money that I have in my pocket against an equal sum on the toss of a coin—heads I win, tails I lose." Jim took the bet. The coin was tossed and the money paid. Joe repeated this offer again and again, always betting one-half the money then in his possession. After a number of bets Joe found that he had won just as many times as he had lost. Now, did Joe gain or lose on these bets?

▲ EXERCISES

1. Explain why the median or mode is a better measure of the central tendency than the mean when the frequency distribution is skewed.
2. Look over the frequency polygons that you made for the exercises following Sec. 5 and the review exercises of Chap. 18. Determine which distributions are symmetrical and which are skewed.
3. Erect vertical lines at the points of the mode, median, and mean on each of the frequency polygons that you have made for the distributions in Exercise 2 above.
4. In symmetrical or moderately skewed bell-shaped distributions having to do with a rather large number of cases, the standard deviation of the distribution is roughly about one-sixth of the distance between the lowest and the highest class mark. This fact is useful in estimating the value of σ before it is actually computed. See how good this approximation is for the values of the standard deviations you found for the frequency distributions in Exercise 2 above.

The Normal Frequency Curve

While the plotted frequency polygon of data may sometimes have more than one mode and may be skewed, it has been observed that more often data plot into frequency polygons which have a single hump and which are symmetrical about the mean. This is particularly true of data resulting from measurements in the physical, social, or biological sciences. These distributions have the form of a certain mathematical curve. Hence this curve has been extensively used as a model in studying distributions. This curve has been named the *normal curve,* and a distribution that has a frequency polygon which has

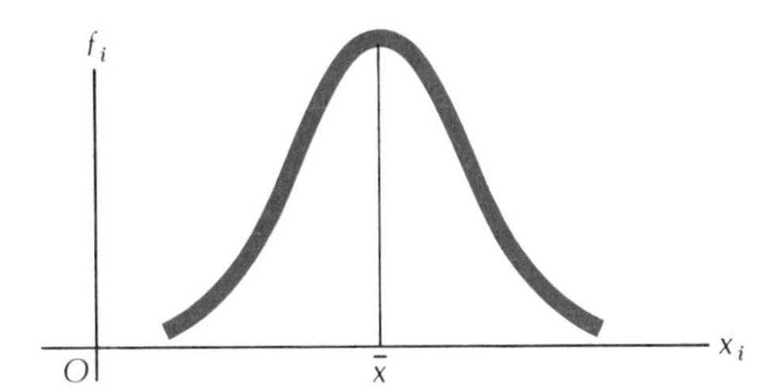

FIG. 92

this form is called a *normal distribution.* In Fig. 92 we show a normal distribution in which the frequency polygon has been plotted with such short class intervals that the polygon resembles a smooth curve.

3 The Equation of the Normal Curve

Let us plot the graph for the curve

$$y = e^{-t^2/2}$$

TABLE 58

t	0	$\pm.5$	$\pm.7$	$\pm.9$	±1	±1.5	±2	±3
y	1	.88	.78	.67	.61	.32	.14	.01

We obtain a curve (Fig. 93) which is very similar to the shape of the distribution plotted in Fig. 92, and statisticians have used this curve to represent the idealized distribution of data.

However, we must modify this equation to make it fit the practical distributions. We know that the normal distribution centers about its mean $\bar{x}$, while the curve we have just plotted is symmetrical about the y axis, that is, the line $t=0$. To change the symmetry to the proper line, we will replace t by $(x-\bar{x})/\sigma_x$. The factor σ_x is introduced

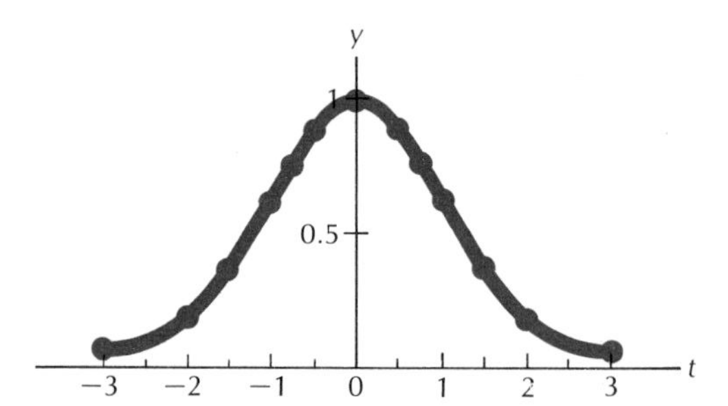

FIG. 93

so that the width of the hump will change with the variability of the distribution. If σ_x is small, the hump in the curve will be steep and narrow; if σ_x is large, the hump will be broad and flat. Our equation now becomes

$$y = e^{-(x-\bar{x})^2/2\sigma_x{}^2}$$

If we plot this curve for $x=0$ and $\sigma_x = 1$, we get the graph of Fig. 93. Other values σ_x will make the hump steeper or flatter but will not change the general shape of the curve.

We notice that the curve flattens out very near the x axis but never touches it, no matter how far out we plot it. The curve and the x axis bound an area which extends to infinity in both directions. While it extends to infinity it becomes narrow so rapidly that the area turns out to be finite. Mathematicians have computed the area bounded by this curve and the x axis and have found it to be $\sqrt{2\pi}\sigma_x$. The applications we shall make of the normal curve have to do with questions of probability. Portions of the area will represent probabilities, which we remember are always fractions from 0 to 1, and thus it will be desirable for the total area to equal 1, which is the largest probability number. We can do this by introducing a factor $\sqrt{2\pi}\sigma_x$ in the height of the curve. With this final modification we take as the equation of the *normal curve*

$$y = \frac{1}{\sqrt{2\pi}\sigma_x} e^{-1/2[(x-\bar{x})/\sigma_x]^2} \tag{1}$$

In this equation, if $(x - \bar{x})/\sigma_x$ is replaced by t and $\sigma_x = 1$, the normal curve is said to be in standard form. So the *standard normal curve* has the equation

$$y = \frac{1}{\sqrt{2\pi}} e^{-t^2/2} \tag{2}$$

To find areas under parts of the normal curve of equation (1) would require different tables for all possible values of $\bar{x}$ and σ_x, while only one table is required for finding areas under parts of the standard curve of equation (2). A frequency distribution is said to be *standardized* when the variable x is changed to the variable $t = (x - \bar{x})/\sigma_x$. Notice that t tells us how many standard deviations a given x is from its mean $\bar{x}$. If x is one standard deviation σ_x away from its mean $\bar{x}$, that is,

$$x - \bar{x} = \pm\sigma_x$$

then the corresponding t is $+1$ or -1 according as x is larger or smaller than $\bar{x}$. For $t = 2$, $x - \bar{x} = 2\sigma_x$, and x is two standard deviations, $2\sigma_x$, above its mean. This means that a distance $x - \bar{x} = 2\sigma_x$ on the x axis of Fig. 92 corresponds to a distance of 2 units on the t axis of Fig. 93. For $t = -2$, $x - \bar{x} = -2\sigma_x$, and x is two standard deviations, $2\sigma_x$, below its mean.

A Property of the Normal Curve

When we were working with histograms we may have noticed that

$$\frac{\text{Area of one rectangle}}{\text{Total area of histogram}} = \frac{\text{class frequency times class width}}{\text{total number of cases times class width}}$$
$$= \frac{\text{number of cases in the one class}}{\text{total number of cases}}$$

From our definition of probability, this last ratio is the probability that an x chosen at random from the distribution will be in the particular class represented by the one rectangle. In general, the probability for an x chosen at random to lie between two limits x_1 and x_2 will be equal to the sum of the areas of the rectangles between x_1 and x_2 divided by the total area of the histogram. Now we may think of the graph of equation (1) as a histogram with extremely narrow rectangles. Then the probability that an x chosen at random from a frequency distribution will lie between two values x_1 and x_2 is simply the area under the curve between these two limits divided by the total area. Here we see the advantage of introducing the factor $\sqrt{2\pi}\sigma_x$ into the equation to make the total area 1. With this total area 1, we do not need to divide, and the area under the curve between x_1 and x_2 is exactly equal to the probability that the random x occurs between x_1 and x_2.

As we have already discussed, it is more convenient to use the standard normal curve, which we do by changing x_1 to t_1 and x_2 to t_2. The equations $t_1 = (x_1 - \overline{x})/\sigma_x$ and $t_2 = (x_2 - \overline{x})/\sigma_x$ will give the desired values for t_1 and t_2. If we have a table that gives the areas under the standard normal curve of equation (2) for different values of t, we shall be able to use this table to find the probability that the variable t lies between t_1 and t_2. This, in turn, will be the probability that a variable x will fall between x_1 and x_2; that is, the probability that an x chosen at random will lie between the values x_1 and x_2 is equal to the

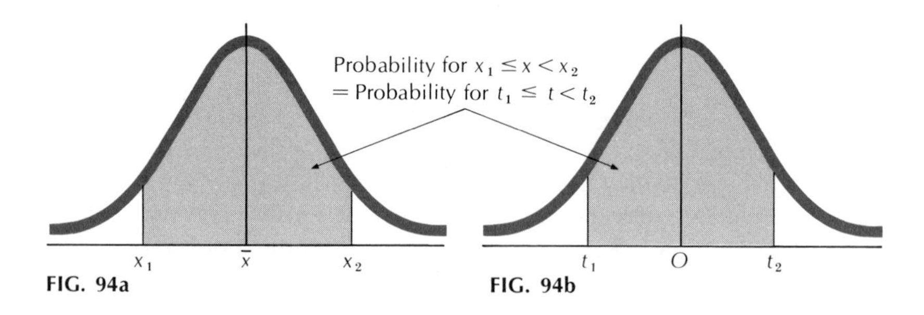

FIG. 94a **FIG. 94b**

shaded area in Fig. 94*a*. This probability is equal to the probability that t lies between $t_1 = (x_1 - \overline{x})/\sigma_x$ and $t_2 = (x_2 - \overline{x})/\sigma_x$, which is equal to the shaded area in Fig. 94*b*.

5 *The Probability That an* x *Differs from* $\overline{x}$ *by Less Than* $t\sigma_x$

Suppose a problem requires that we determine the probability that an x in a normal distribution *differs* from $\overline{x}$ *by less* than $1\frac{1}{2}$ standard deviations. Then $t = (x - \overline{x})/\sigma_x = 1.5$, or $t = (x - \overline{x})/\sigma_x = -1.5$. When

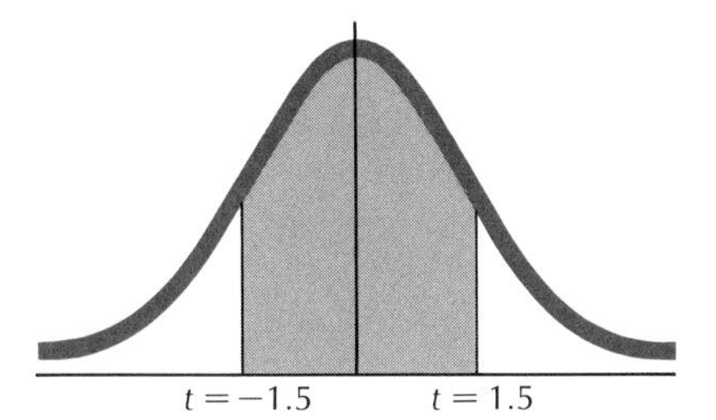

FIG. 95a

we use the word *differ*, we imply that the difference may be either to the left or to the right of the mean. The probability that x *differs* from $\bar{x}$ *by less* than $1.5\sigma_x$ will be the area under the standard normal curve between $t = +1.5$ and $t = -1.5$. This area is shown as the shaded area in Fig. 95*a*.

Table 59 gives us the area under the standard normal curve from $-t$ to $+t$.

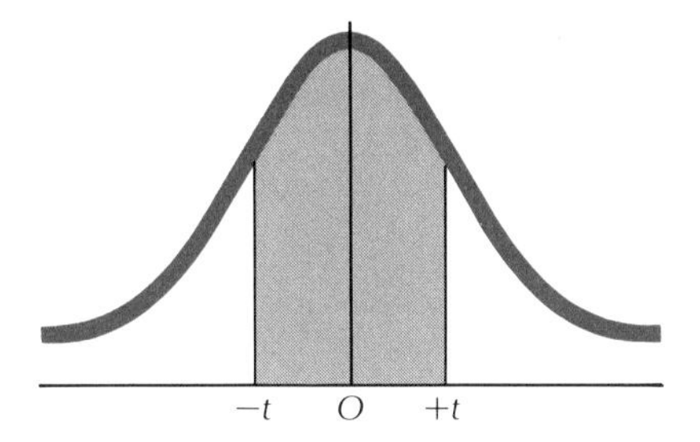

TABLE 59 Probability of x differing from $\bar{x}$ by less than σ_x

t	*Probability*	t	*Probability*	t	*Probability*
0.0	0.000	1.3	0.807	2.5	0.988
0.1	0.080	1.4	0.838	2.58	0.990
0.2	0.159	1.5	0.866	2.6	0.991
0.3	0.236	1.6	0.891	2.7	0.993
0.4	0.311	1.7	0.911	2.8	0.995
0.5	0.383	1.8	0.928	2.9	0.996
0.6	0.451	1.9	0.943	3.0	0.997
0.7	0.516	1.96	0.950	3.1	0.998
0.8	0.575	2.0	0.955	3.2	0.9986
0.9	0.632	2.1	0.964	3.3	0.9990
1.0	0.683	2.2	0.972	3.4	0.9993
1.1	0.729	2.3	0.979	3.5	0.9995
1.2	0.770	2.4	0.984	4.0	0.9999

In the work that follows, we will find it necessary to determine the probability that a value x, in a normal distribution, differs from the mean value $\overline{x}$ in either direction by less than a specified number of standard deviations. We must find how large t needs to be in order that x lie as far from $\overline{x}$ as the given number of σ_x's. Then we use Table

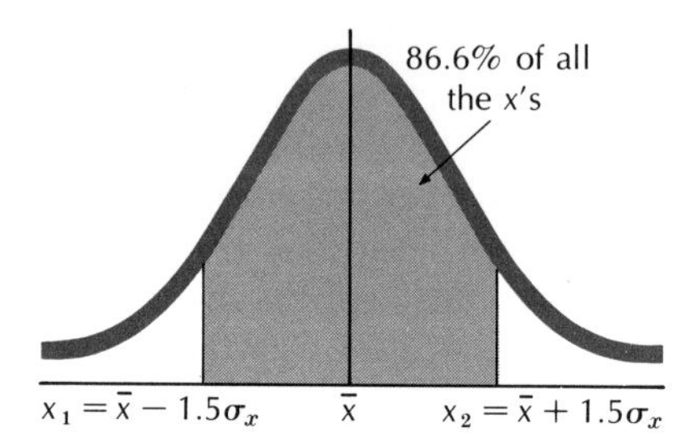

FIG. 95b

59 to find the required probability. From the table, we find that for $t = 1.5$, the required probability for a t to differ from $t = 0$ by less than 1.5 is 0.866 and the shaded area of Fig. 95*b* is 86.6% of the total area under the curve. This tells us that 86.6% of all the x's in a normal distribution are less than 1.5 standard deviations $1.5\sigma_x$ from the mean.

The shaded area in Fig. 95*b* is then 86.6% of the total area under the curve. Since the normal curve is symmetrical about the mean, one-half or 86.6%/2 = 43.3% of all the x's fall between $\overline{x}$ and $\overline{x} + 1.5\sigma_x$. Similarly, 43.3% of all the x's fall between $\overline{x} - 1.5\sigma_x$ and $\overline{x}$.

In all the problems that follow, it is good practice to sketch a normal curve representing the normal distribution being considered and the corresponding standard normal curve. Then on both sketches shade the areas being considered. In this way, errors are avoided that are likely to be made by hurrying through the problems.

■ **Example 1.** It is found that the test grades on a certain test follow a normal curve. If grades of A and F are to be given to the students who are more than 1.6 standard deviations from the mean, what proportion of the students will receive grades of B, C, and D?

□ ***Solution.*** The dividing line for the F and A grades is at

$$t = \frac{(x - \overline{x})}{\sigma_x} = -1.6 \qquad \text{and} \qquad t = \frac{(x - \overline{x})}{\sigma_x} = 1.6$$

respectively. From the table, we find that the proportion of the area lying less than $t = 1.6$ units on either side of the mean is 0.891; that is, the unshaded area (Fig. 96) under the t curve

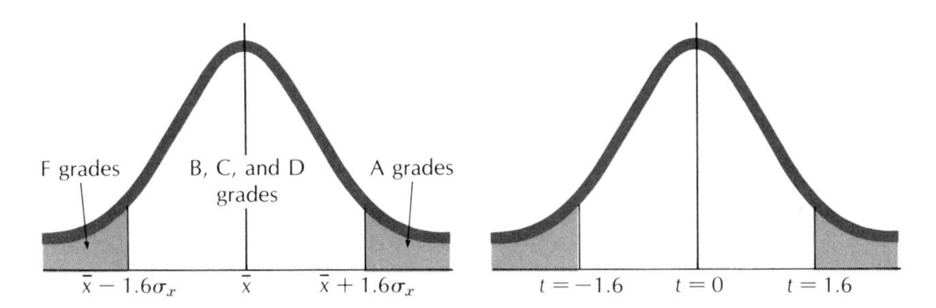

FIG. 96

is 0.891 and the shaded area is 0.109. Hence 89.1% of the grades will be B's, C's, and D's, and 10.9% of the grades will be A's and F's.

■ **Example 2.** The semester grades given by a certain department in a university follow a normal curve. If 83.8% of the grades were B's, C's, or D's, how many standard deviations from the mean were the grades of the students who got A's or F's?

□ ***Solution.*** Here we must use Table 59 in reverse. Since 83.8% of the grades were B's, C's, or D's, the center area in Fig. 97 has an area of 0.838. We look at the entries in Table 59 and see that an area of 0.838 under the curve lies between $t = 1.4$ and $t = -1.4$. Then the x for any grade of A or F is 1.4 or more standard deviations σ_x from the mean of the grades.

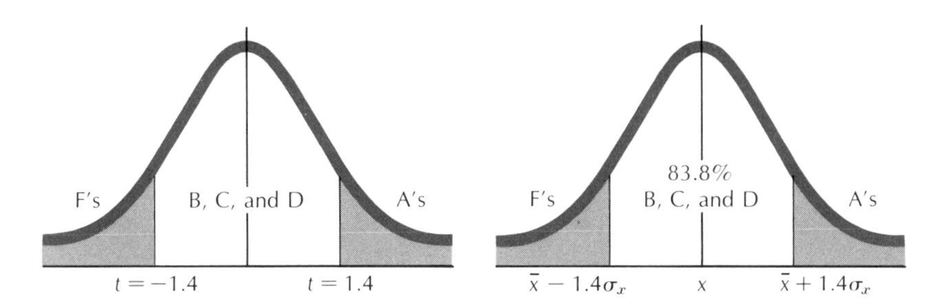

FIG. 97

■ **Example 3.** It is known that the lengths of telephone calls form a distribution which is approximately normal, with a mean of 8 minutes and a standard deviation of $2\frac{1}{2}$ minutes. What is the probability that a telephone call selected at random will last more than 15 minutes?

□ ***Solution.*** We have $\bar{x} = 8$ minutes, $\sigma_x = 2.5$ minutes, $x = 15$ minutes. Then

$$t = \frac{x - \bar{x}}{\sigma_x} = \frac{15 - 8}{2.5} = 2.8$$

From Table 59 we see that the area under the curve from $t = -2.8$ to $t = +2.8$ is 0.995, or that the chance that a randomly chosen t will lie between these limits is 99.5% (Fig. 98). Then $\frac{1}{2}$

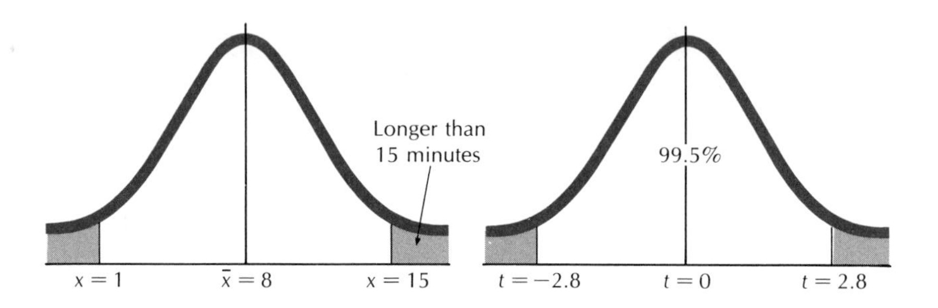

FIG. 98

of 1% of all calls will lie outside these limits, which are $x = 15$ and $x = 1$. The value $x = 1$ corresponds to $t = -2.8$. Since these two "outside" areas are equal, we see that about $\frac{1}{4}$ of 1% of all calls, or 1 out of 400, are longer than 15 minutes. We also see that about $\frac{1}{4}$ of 1% of all calls will last less than 1 minute.

■ **Example 4.** Within what limits do we find 50% of all telephone calls?

☐ ***Solution.*** Given $\bar{x} = 8$ minutes and $\sigma_x = 2.5$ minutes, we need to know within what values of x will be found 50% or 0.500 of the area under the normal curve. Consulting Table 59, we see that this corresponds to t about 0.67. Then

$$x_2 = \bar{x} + t\sigma_x = 8 + (0.67)(2.5) = 9.7 \text{ minutes}$$
$$x_1 = \bar{x} - t\sigma_x = 8 - (0.67)(2.5) = 6.3 \text{ minutes}$$

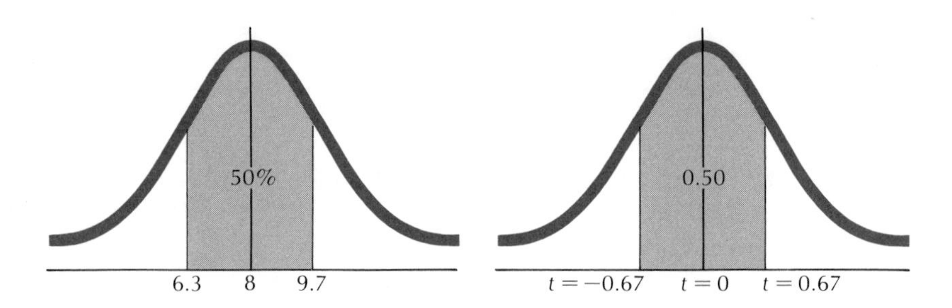

FIG. 99

We see in Fig. 99 that 50% of all telephone calls last between 6.3 and 9.7 minutes.

▲ EXERCISES

In each of the following problems assume that a normal distribution represents the frequencies of the x's. In each case draw the two normal curves, shading the areas under consideration.

1. What proportion of the x's are greater than the mean $\overline{x}$?
2. What proportion of the x's are less than two standard deviations away from $\overline{x}$? More than two standard deviations from $\overline{x}$?
3. What is the probability for an x to lie within $1.8\sigma_x$'s from $\overline{x}$?
4. What is the probability for an x to be *less* than $\overline{x}$ by more than $\frac{1}{2}\sigma_x$? What is the probability for x to deviate from $\overline{x}$ by more than $\frac{1}{2}\sigma_x$? To deviate from $\overline{x}$ by less than $\frac{1}{2}\sigma_x$?
5. A normal distribution of 500 weights has $\overline{x} = 145$ pounds and $\sigma_x = 4$ pounds.

   a. What proportion of these weights are within 6 pounds of 145 pounds? How *many* of the weights are within this range?
   b. If a weight is selected at random, what is the probability that it is more than 12 pounds away from 145 pounds?
   c. What is the probability for a weight to lie between 140 pounds and 150 pounds?
   d. What range of weights will include the middle 50% of the cases?
   e. What weight is exceeded by only 10% of all the cases?

6. A group of tobacco farms in 1926 had a mean income of \$905 with a standard deviation of \$1,409. What proportion of the farms lost money if the incomes followed a normal distribution?
7. For a random sample of 100 people in the home-coming crowd, given a normal distribution in which the mean age is 32 years with a standard deviation of 17.2 years:

   a. Find the percentage of this sample whose ages lie within 1.2 standard deviations of the mean.
   b. Between what age limits would one expect to find 52% of the people in this sample of the home-coming crowd?

8. If 12% of the population in a normal distribution exceeds the mean by more than t standard deviations, what is the value of t?
9. Suppose that a distribution of heights of 1,000 college freshmen is normal with a mean of 66.5 inches and a standard deviation of 3.5 inches.

   a. How many students have heights between 70 and 63 inches?
   b. What is the probability that a student is less than 64.4 inches in height?
   c. Between what heights would you find 98% of the students?
   d. What is the height that only 25 students will exceed?

Problems just for fun

Three men raced up a staircase, and the finish of the race is shown in the sketch. Allen, who is leading, went up three risers at a time; Jones, the second man, went up four risers at a time; and Smith, who is last, went up five risers at a time. Of course, Allen wins the race; but what is the smallest number of risers, counting the top landing as a riser, in the staircase? (The entire staircase is not shown.)

6 *Problems Involving Sampling*

We often find ourselves desiring to make statements about averages: the average grade of a class of students on a botany test, the average salary of department-store clerks, the average weight of football players, the average life of an automobile tire. If the number of objects involved is small, this is a simple computation. For example, if we wished to know the average weight of the players on the Georgia Tech football line, it would not require too much time to write down the weight of each man and compute the mean. But if we should wish to know the average weight of all the college football linesmen in the United States, the problem would be of quite a different magnitude. There are more than one thousand teams, and even if we took the time and expense of writing to all of them, we should probably not get replies from all. And even if we received replies from every one of them, we should have such a mass of figures that it would require many hours to compute the mean.

Common sense would lead us to sampling. We should collect the weights of linesmen from a group of colleges, being careful to include both large and small colleges. If we chose the colleges to be reasonably representative of the entire group, we should reason that the mean figure obtained is quite satisfactorily near to the mean of the entire group. As indicated above, the mean of the entire group would require too much work to compute or might be impossible to compute (if some colleges did not reply to letters).

If a tire manufacturer wishes to state exactly the average life of his tires, he can do so only if he tests every tire he manufactures. Obviously he cannot afford to do this, for he would use every tire in his road tests and have no tires to sell. So he must choose a sample to test and expect that the average of his sample is close to the average of his entire production. But how many shall he use for testing? One? Ten? One hundred? And having made his test, how close can he expect his sample average to be to his total production average? This is the problem of sampling.

Suppose a tire manufacturer chooses a random sample of 100 tires and road tests them. Suppose he finds that the mean life of these tires is 30,000 miles. Should he claim that 30,000 miles is the mean life of all his tires? Obviously not, for a different sample of 100 tires may give another result just by chance variation. It is evident that the manufacturer can never know the true mean of his production and must be satisfied with an estimate. This estimate is made by the method of *confidence intervals.* He claims that the true mean of his production is somewhere between $\bar{x} - d$ and $\bar{x} + d$ where $\bar{x}$ is the mean of his tested sample. Our problem is to give a reasonable value to d. For example, our tire manufacturer may be certain that his tires will wear out between 0 and 60,000 miles, i.e., $d = 30{,}000$, but this range is so wide as to be of little interest. On the other hand, if he takes $d = 5$ and says the average tire will wear out between 29,995 and 30,005 miles, we may not believe him.

Since we can never be certain what the true average of his production is, we must state our conclusions in *confidence levels.* Suppose $\overline{x}$ is the mean of one sample. Other samples may give different means by chance variation. What percentage of them lie between $\overline{x} - d$ and $\overline{x} + d$ with 90% confidence that we are correct. We can never be 100% certain unless we test every tire and have none to sell. While we like certainty we are not willing to pay this price (bankruptcy) for it!

Chance Variation in Sample Means

The problem is to estimate the mean of a population of x values. These x values might be tire lives, test grades, girls' weights, etc., and the *population* consists of all such values under consideration. If we were able or willing to observe each and every x in the population, there would be no sampling problem and the population mean could be computed precisely. But if sampling is to be used, we can only estimate the population mean with a specified degree of confidence, as has been explained above. The tire manufacturer would certainly not test every tire produced, since he would have no tires left to sell—an unpleasant thought for the stockholders in his company. We shall use the symbol $\tilde{x}$, called *x tilde,* to represent the unknown population mean, and the symbol $\overline{x}$ will stand for the mean of our sample. Now there will be a standard deviation for all the x's in the population, as well as a standard deviation for the relatively few x's in the sample. We shall use σ_x for the standard deviation of the population and s_x for the standard deviation of our sample. Usually we have no more knowledge about the value of σ_x than we do about the value of $\tilde{x}$. As we shall see, our estimate of $\tilde{x}$ will require that we have a value for σ_x. It can be shown that if the sample has a sizable number of cases, say, over 20 or 30, we may take s_x to be the same as σ_x. This replacement will bring about no appreciable difference in the desired estimate of $\tilde{x}$. For example, the tire manufacturer does not know σ_x, the standard deviation of all his tires in production. However, if his sample of 100 tires gives $s_x = 2{,}000$, he may use this for σ_x since the sample is fairly large.

To admit that different samples will give different values for $\overline{x}$'s just by chance alone is to admit the existence of a frequency distribution of the sample means. If all possible samples, each having the same number of cases, are drawn from a population of x's, the sample means $\overline{x}$'s could be tallied into a frequency table just as we have done before with other statistical data. This distribution of sample means would have its own mean and standard deviation, as does any frequency distribution. The histogram or frequency polygon could be plotted in order to observe the shape of the distribution of means. Fortunately we do not have to do this for the populations we study, because mathematicians have proved that the distribution of all possible sample means, each having N cases, drawn from a large population of x's, will have the following properties:

1. The distribution will be normal, or very nearly so, even if the original population does not have a normal distribution.
2. The mean of all the sample means will be the population mean $\tilde{x}$.
3. The standard deviation of the distribution of sample means will be the standard deviation of the population divided by the square root of N, that is,

$$\sigma_{\overline{x}} = \frac{\sigma_x}{\sqrt{N}}$$

Figure 100a and b illustrates what we have just said. If Fig. 100a

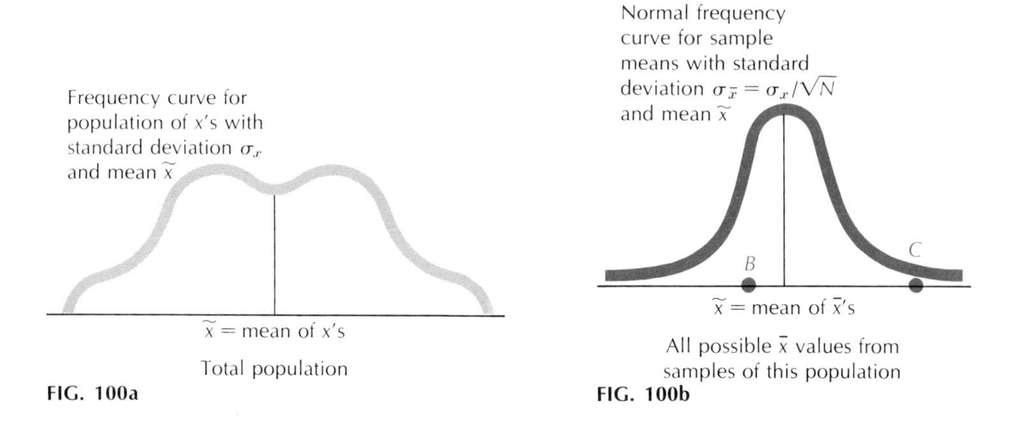

FIG. 100a

FIG. 100b

is the frequency polygon or frequency distribution of a population, Fig. 100b is the frequency distribution of means of all possible samples of a fixed size N drawn from the population. The mean of these two distributions is the same, namely x.

In order to illustrate our previous discussion, we shall relate the following actual experiment: A class of students drew 360 samples of 10 counters each from a bag. The x's were numbers stamped on the counters. The mean of all the x's in the counter population was $\tilde{x}=2.00$, and their standard deviation was $\sigma_x=1.715$. When the 360 sample means were computed and tallied, the distribution of means obtained by the students was as shown in Table 60. The mean of this distribution of $\overline{x}$'s is 1.98, and its standard deviation $\sigma_{\overline{x}}$ is 0.52.

TABLE 60

$\overline{x}$ = *mean of a sample of 10 counters*	*Number of samples having this mean*
0–0.5	2
0.5–1.0	9
1.0–1.5	47
1.5–2.0	122
2.0–2.5	134
2.5–3.0	38
3.0–3.5	7
3.5–4.0	1

Question. According to the previous theory, what would have been the mean and standard deviation in this experiment if *all possible* samples of 10 counters had been drawn?

Estimating the Population Mean from a Sample Mean

Now, in a practical situation, we have drawn only one sample and have only one mean. Figure 100*b* shows how all possible means will behave under chance variation, but we have no way of knowing whether our sample mean is at a point like B, at a point like C, or at any other point along the $\overline{x}$ scale. A mean such as B is below the true mean and is relatively near $\tilde{x}$, while a mean like C is considerably above $\tilde{x}$. As we have said, we shall estimate the location of $\tilde{x}$ by going out a distance d on either side of $\overline{x}$ and then claim that x is in this interval. It should be clear that, if our claim is correct, the distance d will depend upon the width of the hump in Fig. 100*b*; that is, d depends on the size of the standard deviation $\sigma_{\overline{x}}$ of all the sample means. If $\sigma_{\overline{x}}$ is small, the dispersion of all possible sample means is small and the distance d does not need to be very large in order to ensure that $\tilde{x}$ is between $\overline{x}-d$ and $\overline{x}+d$. If $\sigma_{\overline{x}}$ is large, the sample means are more scattered and a larger d will be necessary for an accurate estimate of $\tilde{x}$. The size of σ_x measures the reliability of the mean, or the extent to which $\bar{x}$ is expected to be in error from $\tilde{x}$ just by chance variation. When we observe the formula for $\sigma_{\bar{x}}$, we see that the mean of the sample becomes more reliable as the number of cases in the sample is increased. For example, if the tire manufacturer had selected a sample of 25 tires, $\sigma_{\bar{x}} = \sigma_x/5$ instead of $\sigma_{\bar{x}} = \sigma_x/10$. That is, a sample of 100 tires gives twice the reliability that is given by a sample of 25 tires. Now even if $\sigma_{\bar{x}}$ is relatively small, there can still be some sample means as far away from $\tilde{x}$ as the point C in Fig. 101, just by chance variation alone. And although we take d large enough so that the intervals $A \pm d$ and $B \pm d$ do include $\tilde{x}$, i.e., they give a correct claim as to the location of $\tilde{x}$, the same d is not large enough to make correct the claim that $\tilde{x}$ is in the interval $C \pm d$. But remember that we are willing to run a specified risk of making an inaccurate claim. Let us agree that we need to be only 90% confident that our claim about $\tilde{x}$ is true; that is, we would expect only 9 of 10 such claims to be correct. This means that we shall take d large enough that the claim would be correct for 90% of all possible sample means, and we expect to be incorrect for 10% of all the possible sample means. But our normal-curve-area table tells us that 90% of all the cases in a normal distribution are no more than 1.6 standard deviations from the mean of the distribution. Since the standard deviation of the normal curve in Fig. 101 is $\sigma_{\bar{x}}$, then 90% of all sample means are within $1.6\sigma_{\bar{x}}$ of $\tilde{x}$.

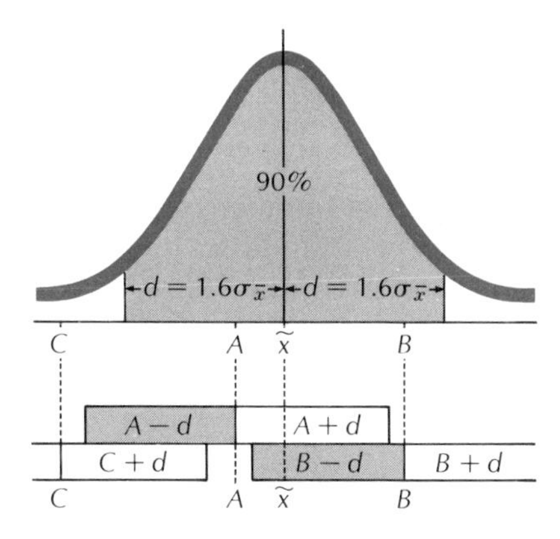

FIG. 101

So if we make the claim that $\tilde{x}$ is in the interval from $\bar{x} - 1.6\sigma_{\bar{x}}$ to $\bar{x} + 1.6\sigma_{\bar{x}}$, we can be 90% certain that our claim is correct. This is because only 10% of all possible sample means are like C, which is farther away from $\tilde{x}$ than $1.6\sigma_{\bar{x}}$.

Thus we have found how to determine the value of d for a given confidence level. Returning to the tire problem, suppose the manufacturer wants a 90% confidence interval for the true mean life of all his tires. From Table 59 we find that d will have to be 1.6 standard deviations. So $d = 1.6\sigma_{\bar{x}} = \dfrac{1.6\sigma_x}{\sqrt{100}} = 1.6(2{,}000) \div (10) = 320$ miles. Thus the manufacturer may have 90% confidence in his estimate that $\tilde{x}$ is between 29,680 and 30,320.

■ **Example 1.** Suppose the tire manufacturer desires more confidence in his statement, say 95%. We know that 100% is impossible unless the range is so large as to be of no value. We look at Table 59 and find that when $t = 1.96$, 95% of the sample means are within the range $\pm t\sigma_{\bar{x}}$ of the true mean $\tilde{x}$. Then

$$d = t\sigma_{\bar{x}} = \frac{t\sigma_x}{\sqrt{N}} = \frac{ts_x}{\sqrt{N}} = 1.96 \times \frac{2{,}000}{\sqrt{100}} = 390 \text{ miles approximately}$$

Then we may state that the average life of the tires lies between 29,610 and 30,390 miles with 95% confidence. This means that only 1 out of 20 samples of 100 tires will test out of this range.

■ **Example 2.** If our tire manufacturer wants 99% confidence, we find from Table 59 that $t = 2.58$ and $d = 2.58 \times \dfrac{2{,}000}{\sqrt{100}} = 516$ miles. The manufacturer may state with 99% confidence (one chance in 100 of being wrong) that the average life of his tires is between 29,484 and 30,516 miles. It is on the basis of this type of information that the tire manufacturer may write guarantees.

■ **Example 3.** What is the average number of hours of study per week of a college student? Suppose a random sample of 50 students shows a mean of 20.6 hours and a standard deviation of 2.2 hours. Suppose we desire 95% confidence. Then $t = 1.96$, and

$$d = t\frac{s_x}{\sqrt{N}} = 1.96 \times \frac{2.2}{\sqrt{50}} = 0.6 \text{ hour}$$

We may state that the true mean $\tilde{x}$ lies between 20 and 21.2 hours with only 1 chance in 20 of being wrong.

9 *An Experiment in Sampling*

Place the numbers 0 to 10 on 200 counters in accordance with the frequencies given in Table 61. The mean of this total counter population is obviously $\tilde{x} = 5.0$. Have each member of the class select a random sample of 16 counters. This is to be done with replacement; that is, a counter is drawn and its number recorded, then it is replaced and the counters mixed before the next counter is drawn. Each student is to compute $\overline{x}$ and s_x for his sample and from this compute his estimate of $\tilde{x}$ with confidence interval at the 90% level. Find the percentage of students whose intervals contain the true population mean $\tilde{x} = 5.0$, and compare this with the percentage expected by the theory of sampling, which is 90%. Use the same samples to compute 95 and 99% confidence intervals, comparing the results obtained in the class with the theory of sampling.

TABLE 61

Mark printed on counter	0	1	2	3	4	5	6	7	8	9	10
Number of counters with this mark	1	3	10	23	39	48	39	23	10	3	1

For this population, the true standard deviation $\sigma_x = 1.715$. Use this figure in place of the standard deviation s_x of the sample in computing the 90, 95, and 99% confidence intervals. To what extent does this change the results?

▲ EXERCISES

1. Suppose we are given a sample and are asked to locate the true mean of of the entire population. What sacrifice is made as a higher and higher degree of confidence in our answer is required?
2. We can obtain more accurate information about the location of the true mean by increasing the size of our sample. Suppose the tire manufacturer had used 200 tires in the test which gave $\overline{x} = 30{,}000$ miles and $s_x = 2{,}000$. Compute d for 90% confidence. Do you think this increased accuracy would justify the doubled expense of the testing?
3. Compute d for 90% confidence if the tire manufacturer had used only 10 tires in the test.
4. In a certain town, a random sample of 25 families showed a mean yearly income of \$4,200, with a standard deviation of \$400. Make a 90% confidence statement about the location of the mean yearly income of all the families in this town.
5. In a breeding experiment, the average height of a certain type of cornstalk was theoretically 60 inches. A random sample of 25 stalks had a mean height of 62 inches with a standard deviation of 6 inches. Decide whether the theoretical height of 60 inches should be revised.
6. Suppose it is found from a random sample of 400 students that the average semester expense of going to college is \$1,144, with a standard

deviation of \$300. You would like to be able to say that the average semester expense of the entire student body is no more than D dollars and be quite certain that the number D is quoted correctly. What will you give as a value for D?

7. A sample of 100 students enrolled in English 185 read an average of 600 words per minute with a standard deviation of 60 words per minute. Give 95% confidence limits for the average number of words read per minute by all the students enrolled in English 185.
8. A random sample of nine breakfast checks, in cents, at a student cafeteria gave 35, 42, 40, 57, 62, 71, 48, 54, and 68.

   a. Between what limits would you expect to find 52% of the breakfast checks?
   b. Estimate the mean breakfast check for the entire number of breakfasts served with 90% confidence.

9. For another sample of 100 checks, the mean was 56 cents, with a standard deviation of 8 cents.

   a. Find the percentage of this sample whose checks lie within 1.3 standard deviations of the mean.
   b. Estimate the mean breakfast check for the entire number of breakfasts served with 90% confidence.
   c. Is the difference between the results in Exercises 8 and 9 real or is it due to random samples? Discuss.

10. The average grade index of 10,000 students at a state university was 2.23 with a standard deviation of 0.63. The average grade index of 100 freshmen students was 2.47 with a standard deviation of 0.56. Is the difference between these indices real or is it due to sampling errors? Discuss.

Factorial 10,000 = 10,000! is an enormous number. Show that it ends in exactly 2,499 zeros.

Review Exercises

1. It is known that scores obtained on a test followed a normal distribution. The mean score was 80, and the standard deviation was equal to 8.

a. What is the probability that an individual will get a grade which is greater than 90?
b. What is the probability that an individual will get a grade between 88 and 92?
c. What is the probability that an individual will get a grade less than 88?
d. Between what two numbers will 95% of the scores fall?

2. A coin was tossed 500 times and heads turned up 275 times. Would this be considered a reasonable result?
3. The average number of rainy days on a certain island was 120 days with a standard deviation of 10 days. Assuming a normal distribution, find the probability that in a given year:

   a. There will be less than 100 rainy days.
   b. There will be more than 130 rainy days.

4. The heights of 100 men were measured. These 100 heights showed an arithmetic mean of 67.2 inches and a standard deviation of 2.3 inches. Make a 95% confidence statement about the mean of the heights of the population from which the sample was drawn, assuming the sample to be random.
5. In a random sample of 100 women on a certain campus, 58 are found to be sorority members. Make a 95% confidence statement about the proportion of women on that campus who belong to sororities.

John Holiday bought some gifts for relatives and friends at the Gift Shoppe. He bought some one dollar gifts, ten times as many two dollar gifts, and finally, some five dollar gifts for special persons. His total bill was $200. How many of each price gift did he buy?

CHAPTER TWENTY

The Basis of Our Number System

God created the natural numbers; everything else is man's handiwork. LEOPOLD KRONECKER

1 *What Is a Number?*

At the center of all mathematics is the concept of number. But what is number? What do we mean when we write $\frac{1}{2} + \frac{1}{4} = \frac{3}{4}$, $2 \cdot 4 = 8$, $\sqrt{3} \cdot \sqrt{3} = 3$? We learn the rules of the game and can solve problems with numbers, but how many of us have ever thought about "What is a number?" In order to examine this question, let us turn to the simplest of all numbers, the whole numbers, or integers, or natural numbers, 1, 2, 3, 4, 5, ...

The integers were created in order to count various collections of things, and they have no physical being. The number 7 applies whether it is seven eggs, seven cars, seven words, or seven ideas. In fact, the number 7 may be defined as that quality which these four groups of objects, as well as other similar-sized groups, have in common. The number 7 answers the question "How many?" but gives us no information on size, shape, or nature. We use it so commonly and with such certainty that most of us are immediately shocked that anyone should even ask, "But what do you mean by the number 7?" To children, numbers are always connected with definite objects, such as fingers or beads. But clearly, the number 7 is not tied up with fingers or beads; it has a meaning independent of these.

Also, it is clear that there is nothing magic in the form of 7.

While it might confuse us at first, we could soon learn to compute if 7 were replaced by $\exists$ or Λ or some other symbol. Once we became accustomed to it, we could add and subtract and multiply just as rapidly as before. So 7 is just a symbol which we all understand and which has only this importance.

The Roman System

Roman numerals, a product of the number system of the Latin language, are used today for numbering clock faces, for marking books (particularly the pages of prefaces, the chapters, and the volumes of a series), for dates on monuments and public buildings, and for numerous less important things. In the beginning it was not a letter system, although it is now written with the capital letters of the Latin alphabet. The Roman numerals consist of seven symbols as follows (note that there is no zero):

I = 1
V = 5
X = 10
L = 50
C = 100
D = 500
M = 1,000

Numbers are written from left to right and are made thus: an M is written down for every separate thousand; then 500 is taken, and D is written for it; next, as many hundreds are taken as possible, and a C is written for each; 50 is then taken, and L is written for it; as many 10's as possible are next taken, and X is written for each; 5 is then taken, and V is written for it; and finally, I is put down as many times as there are units left over. Thus 3,500 is written MMMD (and means 1,000 plus 1,000 plus 1,000 plus 500); 1,550 is written as MDL; and 1,883 as MDCCCLXXXIII.

At a later date, subtraction by changing the position of the symbols came into use. Instead of writing IIII, the I is placed before the V and we have IV, which means 1 subtracted from 5. Thus 9 is written as IX instead of VIIII, 40 as XL instead of XXXX, 90 as XC, and 900 as CM. Thus, 1,949 would be written as MCMXLIX and 1,950 as MCML.

Addition is not too difficult, and we might do a problem in this way:

```
CXXXVII
CXX  III
-------------
CCXXXXXVIIIII
    L     V  =CCLX
```

Remember to think entirely in Roman symbols—it is cheating to translate into Arabic figures, do the problem, and translate back.

Multiplication is more complicated, and we may consider multiplying XVI by XII as follows:

```
XV I
XI I
-----------
XXXXXXXXXX
VVVVVVVVVV
I I I I I I I I I I
XV I    XV I
-----------
CLX
XXXI I
-----------
CLXXXXI I
-----------
CXCI I
```

We are not sure that this is exactly the way the Roman schoolboy did this problem, but we can see that his problem was not as easy as ours, and we can appreciate the difficulty involved in using large numbers. Also, the very contemplation of the problem of dividing MMCCLIX by XVI is enough to produce cold shivers. This Roman symbolism, which is a good example of an additive number system, is cumbersome, and, as we pass to larger numbers, new symbols must be added at each stage to keep the figures within even reasonable bounds. Imagine the difficulty of expressing some of the numbers of astronomy or atomic physics in this system! The distance that light travels in 1 year is 5.87×10^{12} miles (see page 165). In Roman numerals this number would probably be 20 feet long. At this point, we can begin to understand why all computing in Roman times was done by a few specially trained persons.

▲ EXERCISES

Solve the following problems in addition and multiplication. Remember that translating into Arabic symbols is against the law!

1. XXVII + XVIII
2. CXXII + LXXXI
3. CCCLXXIII + CLXVI
4. XIX + XCIV
5. MCMLXVIII + CXXV
6. MCMV + XCV
7. VIII · V
8. VI · XIII
9. XII · XIII
10. CXI · XI
11. IX · IV
12. XLII · XIV

Problems just for fun

A cylindrical glass is 6 inches high and 6 inches in circumference. On the inside of the glass, 1 inch from the top, is a drop of honey, and on the opposite side of the vessel, 1 inch from the bottom on the outside, is a fly. How far must the fly walk to reach the honey?

3 *Positional Notation*

Having floundered through these problems in Roman notation, we begin to appreciate the simplicity of our modern notation. The invention of positional notation was one of the great inventions of all time. Its discovery is attributed to the Babylonians, and it was developed by the Hindus. It was introduced into medieval Europe by the Italian merchants, who had learned it in their travels to the East. For the first time, computation was no longer entirely in the hands of trained specialists. Now any school child could do problems in multiplication and division which had been mysteries to all but these specialists, and the computation time was cut from minutes to seconds.

Positional notation derives its name and its great simplicity from the fact that we use only 10 symbols, 0, 1, 2, 3, 4, 5, 6, 7, 8, and 9, and may denote any number by the position in which these symbols appear. Thus 354 means something quite different from 543. In the symbol 543, we mean 5 hundreds, 4 tens, and 3 units. We may express larger numbers with the same 10 symbols by using more positions:

$$\begin{aligned} 3{,}642 &= 3 \cdot 1{,}000 + 6 \cdot 100 + 4 \cdot 10 + 2 \\ &= 3 \cdot 10^3 + 6 \cdot 10^2 + 4 \cdot 10 + 2 \end{aligned}$$

We see that any number from 100 to 999 may be expressed by

$$abc = a10^2 + b10^1 + c$$

where a, b, and c are chosen from the 10 original number symbols, 0, 1, 2, 3, 4, 5, 6, 7, 8, and 9. Similarly, any number from 1,000 to 9,999 may be expressed by

$$abcd = a10^3 + b10^2 + c10^1 + d$$

and any number from 10,000 to 99,999 may be expressed by

$$abcde = a10^4 + b10^3 + c10^2 + d10^1 + e$$

Thus the expressing of any number is equivalent to writing an expression such as

$$a10^3 + b10^2 + c10^1 + d$$

in powers of 10.

But why 10? Why not 6 or 8 or 12 instead of 10? The earliest form of counting was done on the fingers. Small children and primitive tribes still use them for this purpose. Note that the English word digit can mean a finger or a numeral. In some languages the names of the numbers show this universal finger counting. For example, the Tamanacas Indian tribe have words for the first four numbers. For five their word means "a whole hand." For six their phrase means "one on the other hand." Ten is "both hands." Eleven is "one on the foot" meaning both hands and one toe. Fifteen is "a whole foot." For 20 they say "one Indian." For 21 it is "one on the hand of the other Indian." For 40 it is "two Indians" and so on.

Thus there is abundant evidence that our number system is based on 10 because it grew out of finger counting. Had man been born with 4 fingers on each hand, it is almost certain that our present number system would be based on 8.

Number System Based on 8

Let us consider what arithmetic would be under a number system based on 8. We would use only the eight symbols 0, 1, 2, 3, 4, 5, 6, and 7; 8 would appear as 10, 9 as 11, and 10 as 12 in this arithmetic.

We may see how the smaller numbers appear in the scale of 8 in Table 62.

TABLE 62

Scale of 10	1 2 3 4 5 6 7 8 9 10 11 12 13 14 15 16 17 18 19 20
Scale of 8	1 2 3 4 5 6 7 10 11 12 13 14 15 16 17 20 21 22 23 24
Scale of 10	21 22 23 24 25 26 27 28 29 30 31 32 33 34 35
Scale of 8	25 26 27 30 31 32 33 34 35 36 37 40 41 42 43

We see that $28 = 3 \cdot 8 + 4$ and so appears as 34 in the scale of 8. Similarly, $35 = 4 \cdot 8 + 3$ and appears as 43.

To express any number x in this system, we must determine the letters a, b, c, and d in the equation

$$x = a8^3 + b8^2 + c8^1 + d$$

just as we did before with the equation

$$x = a10^3 + b10^2 + c10^1 + d$$

Thus the number commonly written as 371 becomes 563 in the system based on 8, since

$$371 = 5 \cdot 8^2 + 6 \cdot 8^1 + 3$$

Do you see how the numbers 5, 6, and 3 in the equation above are obtained? One simple way is to find the largest power of 8 which is less than 371. This is 64 and we divide it into 371:

$$\begin{array}{r|r|l} 64 & 371 & 5 \\ & 320 & \\ \cline{2-2} & 51 & \end{array}$$

Then $371 = 5 \cdot 8^2 + 51$. Now divide the remainder 51 by 8.

$$\begin{array}{r|r|l} 8 & 51 & 6 \\ & 48 & \\ \cline{2-2} & 3 & \end{array}$$

Then $51 = 6 \cdot 8 + 3$ and

$$371 = 5 \cdot 8^2 + 6 \cdot 8^1 + 3$$

This may seem complicated and queer to us at first, but our system would seem just as queer to a man from Mars where each man had four fingers on a hand. Where we write 371 and say "three hundred and seventy-one," our man from Mars would write 563 and say— we are not sure what—but certainly not "five hundred." He might say "five sixty-fours" or he might have some other name for "sixty-four." Since we do not know, we will just call the digits in sequence and read 563 as "five six three."

In an arithmetic based on 8, the rules are the same, but the child would learn a different set of addition and multiplication tables. Just as our children parrot "5 times 4 is 20," the children of the four-fingered race would parrot "5 times 4 is 24" since

$$20 = 2 \cdot 8 + 4$$

Let us try a few simple problems in this four-fingered arithmetic. But first let us write down the addition and multiplication tables which the four-fingered child has memorized just as we have memorized our multiplication tables.

Addition

	1	2	3	4	5	6	7
1	2	3	4	5	6	7	10
2		4	5	6	7	10	11
3			6	7	10	11	12
4				10	11	12	13
5					12	13	14
6						14	15
7							16

Multiplication

	1	2	3	4	5	6	7
1	1	2	3	4	5	6	7
2		4	6	10	12	14	16
3			11	14	17	22	25
4				20	24	30	34
5					31	36	43
6						44	52
7							61

Now let us try a few problems in base 8 and transfer to base 10 to check our work.

■ Example 1

Scale of 8

$$\begin{array}{r} 25 \\ +\ 46 \\ \hline 13 \\ 6\ \\ \hline 73 \end{array}$$

Scale of 10

$$\begin{array}{r} 21 \\ +\ 38 \\ \hline 59 \end{array}$$

□ ***Check***

73 (scale of 8) $= 7 \cdot 8 + 3 = 59$ (scale of 10)

■ Example 2

$$\begin{array}{r} 64 \\ +56 \\ \hline 12 \\ 13\ \\ \hline 142 \end{array}$$

$$\begin{array}{r} 52 \\ +46 \\ \hline 98 \end{array}$$

□ ***Check***

142 (scale of 8) $= 1 \cdot 8^2 + 4 \cdot 8 + 2 = 98$ (scale of 10)

■ Example 3

$$\begin{array}{r} 24 \\ \times 13 \\ \hline 14 \\ 6\ \\ 24\ \\ \hline 334 \end{array}$$

$$\begin{array}{r} 20 \\ \times 11 \\ \hline 20 \\ 20\ \\ \hline 220 \end{array}$$

□ ***Check***

$$334 \text{ (scale of 8)} = 3 \cdot 8^2 + 3 \cdot 8 + 4 = 220 \text{ (scale of 10)}$$

■ **Example 4**

```
   45                 37
  ×63                ×51
  ---                ---
   17                 37
   14                 35
   36                15
  30                 -----
 -----               1,887
 3,537
```

□ ***Check***

$$3{,}537 \text{ (scale of 8)} = 3 \cdot 8^3 + 5 \cdot 8^2 + 3 \cdot 8 + 7$$
$$= 1{,}887 \text{ (scale of 10)}$$

▲ **EXERCISES**

The numbers in these problems are given in four-fingered arithmetic. Do these problems using the scale of 8 addition and multiplication tables, and then check your work by translating into the scale of 10 and solving.

1. $34+31$
2. $34+46$
3. $57+66$
4. $263+345$
5. $637+146$
6. $675+753$
7. 24×4
8. 36×4
9. 23×35
10. 65×27
11. 326×62
12. 547×64

Problems just for fun

The following is a problem in addition in which each letter represents a number and two different letters cannot be the same number.

```
  S E N D
  M O R E
---------
M O N E Y
```

Can you figure out what numbers the letters represent?

Other Number Systems

Having tried our hand at four-fingered arithmetic, we may ask whether systems other than 10 have ever been used or have any value as possibilities. There has been some argument in favor of a number system based on 12. The points in favor of such a system are its ease in division and handling of fractions. In the scale of 10, only the fractions $\frac{1}{2}$, $\frac{1}{5}$, and $\frac{1}{10}$ have simple equivalents 0.5, 0.2, 0.1. On the other hand, the scale of 12 would give this same simplicity to $\frac{1}{2}$, $\frac{1}{3}$, $\frac{1}{4}$, $\frac{1}{6}$, $\frac{1}{12}$. These fractions would appear as 0.6, 0.4, 0.3, 0.2, and 0.1. To change to a scale of 12 would require the introduction of two new symbols for 10 and 11 and the learning of different tables of addition and multiplication.

There are some vestiges of number systems based on 12 in the history of civilization. There is evidence that the Mayans of Central America had such a system. Also, both the English and German languages have special words for 11 and 12 but revert to words based on 10 at 13, 14, etc. This is exactly what we should expect to find in a number system based on 12.

The Babylonians had a number system based on 60. They made extensive studies of the movements of the sun and stars and our modern use of $12 = \frac{60}{5}$ for hours on a clock can be traced back to this system. The ancient Greeks, who wore open sandals, used a number system based on 20, the visible fingers and toes. The French word for 80 is "quatre-vingts," that is, 4 times 20, and it is likely that this is a modern vestige of such a number system.

The Dyadic System

If the number system has a large base, we have too many symbols for the digits and too complicated a multiplication table. It might seem desirable, then, to change to the dyadic system, or scale of 2. This uses exactly two symbols, 0 and 1, and has the following simple tables:

Addition

	0	1
0	0	1
1	1	10

Multiplication

	0	1
0	0	0
1	0	1

This would appear to be a paradise for the grade-school child, with only two symbols and almost no multiplication table to learn. However, even relatively small numbers become formidable in the dyadic system. Thus 87 becomes 1010111, since

$$87 = 1 \cdot 2^6 + 0 \cdot 2^5 + 1 \cdot 2^4 + 0 \cdot 2^3 + 1 \cdot 2^2 + 1 \cdot 2 + 1$$

Multiplication is very simple in the dyadic system, since we need remember only that $1 \times 1 = 1$ and $1 + 1 = 10$. As an example, let us multiply 7 and 11, which are 111 and 1011.

$$\begin{array}{r} 1011 \\ \underline{111} \\ 1011 \\ 1011 \\ \underline{1011} \\ 1001101 \end{array} = 2^6 + 2^3 + 2^2 + 1$$

This gives us 77, as it should.

While individual numbers have complicated representations, the dyadic system is the most important of all bases except 10, for it is the system used by all high-speed digital computers. Without the dyadic system IBM, Control Data, and the other manufacturers of computers would be out of business. A light is either on or off, electric current is either flowing or not, a switch is either open or closed. While we may not fully understand the complexity of a computer, we can see that a number system using only 1 or 0 fits the situation perfectly.

Computers

As soon as man began using mathematics to solve problems, he looked for ways to shorten the time spent in carrying out the computations. Many centuries ago the Chinese invented the abacus, which consists of beads strung on rods in a frame. This simplified addition and subtraction. Also multiplication can be done by a series of additions and division by a sequence of subtractions. The abacus is widely used even today in the Orient and a clever operator can do additions and subtractions faster than an electric calculator.

Shortly after the discovery of logarithms, the first slide rule was devised. The slide rule, which was the trademark of an engineer for many years, has been improved but is still limited to a few operations such as multiplication, division, raising to powers, or extracting roots of numbers. Also slide rules are limited in the number of significant figures that can be found. Nevertheless, they are still widely used to give a quick approximation to a numerical problem.

There are many forms of adding machines other than the fundamental type which adds a column of figures. In the check-out lanes of supermarkets are machines which will give a daily total for each type of product: meat, produce, dairy products, staples, etc. They give a subtotal for each customer and will calculate the amount of change to give the customer. Also they keep a record of the sales taxes collected.

Another type of adding machine is called the desk calculator and will add, subtract, multiply, and divide. The early desk calculators were run by hand, later models by electricity; recent advances in solid-state physics have produced very high speeds.

Designs for high-speed digital computers existed on paper before World War II but the state of engineering did not make one feasible. But shortly after the war advances in solid-state physics and the transistor enabled computers to be built. The first ones used hundreds of radio tubes. These were space filling, gave off large quantities of heat requiring extensive air conditioning, and needed constant replacement of bad tubes. But with transistors and other miniature electronic elements these difficulties have been overcome and the speed increased enormously.

A computer is a very complicated machine but fundamentally it merely adds and subtracts in the dyadic system. But it does these simple bits so fantastically fast that computers are taking over much of the work of our modern society. They calculate and print out our paychecks, keep track of our bank accounts and charge accounts, maintain business records, print out our semester grade reports and compute our grade averages, and much more. Manufacturing plants use them to control processes. The space agency uses them to track and guide space vehicles. And the Weather Bureau can feed in weather data from many stations and summarize them for better predictions.

These machines will do exactly what the operator, called a *programmer,* tells them to do. The programmer must prepare a series of orders for the machine before it can do any problem. This is called a *program* and its preparation can take considerable time. The program must be detailed and exact. Machines don't make mistakes, but programmers do. Program preparation varies from computer to computer and special languages have been developed to assist in this task.

8 *How Much Is 6 and 5?*

Now turn to the problem, "How much is 6 and 5?" There is only one answer, "11," if the symbols 6 and 5 have their usual meaning. But how do we write 11? Before studying this chapter, we could have answered easily and confidently, "11." But now we have learned to be more cautious, and we ask, "In what number base?" Six plus five equals 11 in base 10, 13 in base 8, 1011 in a dyadic system, and whatever the single symbol for 11 is if the base is 12.

A schoolboy on Neptune solves the quadratic equation

$$x^2 - 10x + 31 = 0$$

and finds that the solution is $x = 5$ or 8. What is the basis of his number system?

▲ EXERCISES

1. Express 127 in the base of 6.
2. Express 1,425 in the base of 12.
3. Express 12 in the bases of 2, 3, 4, and 5.
4. The following statements contain numbers in the dyadic system. Translate them into the usual form.

   a. In one season Babe Ruth hit 111100 home runs.
   b. Siwash football team beat Falling Rain College 11011 to 1101.
   c. Henry asked to drive Mary home from the pep rally. She said, "But I live at 10001101 Main Street. Do you want to drive that far?"
   d. Mrs. Housewife asked for a dozen oranges and received 1011.

5. Fill in the missing numbers in dyadic form:

   a. Christmas comes on December ______.
   b. In Leap Year, February has ______ days.
   c. Each day contains ______ hours.
   d. The disaster of Pearl Harbor happened in the year ______.

   (Just in case your memory is poor, it was A.D. 1941.)

6. Using the dyadic base, add the numbers 1011, 1101, 110, and 1110. Check by translating to base 10.

7. Using the dyadic base, add the numbers 11111, 1001, 101, and 11010. Check by translating to base 10.
8. Multiply in dyadic base the numbers 1101 and 101. Check by translating to base 10.
9. Multiply in dyadic base the numbers 1011 and 1001. Check by translating to base 10.
10. Construct the addition and multiplication table for base 5. Use this to multiply the numbers 123 and 43 in base 5. Check the result by translating to base 10.
11. In what base is the following addition problem?

$$\begin{array}{r} 26 \\ \underline{35} \\ 62 \end{array}$$

12. Professor Digit's son found a sheet of paper in his father's study and thought his father had gone senile. What system was his father using and what is the division in base 10?

$$\begin{array}{r} 34 \\ 32\overline{)2243} \\ \underline{201} \\ 233 \\ \underline{233} \end{array}$$

9 *Properties of Number Systems*

We use many properties of our numbers without thinking about them. We accept without question the fact that $2 + 5$ is the same as $5 + 2$. This may seem obvious, but we recognize the importance of order in most other life situations. For example, "take five steps and bend over" is quite different from "bend over and take five steps," and "drive 100 yards and turn right" instead of "turn right and drive 100 yards" may produce an accident. Mathematicians use the word *commutative* whenever the order does not affect the result. Thus $2 + 5$ is the same as $5 + 2$ and is the commutative property of our number system.

We might also ask whether the multiplication of two numbers is a commutative process? If $(a)(b) = (b)(a)$, i.e., if the order in which the operation of multiplication is performed is not important, then multiplication is also a commutative process.

There are four properties of addition and multiplication that are satisfied by all the number systems discussed so far in this chapter. If a, b, and c are numbers in a number system, we state these four properties in symbolic form:

1. Commutative property: $a+b=b+a$, $ab=ba$
2. Associative property: $(a+b)+c=a+(b+c)$
   $(ab)c=a(bc)$
3. Distributive property: $a(b+c)=ab+ac$
4. Property of closure: $a+b$ and ab are always numbers belonging to the system

Property (2) tells us that we need not worry about the order in which we perform the operations. Property (3) tells us that we can add b and c and then multiply or that we can add the products ab and ac. Property (4) seems like an obvious property, but there are some queer number systems for which this property does not always hold. Thus we see that the four properties given above are desirable properties for a number system, because when they hold we need not worry about the order in which we do the usual operations.

A fifth property which seems desirable is that the equation $a + x = b$ shall always have a solution; that is, that the difference of any two numbers $b - a$ can always be found. If we restrict our number system to the positive numbers, then this property is no longer true because $8 + x = 6$ has no solution in positive numbers. In order to have this property hold, we must add zero and the negative numbers to our number system of positive numbers.

We are accustomed to zero and negative numbers and accept them without question, but it is interesting to note that mankind struggled for centuries before reaching these ideas. Mathematics, including some fairly complex methods, dates back to the very dawn of history. Yet it was not until the thirteenth century that zero came into general use. Negative numbers were rejected until the beginning of the seventeenth century, although we find them in some Hindu writings of the twelfth century associated with the idea of debt. The ideas of negative direction along lines first appear in the writings of L. N. M Carnot (1753–1823) and A. F. Möbius (1790–1868).

Each number system, and there are many, satisfying the above four properties plus the property that the equation $a+x=b$ always has a solution is called a *commutative ring*.

Another property of our number system is that $ab = 0$ only if $a = 0$, $b = 0$, or both. If this property holds, in addition to the preceding five properties, we say that our number system is an *integral domain*.

Provided that we allow zero and the negative numbers, all of the number systems discussed so far in this chapter are integral domains. However, let us look at some different systems.

Finite Number Systems

The number systems that we have discussed so far in this chapter have many many numbers, in fact, infinitely many. However there are interesting finite systems of numbers which obey the same properties.

For example, consider the days of a week. Four days after Monday is Friday and six days after Sunday is Saturday. If we assign 0 to Sunday, 1 to Monday, 2 to Tuesday, etc., we may interpret "Four days after Monday" as $1+4=5=$ Friday and "Six days after Tuesday" as $2+6=1=$ Monday. Then we have a number system of seven symbols, 0, 1, 2, 3, 4, 5, 6, in which we have the following addition table:

TABLE 63

+	0	1	2	3	4	5	6
0	0	1	2	3	4	5	6
1	1	2	3	4	5	6	0
2	2	3	4	5	6	0	1
3	3	4	5	6	0	1	2
4	4	5	6	0	1	2	3
5	5	6	0	1	2	3	4
6	6	0	1	2	3	4	5

Notice that the result of any addition in this system is the ordinary sum reduced by a multiple of 7, that is, $6+3=9-(1)(7)=2$, $5+6=11-(1)(7)=4$. A number system in which addition is defined in this way is called a *system modulo* 7. Multiplication in a modulo 7 system is also the ordinary multiplication reduced by a multiple of 7, that is, $4 \cdot 5=20-(2)(7)=6$.

We may check and see that this system of seven symbols satisfies all the properties of the number systems listed in the previous section. It satisfies the commutative, associative, and distributive properties. Also the system is closed with respect to addition and multiplication as defined. From the table for addition, it is easy to verify the fact that the equation $a+x=b$ always has a solution. Finally, $ab=0$ if and only if $a=0$, $b=0$, or both. In other words, this set of seven symbols satisfies all the properties of a number system mentioned, and therefore is an integral domain.

But there is nothing sacred about the number 7. Let us consider a number system modulo 12. Here we have 12 numbers, 0, 1, 2, 3, 4, 5, 6, 7, 8, 9, 10, 11, and the practical application is the months of the year. The symbol $8+7$ means 7 months after August (8th month), which would be March (3rd month). This agrees with the definition of addition for such systems for $8+7=15-(1)(12)=3$. On checking, we can see that this new system modulo 12 satisfies the six properties of a number system with one exception. This is that $ab=0$ without either $a=0$ or $b=0$, as we may see from the equation $3 \cdot 4=0$, and certainly neither 3 nor 4 is zero. Then the number system of addition modulo 12 is a commutative ring, but is not an integral domain.

These two finite number systems have another surprising property. All of the earlier number systems have an order relationship; that is, for any two numbers we may say that one of them is the larger. For example, 9 is larger than 7 since $7+2=9$. However this is not possible in these last two finite systems. Consider the system modulo 7. Since $3+1=4$, it would be natural to say that 4 is greater than 3. However, $4+6=3$, so 3 is greater than 4. Clearly these two statements are contradictory, so we must drop the idea of "greater than," and we cannot say that any one of the numbers is larger than another in the two finite systems discussed here.

▲ EXERCISES

1. In the number system modulo 7, solve the following problems:

   a. $4+2$　　b. $5+4$
   c. $6+4$　　d. $6+5$
   e. $6+6$　　f. $5+5$

2. Construct the multiplication table for the system modulo 7 and solve the following problems:

   a. $3 \cdot 2$　　b. $1 \cdot 6$
   c. $4 \cdot 5$　　d. $5 \cdot 2$
   e. $5 \cdot 6$　　f. $3 \cdot 5$

3. In the number system modulo 12, solve the following problems:

   a. $5+4$　　b. $8+3$
   c. $9+7$　　d. $11+7$
   e. $7+8$　　f. $10+10$

4. Construct the multiplication table for the number system modulo 12 and solve the following problems:

   a. $3 \cdot 3$　　b. $2 \cdot 6$
   c. $4 \cdot 6$　　d. $3 \cdot 9$
   e. $5 \cdot 7$　　f. $11 \cdot 6$

5. In addition to 12, for what other values of K will we have a system modulo K in which $ab=0$ when neither $a=0$ nor $b=0$?

Problems just for fun

The persons attending a picnic paired up for a game and one person was left out. For another game, they formed groups of three and again one person was left out. In groups of four, one was left out. In groups of five, again one was left out. In groups of six, again one. But when they played in groups of seven, no one was left out. How many persons were at the picnic?

CHAPTER TWENTY-ONE

Linear Programming

Mathematics is the predominant science of our time; its conquests grow daily, though without noise; he who does not employ it *for* himself, will some day find it employed *against* himself. J. F. HERBART

1 *History*

During World War II groups of scientists were assembled in Britain, as a last ditch desperation measure, to provide assistance with several critical war problems. Most notable of these were the German air attacks and the U-boat menace. It is generally conceded that their work was a substantial contribution to the strategies which proved successful. Their studies showed that improvement in the use of existing weapons and equipment could accomplish more in a short time than could improvement in the weapons and equipment themselves. Hence they devised methods of obtaining most effective use of existing weapons and equipment. These techniques were the beginnings of what we now call *linear programming*. It was quickly realized that the business executive or manufacturer faced similar problems of achieving best results with existing equipment, space, manpower, etc. Following the war George B. Dantzig, Abraham Charnes, and many others improved and extended these British ideas to fit many different situations in business affairs. In the quarter century since World War II these methods have been studied and applied extensively by executives and managers.

What Is Linear Programming?

To the mathematician, linear programming means finding the maximum (or minimum) of a linear function, hence the name, subject to a number of restrictions.

To the business executive, it is a device for allocating limited space, dollars, manpower, etc.; for producing maximum profits or minimum costs. Some of the problems which yield solutions from linear programming are:

1. Determining most profitable product mix to be obtained from existing manufacturing facilities.
2. Determining which parts to make and which parts to buy for best profits.
3. Locating warehouses properly to minimize transportation costs.
4. Determining the feed mix to provide proper nutrition and reduce the cost of raising livestock and fowl.
5. Supplying a varying sales demand with minimum inventory cost.

In actual business or manufacturing situations, the number of variable factors is usually so large that extensive clerical work or a high-speed computer is necessary to arrive at a solution. However, we can learn what linear programming means by looking at some simple examples.

The Graphical Method

■ **Example 1.** A dealer wishes to stock at most 100 refrigerators and stoves. A refrigerator weighs 200 pounds and a stove 100 pounds and he is limited to a total weight of 12,000 pounds. If he makes a profit of \$35 per refrigerator and \$20 per stove, how many of each should he stock for maximum profit?

□ ***Solution.*** Let R be the number of refrigerators and S be the number of stoves. We have

$$200R + 100S \leqq 12{,}000$$

or

$$2R + S \leqq 120$$

Also

$$R + S \leqq 100$$

We plot the lines $R + S = 100$ and $2R + S = 120$ in Fig. 102 and see that the solution must lie in the shaded area.

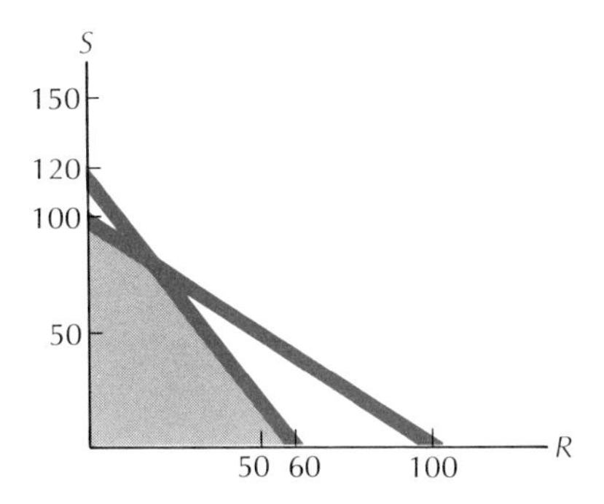

FIG. 102

The profit in dollars $P = 35R + 20S = f(R,S)$. We note that this function $f(R,S)$ is of first degree or linear in R and S. We wish to determine the maximum of this linear function over the shaded area of Fig. 102, hence the name linear programming. If we plot P as a vertical height over each point of the shaded area, we form a flat but slanted roof over the area. No matter how you hold your book, you see that the highest (and lowest) point of the shaded area occurs at one of the four corners. Thus the maximum profit P must occur at one of the four corners of the shaded area. We compute:

$$f(0,0) = 0 \qquad f(60,0) = \$2{,}100$$
$$f(20,80) = \$2{,}300 \qquad f(0,100) = \$2{,}000$$

Hence, under the restrictions on number and weight, the most profitable mix is 20 refrigerators and 80 stoves.

Example 2. Determine the most profitable mix of two products A and B which have to be processed through three machines to manufacture a salable item. The conditions of the problem are given in Table 64.

TABLE 64

Machine	*Hours per piece* Product A	*Hours per piece* Product B	*Hours available*
I	2	6	Up to 50
II	4	4	Up to 60
III	3	0	Up to 42
Profit per piece	\$6	\$4	

☐ ***Solution.*** Let A and B represent the numbers of Products A and B. Then Machine III will produce $3A \leqq 42$. Machine II will produce $4A + 4B \leqq 60$. Machine I will produce $2A + 6B \leqq 50$. Since we must stay within the weekly capacity of the three machines, we are confined to the shaded area in Fig. 103.

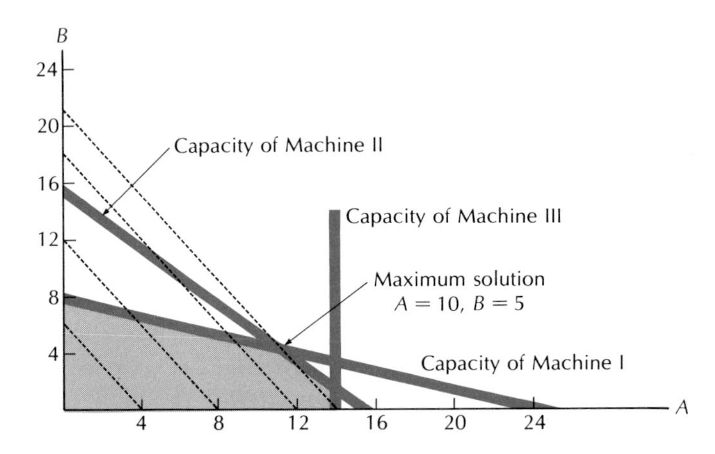

FIG. 103

Our profit is $6A + 4B$, which we wish to make as large as possible within the capacity of the three machines. The family of parallel lines $6A + 4B = k$ is shown by dashed lines in Fig. 103 and we see that the maximum k occurs at the corner of the shaded area where $A = 10$ and $B = 5$. Note that this does not use the total capacity of Machine III.

The Simplex or Transportation Method

As more variables enter the problem, the graphical method becomes too complicated. It requires graphing in spaces of higher dimensions. This is difficult for the professional mathematician to visualize and just about impossible for the rest of us. To handle such problems methods other than graphic have been devised. In essence they involve computing a possible solution and then improving it step by step until the best (maximum or minimum) solution is reached. This may involve many steps and partly explains why actual business problems require much clerical computing or a computer. However, let us look at two simple examples.

■ **Example 3.** To handle rush-hour traffic, a bus company dispatcher must order extra buses from three garages to three different downtown starting points of bus routes. He has six buses at garage G_1, three at garage G_2, and five at garage G_3. He needs six at starting point A, three at B, and five at C. One solution is obvious. But he wishes to save as much time and money as possible and wishes to determine the distribution which will give the least total travel time. The travel time in minutes from each garage to each starting point is given in Table 65.

TABLE 65

	A	B	C
From garage G_1	15	18	12
From garage G_2	17	13	14
From garage G_3	20	15	22

If he chooses the lazy, obvious solution of sending all buses at each garage to the starting point needing that number of buses, the total travel time will be 239 minutes. To improve on this, suppose we select the lowest possible travel time, 12 minutes, in Table 65 and allot the 5 buses needed there from G_1. Then consider the next lowest travel time, 13 minutes, and allot the 3 buses needed there. Continuing, we reach the allotment shown in Table 66.

TABLE 66

		A 6	B 3	C 5
From G_1	6	1		5
From G_2	3		3	
From G_3	5	5		

This gives a total travel time of 214 minutes, which is quite a saving over the lazy, easy method requiring 239 minutes. But can this be improved? To do this we must examine the empty spaces in Table 66 and see what the net change would be if we divert a bus here with correspondent change to balance the situation. We see that diverting one of the buses from $G_2 - B$ to $G_2 - A$ would cost four minutes, but the accompanying change from $G_3 - A$ to $G_3 - B$ would gain five minutes. This would reduce the total time one minute. If the dispatcher decides to shift one bus from $G_2 - B$ to $G_2 - C$, it raises the total time one minute and the corresponding change of $G_1 - C$ to $G_1 - B$ costs six more minutes, so the total result of such a shift would add seven minutes to the total time. We could do this for each of the vacant spaces in Table 66 to show the net result on the total time of putting a bus in the vacant spot. The result of each such shift in the total time is shown in Table 67.

TABLE 67

	A	B	C
G_1		+1	
G_2	−1		+7
G_3		−1	+5

Here we see that the first shift mentioned is the only one which gives an improvement in total time. In more complicated problems, it is very time consuming to compute this for each vacant space in the table. Whenever mathematics runs into a time consuming problem it looks for an easier way; therefore, the method of *fictitious costs* has been devised.

From Table 66 we enter the time of the buses being used into Table 68. We border the table with vertical and horizontal fictitious costs as indicated and wish to choose *f.c.*'s so that each entry is the sum of its two fictitious costs. We may arbitrarily assign zero as one fictitious cost, say the middle space in the vertical column. This forces us to place 13 in the *f.c.* space under B. No other number is affected, so we may again choose a zero, say under C. Under the restriction that each entry in the table is the sum of its two *f.c.*'s, the entries of 12, 3, and 17 as *f.c.*'s are determined as shown in Table 68.

TABLE 68

	A	B	C	*f.c.*
G_1	15		12	12
G_2		13		0
G_3	20			17
f.c.	3	13	0	

Now we compute the *fictitious cost* of the blank spaces in Table 68 by adding the two border figures as shown in Table 69.

TABLE 69

	A	B	C	*f.c.*
G_1		25		12
G_2	3		0	0
G_3		30	17	17
f.c.	3	13	0	

If any of these fictitious costs are greater than the actual costs, in our problem time, we can improve the solution. We select the largest difference and shift as many buses as fit the problem. This gives the allotment shown in Table 70.

TABLE 70

		A 6	B 3	C 5
G_1	6	1		5
G_2	3	3		
G_3	5	2	3	

If now we construct a fictitious cost table for this allotment, we have the results shown in Table 71, in which the boldface numbers are the computed fictitious costs.

TABLE 71

	A	B	C	*f.c.*
G_1	15	**10**	12	0
G_2	17	**12**	**14**	2
G_3	20	15	**17**	5
f.c.	15	10	12	

Here no fictitious cost is greater than the actual cost (time) in Table 65, so the distribution of buses given in Table 70 is the best possible. The total travel time is 211 minutes.

Comment. The fictitious cost 14 in Table 71 for $G_2 - C$ is the same as the actual time shown in Table 65. This indicates that there is another solution which gives the same minimum time. We observe that we could change one or more buses from $G_1 - C$ to $G_1 - A$ with corresponding shift from $G_2 - A$ to $G_2 - C$. Each such change would gain and lose 3 minutes so that the total time of 211 minutes would not be affected.

■ **Example 4.** The buyer for a chain of department stores takes bids for 1,000 light fixtures; 300 to be delivered to warehouse A, 400 to warehouse B, and 300 to warehouse C, located in the East, Midwest, and Far West. Manufacturer Brown bids \$12, \$15, and \$15 for fixtures delivered to the 3 warehouses, up to his manufacturing capacity of 500. Manufacturer Jones bids \$13, \$14, and \$16, up to his limit of 300 fixtures. Manufacturer Smith bids \$9, \$11, and \$13, up to a total of 200 fixtures; but also bids \$10, \$12, and \$14, up to an additional 600 (this will cost him some overtime pay). How does our buyer allot his order so as to produce the least cost?

□ ***Solution.*** Let us assume that each manufacturer produces his maximum number but some of them may go to a dummy warehouse D at zero cost, that is, this part is not really produced at all. Furthermore, we handle the Smith bid by dividing it into two theoretical manufacturers, Smith Regular (R) and Smith Overtime (O). We have the cost situation shown in Table 72.

TABLE 72

		A 300	*B* 400	*C* 300	*D* 600
Brown	500	12	15	15	0
Jones	300	13	14	16	0
Smith (R)	200	9	11	13	0
Smith (O)	600	10	12	14	0

Using the initial allotment method of first filling the square of lowest cost, then next lowest, etc., in units of 100 fixtures, we have the first solution shown in Table 73.

TABLE 73

		A 3	*B* 4	*C* 3	*D* 6
Brown	5			2	3
Jones	3				3
Smith (R)	2	2			
Smith (O)	6	1	4	1	

This gives a total cost of $12,000. We must now investigate how this may be improved, using the method of bordering with fictitious costs, arbitrarily choosing one fictitious cost as zero. This gives Table 74.

TABLE 74

	A	*B*	*C*	*D*	*f.c.*
Brown			15	0	2
Jones				0	2
Smith (R)	9				0
Smith (O)	10	12	14		1
f.c.	9	11	13	−2	

Now we fill in the blank spaces of Table 74 by adding the two bordering fictitious costs and we obtain the figures shown in Table 75.

TABLE 75

	A	*B*	*C*	*D*	*f.c.*
Brown	11	13			2
Jones	11	13	15		2
Smith (R)		11	13	−2	0
Smith (O)				−1	1
f.c.	9	11	13	−2	

Since no one of these figures is higher than the actual costs seen in Table 72, the allotment given in Table 73 with a total cost of $12,000 cannot be improved. Our buyer should reject the bid of Jones, instruct Brown to deliver 200 fixtures to warehouse *C*, accept Smith's bid of 200 to go to warehouse *A*, and also accept Smith's *overtime* bid of 600—400 to go to *B* and 100 to each of *A* and *C*.

Comment. Notice that the fictitious cost figures of Smith (R) of 11 and 13 in columns *B* and *C* are the same as the true cost figures. This indicates that our solution is not unique and other solutions exist at the same total cost. For example, changing 100 fixtures of Smith (R) from *A* to *B* will increase the cost by $200, but the accompanying switch of 100 fixtures of Smith (O) from *B* to *A* will save $200.

▲ EXERCISES

1. Suppose the dealer in Example 1 knows that he cannot sell more than 50 stoves. What distribution should he choose for maximum profit?
2. Suppose this dealer can sell any number of stoves but cannot sell more than 15 refrigerators. What distribution should he choose and what is his maximum profit?
3. A dealer has space for at most 10 cars in his showroom. Car A retails for $4,000 and has a gross profit of $750; car B $3,000 and $600; car C $2,500 and $550. If he can afford a gross inventory of $32,500, what combination of cars will give him the largest gross profit?
4. A dealer wishes to stock 40 radios and TV sets, some cabinet and some portable. He wants at most 30 cabinet TVs and wants more cabinet TVs than radios. He wants at least 5 radios and knows that he cannot sell more than 15. If he makes a profit of $40 on a cabinet TV, $20 on a portable TV, and $10 on a radio, what combination shall he buy to give him a maximum profit?
5. The rush-hour dispatcher needs four buses at starting point A and four at B. He has three buses at garage G_1 and five at garage G_2. To give the least total travel time, how should he order them if the travel time in minutes between points is given by Table 76?

TABLE 76

		A 4	B 4
From G_1	3	12	11
From G_2	5	18	14

6. How should he dispatch them if we have the situation shown in Table 77?

TABLE 77

		A 5	B 5	C 5
From G_1	7	20	15	12
From G_2	8	14	11	14

7. How should the dispatcher send the buses if we have the situation shown by Table 78?

TABLE 78

		A	*B*	*C*
		2	7	4
From G_1	3	10	12	14
From G_2	4	16	15	12
From G_3	6	12	18	11

8. How should the dispatcher send the buses if we have the situation shown by Table 79?

TABLE 79

		A	*B*	*C*	*D*
		3	3	5	4
From G_1	7	18	18	12	15
From G_2	6	14	17	13	12
From G_3	2	11	13	20	15

9. Three customers require 30, 50, and 60 tons of oil. Three refineries have 65, 40, and 35 tons available. How shall we make trucking costs least, if their cost in dollars per ton is given by Table 80?

TABLE 80

	Customers		
	A	*B*	*C*
	30	50	60
From Refinery 1 65 tons	4	4	1
From Refinery 2 40 tons	2	3	3
From Refinery 3 35 tons	3	4	5

10. In Example 4 suppose manufacturer Brown's bid was based on a minimum order of 400 fixtures, that is, Brown offers to deliver 400–500 fixtures or none. Under this additional condition, what is the best solution for the department store buyer?

Problems just for fun

Long, long ago silver dollars were the common currency in the gambling casinos of Nevada, and counterfeiters were active. A roulette operator has 10 stacks, each containing exactly 10 dollars, and he has reason to suspect that one entire stack is counterfeit. A counterfeit dollar weighs one gram less than the genuine article. If he has a pointer scale, i.e., one that shows weight in grams, how many weighings are required to spot the false stack?

CHAPTER TWENTY-TWO

The Theory of Numbers

Mathematics is the queen of sciences and the theory of numbers is the queen of mathematics.

K. F. GAUSS

Since the very earliest days of history, man has been fascinated by the whole numbers and their properties. The numbers have always been associated with superstition and mysticism. Although man's superstition has waned, 3 is still thought to be a lucky number, and 13 is considered unlucky by so many people that most hotels and office buildings omit the thirteenth floor. The interest in integers and their properties is just as alive today as it was centuries ago.

1 *Some Parlor Tricks*

The properties of integers form the basis of many parlor tricks. One very puzzling trick is based on the property that any three-digit number repeated gives a six-digit number which is divisible by 7, 11, and 13. Thus on successive division of the six-digit number by these three numbers, the final quotient is the original three-digit number. For example, choose the number 436 and form the number 436,436. Dividing successively by 7, 11, and 13, we have $436{,}436 \div 7 = 62{,}348$; $62{,}348 \div 11 = 5{,}668$; $5{,}668 \div 13 = 436$!!! To use this as a mystifying trick, we ask a person to choose a three-digit number and repeat it to give the six-digit number. We ask a second person to divide it by 7. Have a third and a fourth person divide the result by 11 and 13. Then we astonish the first person by handing him the result, which will be his original number. But we will be the ones astonished unless our friends can divide without mistakes!!!

A much simpler trick is as follows: Think of any number. Add 2. Double the result. Add 6. Divide by 2. Subtract the original number. The result is always 5. Why?

For a somewhat more baffling trick, ask someone to think of a number. Suppose he selects 237. Ask him to add the digits and subtract from his secret number: $2+3+7=12$ and $237-12=225$. Now ask him to multiply by any number he chooses. Suppose he chooses 13: $225 \times 13 = 2{,}925$. Now ask him to strike out any digit except 0 and tell you the result. Suppose he chooses 5 and tells you 292. You add the remaining digits, subtract mentally from the next highest multiple of 9, and tell him the digit he discarded: $2+9+2=13$ and $18-13=5$.

Why does the trick work? If we subtract the sum of the digits of any number from the number, we get a number whose digits add to a multiple of 9, in our case 225. Any number whose digits add to a multiple of 9 will still do so after being multiplied by any number. Notice that this is true of 2,925. So if we omit any one of these digits, we can spot it by subtracting the remaining sum from a multiple of 9. "Elementary, my dear Watson!"

If he had omitted a 0 or 9, we could not tell which. So we avoid this by telling him to drop any digit except a 0.

Problems just for fun

Ask your friend to think of a number, any number. Tell him to double it. Add 25 (or 1 plus any multiple of 8). Square the result. Now subtract 17 (or 1 plus any multiple of 8). Bet a Coke that the result is divisible by 8. Why does it work?

Magic Squares

A *magic square* is a square array of whole numbers such that the sum of the numbers in any row or column or either of the two main diagonals is always the same. Such curious squares and their construction have intrigued men for generations. One of the earliest is found in Albrecht Dürer's famous engraving *Melancholia*, of the year 1514, and is shown in Fig. 104. Here all rows, all columns, and the two diagonals add to the sum 34.

But this square has many other interesting properties. Note the date 1514 in the bottom row. And in addition to rows, columns, and diagonals, the four corner squares add to 34; also the four center squares and the four squares of each quadrant add to 34. Also note the opposite interior squares: $3 + 2 + 15 + 14 = 5 + 9 + 8 + 12 = 34$. But there is more. The sum of the squares of the first two rows equals the sum of the squares of the third and fourth rows. This is also true of the columns. Finally, if we consider the eight numbers on the two diagonals and the eight not on diagonals, we find that their sums, sums of squares, and sums of cubes are equal! Truly there was lots of magic in the sixteenth century.

The method of construction of magic squares may be found in most books on mathematical tricks or recreations.* You may, however, construct one by trial and error, and without too many trials, from the nine numbers 1 to 9 if we give you the hint that the middle number must be 5 and the rows and columns must add up to 15.

But another type of "magic" square is given in Fig. 105, which has some remarkable properties. Ask one of your friends to choose any number from the array. Suppose he chooses the largest—19. Tell him to strike out all the remaining numbers in the same row and column as 19 and choose any number left, say 11. Again strike out all the other numbers in the same row and column as 11. This leaves four numbers, and he may choose one of these and strike out the two in the same row and column. There remains one number; tell him to add his three chosen numbers and this remaining one. The sum is always 38! Try the choice of numbers several times and see that the result is always 38.

This mysterious square really has a simple secret and is a form of addition. We can construct one of any size and adding up to any number we choose. Take a square of any size and write any numbers you wish along the top and one side (Fig. 106). Now fill in each square by adding the numbers above and at the end of the row. We have filled in a few of the squares to show the method. Now erase the numbers along the top and side and you have your "magic" square. The numbers we have used will produce the array shown earlier (Fig. 105), giving the sum 38, but you may choose other numbers along the top and side and produce as many arrays as you wish.

*Maurice Kraitchik, "Mathematical Recreations," chap. 7, W. W. Norton and Company, Inc., New York, 1942.

16	3	2	13
5	10	11	8
9	6	7	12
4	15	14	1

FIG. 104

14	7	11	6
11	4	8	3
19	12	16	11
12	5	9	4

FIG. 105

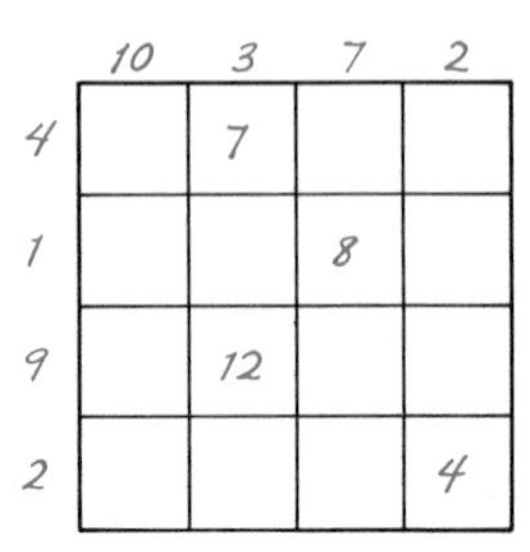

	10	3	7	2
4		7		
1			8	
9		12		
2				4

FIG. 106

The reason for the sum 38 should now be clear. The method of your friend's choice of four numbers always gives him one number from each row and column. And since each number in the square is the sum of the figure at the top of the column and the one on the left, what he is really doing is adding up your secret numbers along the top and side, no matter how he chooses.

3 *The Prime Numbers*

Of foremost interest among the integers are the primes. A whole number greater than 1 is called *prime* if it is divisible only by itself and 1. The smallest prime numbers are 2, 3, 5, 7, 11, 13, 17, etc. The number 1 might be considered prime, but this is not the custom. To do so would force us to state an exception to many of the theorems of prime numbers. For this reason 2 is considered the smallest prime. Information about primes is important, because they are the building blocks of all numbers, and the question of whether a number is prime or not enters into many statements in the theory of numbers.

How many primes are there? More than 2,000 years ago, Euclid, who is most famous for his book on plane geometry, proved that the number of primes is infinite. His proof is so elegant and clear that we shall repeat it here.

First, let us consider the two primes 2 and 3 and form the number $2 \cdot 3 + 1 = 7$. Observe that 7 is not divisible by 2 or 3, for each such division leaves the remainder 1. Thus 7 is a prime, or has a prime factor different from 2 and 3. So we have proved the existence of three primes. Similarly we may show that $2 \cdot 3 \cdot 7 + 1$ is not divisible by 2, 3, or 7 and thus must either be a prime or have as a factor a fourth prime. Euclid's proof follows this line of reasoning and is an elegant example of a *reductio ad absurdum* argument.

Euclid reasons: Suppose the total number of primes is n, and we denote them by $p_1, p_2, p_3, \ldots, p_n$. Now form the integer

$$p_1 \cdot p_2 \cdot p_3 \cdots p_n + 1 = N$$

The number N is not divisible by any of the n primes, since such a division clearly leaves the remainder 1. Then N is either itself a prime or has a prime factor different from any of the n primes, giving a total of $n + 1$ primes. Thus the original assumption that the number of primes is finite leads to a contradiction.

Euclid's elegant proof shows that the sequence of all integers 1, 2, 3, 4, 5, 6, 7, 8, . . . contains infinitely many primes. But we may also show that other sequences, such as 3, 7, 11, 15, 19, . . ., or 5, 11, 17, 23, . . ., contain infinitely many primes. In fact, Dirichlet (1805–1859) proved that every sequence of the form $a, a + b, a + 2b, a + 3b, \ldots$ contains infinitely many primes if a and b have no common factor. However, his proof, unlike that of Euclid, is extremely difficult and complicated.

Now that we know what primes are and that they are scattered among all the numbers, no matter how large, we naturally ask for a formula which will give all the primes. Many such formulas have been given, but someone has always found that each formula gave at least one number which was not a prime. To date, no one has ever found a formula which always yields a prime.

Fermat (1601–1665) made the famous conjecture that $2^{2^n} + 1$ was always a prime. However, Euler (1707–1783) discovered that for $n = 5$, the number given is not prime, for $4{,}294{,}967{,}297 = 641 \times 6{,}700{,}417$.

Another famous formula is $n^2 - n + 41$. This gives a prime number for every $n < 41$. However, for $n = 41$, it gives 41^2, which is certainly not a prime. Incidentally, this formula gives a fine lesson in the danger of drawing conclusions from experiments. Suppose we are asked to test the assertion that "$n^2 - n + 41$ is always a prime." Every try for $n < 41$ will give success. After 10 or 20 such trials, we may be willing to grant the truth of the assertion. Beware of such conclusions!

Some Unsolved Problems about Prime Numbers

Despite the immense amount of knowledge about primes, there are other simple questions, like the formula for primes, whose answers are not as yet known. There remains much to do in mathematics. No one today knows the answer to any of the following simple questions:

1. Given a prime P, what is the next larger prime?
2. Are there an infinite number of pairs of primes differing by 2? Examples are 5 and 7, 11 and 13, 17 and 19, 29 and 31.
3. Given a number N, how many primes are less than N? Or what is

the formula for the number of primes between two numbers M and N?

There is, however, a remarkable theorem about the distribution of primes. Gauss (1777–1855) conjectured that the number of primes less than n is approximately $n/\log n$, where this is the Napierian or natural logarithm to the base $e = 2.718 \ldots$. The proof of Gauss's conjecture was attempted by many mathematicians, but it was nearly 100 years later, in 1896, that Hadamard and de la Vallée Poussin completed the proof, using the most complicated tools of analysis. More precisely, the theorem states that the ratio of P_n, the number of primes less than n, and $n/\log n$ approaches 1 as n becomes large. We may see this for a few values, since P_n is known by counting primes:

$n = 1{,}000$	ratio is 1.16
$n = 1{,}000{,}000$	ratio is 1.08
$n = 1{,}000{,}000{,}000$	ratio is 1.05

4. Is every even number, except 2, the sum of two primes? $4 = 2 + 2$, $6 = 3 + 3$, $8 = 5 + 3$, $10 = 5 + 5$, $12 = 7 + 5$. The famous Goldbach (1690–1764) conjecture affirmed this.

   No one has ever solved this problem, but in 1931 the Russian Schnirelman proved that every integer is the sum of at most 300,000 primes. While 300,000 is so large that it may seem ludicrous to call this a partial solution, Schnirelman's success led others to attempt the problem, and Vinogradoff later proved that all large integers may be expressed as the sum of at most 4 primes. Vinogradoff's proof is of the *reductio ad absurdum* type, so we do not know how large "large integers" are and cannot test the ones below this. What he really proved is that the assumption that the number of integers requiring more than 4 primes is infinite leads to a contradiction.

▲ EXERCISES

1. What are the prime numbers given by the formula $2^{2^n} + 1$ for $n = 0, 1, 2, 3$?
2. Compute the prime numbers given by the formula $n^2 - n + 41$ for $n = 1, 2, 3, 4, 5, 6, 7, 8, 9, 10$.
3. Write down all the prime numbers less than 75.
4. How many pairs that differ by 2 are found among the primes less than 75?
5. Among the ten numbers 1 to 10 are four primes, 2, 3, 5, and 7. Among

the decade 11 to 20 are also four primes, 11, 13, 17, and 19. Find another decade containing four primes. There are three such decades between 21 and 1,000.

6. The Euclid proof shows that $p_1 \cdot p_2 + 1$ is either a prime or contains one different from p_1 and p_2. Similarly, $p_1 \cdot p_2 \cdot p_3 + 1$ finds a fourth prime. Using $p_1 = 2$ and $p_2 = 3$, find three more primes by the Euclid method.
7. Express each of the even numbers from 40 to 50 as the sum of two primes. There are at least three solutions for each number.
8. Show that every odd prime number can be represented as the difference of two perfect squares in one and only one way. Find this representation for 13, 17, and 23.
9. Show that there are no positive integers x and y such that $y^2 = 2x^2$. From this it follows that $\sqrt{2}$ is not a rational number, i.e., the quotient of two integers.

Problems just for fun

A three-digit number is peculiar. It ends with 4. If the 4 is moved to the front, the new number is as much greater than 400 as the original number was less than 400. What is the number?

5 *The Perfect Numbers*

An intriguing group of integers are those which have been named the *perfect numbers.* These are the integers which are exactly equal to the sum of all their divisors except themselves. The smallest perfect number is 6, for all the divisors of 6 are 1, 2, 3, and 6, and $6 = 1 + 2 + 3$. Some numbers, such as 10, have divisors that add to less than themselves, $5 + 2 + 1 = 8$, and some numbers, such as 12, are "too perfect," for $6 + 4 + 3 + 2 + 1 = 16$.

On the surface one sees no connection between prime numbers and perfect numbers. Yet, as in many cases in number theory, there is a very close connection. A theorem states that if $2^n - 1$ is a prime number, than $2^{n-1} (2^n - 1)$ is a perfect number.

Notice that every perfect number given by this formula is an even number, since a power of 2 is one factor. Then what are the odd numbers which are perfect? To date, the mathematicians have not found the answer to this question. No odd perfect number has been found, and it is not known whether one exists.

▲ EXERCISES

1. Test the numbers up to 40 to see which ones are perfect.
2. Using the formula given above, find four perfect numbers and verify that they are perfect.

Find a group of two or more consecutive numbers which add up to 100. Can you show that there are exactly two different solutions?

6 *Pythagorean Numbers*

A second group of interesting numbers are the integers which satisfy the equation $x^2 + y^2 = z^2$. These triples are called *Pythagorean* or *right-triangle numbers*, since it is possible to construct a right triangle with x and y as legs and z as hypotenuse. Thus, if we find all the solutions of $x^2 + y^2 = z^2$ in whole numbers, we have found all the right triangles that we can construct with whole numbers as sides.

In fact, this gives all the shapes of right triangles which have rational numbers as sides, for each right triangle with rational numbers as sides is similar to one with whole numbers as sides. Suppose $\frac{a}{b}, \frac{c}{d}, \frac{e}{f}$ are the sides of a right triangle, where the six letters represent integers. We may write these three sides as $\frac{adf}{bdf}, \frac{cbf}{dbf}, \frac{ebd}{fbd}$. Then the triangle with the whole numbers adf, cbf, ebd as sides is similar to the original triangle. Thus we see that a knowledge of the solutions of

$x^2 + y^2 = z^2$ in whole numbers gives us the knowledge of every type of right triangle possible which does not involve a square root in the length of any side.

One triple, 3, 4, 5, of Pythagorean numbers is well known, for $3^2 + 4^2 = 5^2$. This immediately leads us to a whole series of triples—6, 8, 10; 9, 12, 15; 12, 16, 20; etc., which satisfy the equation $x^2 + y^2 = z^2$. We see that whenever a, b, c is Pythagorean, then ka, kb, kc is Pythagorean for any integer k. If $a^2 + b^2 = c^2$, then

$$\begin{aligned}(ka)^2 + (kb)^2 &= k^2a^2 + k^2b^2 \\ &= k^2(a^2 + b^2) \\ &= k^2c^2 \\ &= (kc)^2\end{aligned}$$

Thus each solution of $x^2 + y^2 = z^2$ leads to a whole infinity of solutions. These are not particularly interesting, for all the triangles formed are similar, differing only in size.

We shall be interested in *primitive* Pythagorean triples, that is, solutions of $x^2 + y^2 = z^2$ such that x, y, and z have no common factor except 1. We see that 3, 4, 5 is a primitive solution, but 6, 8, 10 is not a primitive solution. Two primitive solutions yield triangles which are not similar. For example, 3, 4, 5 and 5, 12, 13 are Pythagorean triples of numbers, and the right triangles produced are quite different in shape.

The problem of finding primitive Pythagorean numbers has been completely studied, and it has been proved that x, y, z will be a triple of primitive Pythagorean numbers if

$$\begin{aligned}x &= 2mn \\ y &= m^2 - n^2 \\ z &= m^2 + n^2\end{aligned}$$

where m and n are any two positive integers such that (a) one integer is even and one integer is odd, (b) the two integers have no common factor except 1, and (c) $m > n$.* Since $m = 2$, $n = 1$ satisfies these conditions, we should expect the resulting x, y, z to be Pythagorean. Substituting into the equations above, we obtain $x = 4$, $y = 3$, $z = 5$, which we already knew was Pythagorean.

▲ EXERCISES

1. Show that $m = 3$, $n = 2$ yields the Pythagorean numbers 5, 12, 13.
2. Find four other sets of primitive Pythagorean numbers by using other

*For proof of this statement see Robert Daniel Carmichael, "Theory of Numbers," p. 85, John Wiley & Sons, Inc., New York, 1914.

small values of m and n. Check to prove that the numbers are Pythagorean.

3. If $x=2mn$, $y=m^2-n^2$, $z=m^2+n^2$, prove that $x^2+y^2=z^2$.

Two ferryboats leave opposite sides of the Hudson River at the same time. They pass each other at a point 700 feet from one shore. Each boat remains at the dock for 10 minutes and returns. On the return trip they pass each other at a point 400 feet from the opposite shore. How wide is the Hudson at this point?

7 *Fermat's Last Theorem*

Having made a complete study of the solutions of $x^2 + y^2 = z^2$, we naturally want to know about the integral solutions of $x^3 + y^3 = z^3$, of $x^4 + y^4 = z^4$, and in general of $x^n + y^n = z^n$ for $n > 2$.

This problem derives its name as the last of the theorems left unsolved by Pierre de Fermat (1601–1665). Fermat owned a copy of Bachet's "Diophantus," and as he read, it was his habit to make marginal notes, frequently stating results but not their proofs. After his death in 1665 this book was found, and various mathematicians supplied the proofs which Fermat did not write down. In time, proofs were devised for all these notes except the one in which Fermat wrote: "On the other hand, it is impossible to separate a cube into two cubes, or a biquadrate into two biquadrates, or generally any power except a square into two powers with the same exponent. I have discovered a truly marvelous proof of this which, however, the margin is not large enough to contain." Thus, Fermat wrote that he had a proof that it was impossible to solve $x^n + y^n = z^n$ in integers for all $n > 2$. It seems very doubtful that Fermat really had a proof, for no one has been able to give a complete proof up to the present time.

In 1908, a bequest of 100,000 marks was made to the Royal Academy of Göttingen to establish a prize for a correct proof of Fermat's last theorem. Since this had a value of about $24,000, interest in the problem was greatly stimulated. Hundreds of solutions poured in, but errors in reasoning were found in all of them. Even some famous mathematicians have thought they had a solution, only to find an error. The inflation following World War I wiped out the monetary value of this prize, but interest in the problem continues. No solution exists today, but some new advance of knowledge may yield a solution tomorrow, next year, next decade, or next century. Who knows?

While no complete solution of Fermat's last theorem is known, many partial solutions are known. Fermat left a proof for $n = 4$, and Euler (1707–1783) proved the theorem for $n = 3$. Dirichlet (1805–1859) proved it for $n = 5$ and $n = 14$. Lamé (1795–1870) proved it for $n = 7$. Kummer (1810–1893) gave a proof which covered all cases except certain ones. The missing cases are all above 100 and are rare among the smaller numbers. Among others, the American mathematicians L. E. Dickson (1874–1954) and H. S. Vandiver (1882–   ) have made significant advances on the theorem. The solution is now known to be impossible for all values of $n < 4{,}000$ and for most values of n to 10,000. In the case where x, y, z, and n are *relatively prime*, that is, no two have a common factor, D. H. and Emma Lehmer proved in 1941 that the solution is impossible for $n < 253{,}747{,}889$!!

8 *Waring's Theorem*

Every integer may be expressed as the sum of perfect squares, since we may write $1^2 + 1^2 + 1^2 + \cdots$ and thus construct any integer n. But the real question is the minimum number of perfect squares necessary to represent any number. Thus,

$$13 = 1+1+1+1+1+1+1+1+1+1+1+1+1$$

but $13 = 9 + 4$ also, so we need only two squares in this case.

▲ EXERCISE

Express each integer less than 100 in as few squares as possible. What is the largest number needed? The general result that every integer may be expressed by not more than this number of squares was stated by Fermat.

Problems just for fun

"Goodbye, my little darlings," said the Sultan of Nabob to his harem. "I'm off to market and won't be back until next week. And girls," said the Sultan, opening two identical boxes of bonbons and chocolates, "I want you to watch your figures. So I am rationing you to 7 bonbons and 11 chocolates apiece."

On his return the Sultan counted the remaining bonbons and chocolates and blew his top. "Somebody has been cheating," he yelled. "There are only 49 bonbons and 17 chocolates left."

"Relax your blood pressure, big boy," replied his favorite wife. "My sister, Grace, paid me a visit, and we gave her the same share."

How many wives did the Sultan have?

"How many children do you have?" asked the census taker. "Plenty," said the housewife, "and we have found that even the smallest children eat as much as my husband and I. After we were blessed with twins, we found that a crate of oranges lasted three days less than before. And if I had 4 more children, the crate would be empty even 4 days earlier." How many children did the census taker write down?

How many cubes are needed to express any integer? Edward Waring (1734–1798) stated without proof that at most nine cubes are needed. This was proved by Wieferich in 1909. Waring also stated that at most 19 fourth powers are needed. Wieferich proved that it could be done with not more than 37, and better results have been found more recently.

▲ EXERCISES

1. Express each integer less than 50 in as few cubes as possible. What is the maximum number needed?
2. Express each integer less than 50 in as few fourth powers as possible. What is the maximum number needed?

CHAPTER TWENTY-THREE

Topology

[Geometry] that held acquaintance with the
stars, And wedded soul to soul in purest bond
Of reason, undisturbed by space or time.

WORDSWORTH

1 *What Is Topology?*

Topology is the branch of geometry which deals with properties which are not affected by changes in size and shape. We may understand its nature if we think of a very elastic rubber sheet on which some geometrical figures have been drawn. We are allowed to stretch and bend the sheet but must not cut or tear it. Distance has no meaning in topology, for two points an inch apart may easily be made 2 inches apart by stretching the sheet. Angle size is meaningless, for by careful stretching we may change an angle of 15° on the sheet to 45°. Even straight lines have no place in topology, for the straight line AB may be changed into any one of the shapes shown in Fig. 107 by proper stretching of the sheet. Perhaps it is not so easy to see how the third one is obtained, but it can be done.

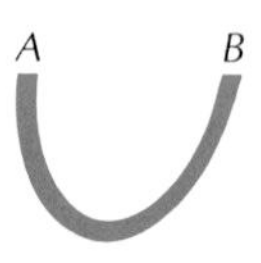

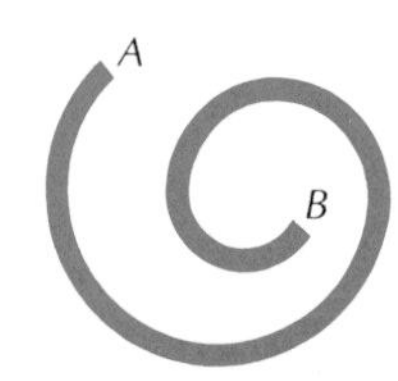

FIG. 107

So many of the properties of geometry become meaningless in this rubber world of topology that it may seem that nothing remains.

But notice that no matter how we stretch or bend, the path from A to B remains a path from A to B which does not cross itself. It may become very crooked, much worse than the three pictures above, but it remains a path from A to B. Thus a path from A to B belongs to the language of topology, and the topologist calls it the *arc* AB.

Now consider a circle on the rubber sheet (Fig. 108). By distorting the sheet it may become an oval (Fig. 109), or something worse (Fig. 110), or even something like Fig. 111. But it remains a path $ABCDA$, which returns to its original point without crossing

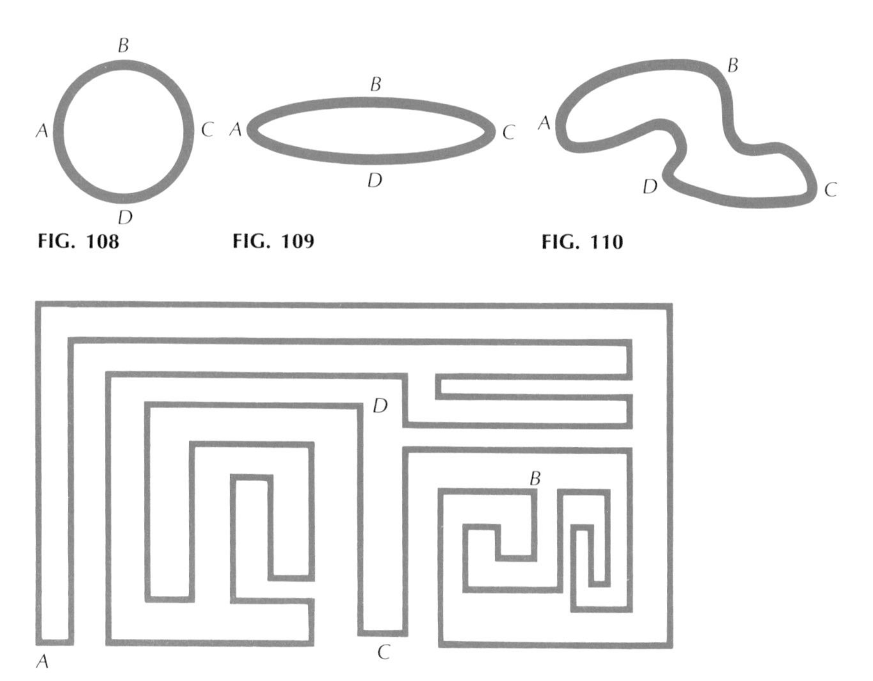

FIG. 108 **FIG. 109** **FIG. 110**

FIG. 111

itself. Since topology ignores changes produced by stretching, all these figures are the same for the topologist, and he calls each of them *a simple closed curve*, or *closed circuit*. Each consists of two arcs ABC and ADC which have only the points A and C in common.

We notice that no matter how the original circle is distorted by stretching, it always divides the sheet into two parts, the outside and the inside regions. This is the Jordan curve theorem, one of the fundamental theorems of topology. This may seem very obvious, but it is not so obvious if we look at Fig. 111, where it is only with some difficulty that we are able to distinguish which is "outside" and which is "inside." Actually, this theorem, which states that two paths from A to C which have no other points in common always divide the plane into two regions, is a very difficult theorem to prove, and the first correct proof was given just at the dawn of the twentieth century.

A Story

But topology is not concerned primarily with proving "obvious" theorems, and we shall begin our brief visit to this fascinating land with a story reprinted with permission from the *Ford Times* (July, 1949).

PAUL BUNYAN VERSUS THE CONVEYOR BELT

by William Hazlett Upson

One of Paul Bunyan's most brilliant successes came about not because of brilliant thinking; but because of Paul's caution and carefulness. This was the famous affair of the conveyor belt.

Paul and his mechanic, Ford Fordsen, had started to work a uranium mine in Colorado. The ore was brought out on an endless belt which ran half a mile going into the mine and another half mile coming out—giving it a total length of one mile. It was four feet wide. It ran on a series of rollers, and was driven by a pulley mounted on the transmission of Paul's big blue truck "Babe." The manufacturers of the belt had made it all in one piece, without any splice or lacing, and they had put a half twist in the return part so that the wear would be the same on both sides.

After several months' operation, the mine gallery had become twice as long, but the amount of material coming out was less. Paul decided he needed a belt twice as long and half as wide. He told Ford Fordsen to take his chain saw and cut the belt in two lengthwise.

"That will give us two belts," said Ford Fordsen. "We'll have to cut them in two crosswise and splice them together. That means I'll have to go to town and buy the materials for two splices."

"No," said Paul. "This belt has a half twist—which makes it what is known in geometry as a Möbius strip."

"What difference does that make?" asked Ford Fordsen.

"A Möbius strip," said Paul Bunyan, "has only one side, and one edge, and if we cut it in two lengthwise, it will still be in one piece. We'll have one belt twice as long and half as wide."

"How can you cut something in two and have it still one piece?" asked Ford Fordsen.

Paul was modest. He was never opinionated. "Let's try this thing out," he said.

They went into Paul's office. Paul took a strip of gummed paper about two inches wide and a yard long. He laid it on his desk with the gummed side up. He lifted the two ends and brought them together in front of him with the gummed sides down. Then he turned one of the ends over, licked it, slid it under the other end, and stuck the two gummed sides together. He had made himself an endless paper belt with a half-twist in it just like the big belt on the conveyor.

"This," said Paul "is a Möbius strip. It will perform just the way I said—I hope."

Paul took a pair of scissors, dug the point in the center of the paper and cut the paper strip in two lengthwise. And when he had finished—sure enough—he had one strip twice as long, half as wide, and with a double twist in it.

Ford Fordsen was convinced. He went out and started cutting the big belt in two. And, at this point, a man called Loud Mouth Johnson arrived to see how Paul's enterprise was coming along, and to offer any destructive criticism that might occur to him. Loud Mouth Johnson, being Public Blow-Hard Number One, found plenty to find fault with.

"If you cut that belt in two lengthwise, you will end up with two belts, each the same length as the original belt, but only half as wide."

"No," said Ford Fordsen, "this is a very special belt known as a Möbius strip. If I cut it in two lengthwise, I will end up with one belt twice as long and half as wide."

"Want to bet?" said Loud Mouth Johnson.

"Sure," said Ford Fordsen.

They bet a thousand dollars. And, of course, Ford Fordsen won. Loud Mouth Johnson was so astounded that he slunk off and stayed away for six months. When he finally came back he found Paul Bunyan just starting to cut the belt in two lengthwise for the second time.

"What's the idea?" asked Loud Mouth Johnson.

Paul Bunyan said, "The tunnel has progressed much farther and the material coming out is not as bulky as it was. So I am lengthening the belt again and making it narrower."

"Where is Ford Fordsen?"

Paul Bunyan said, "I have sent him to town to get some materials to splice the belt. When I get through cutting it in two lengthwise I will have two belts of the same length but only half the width of this one. So I will have to do some splicing."

Loud Mouth Johnson could hardly believe his ears. Here was a chance to get his thousand dollars back and show up Paul Bunyan as a boob besides. "Listen," said Loud Mouth Johnson, "when you get through you will have only one belt twice as long and half as wide."

"Want to bet?"

"Sure."

So they bet a thousand dollars and, of course, Loud Mouth Johnson lost again. It wasn't so much that Paul Bunyan was brilliant. It was just that he was methodical. He had tried it out with that strip of gummed paper, and he knew that the second time you slice a Möbius strip you get two pieces—linked together like an old fashioned watch chain.

3 *The Möbius Band*

Take a long narrow strip of paper (Fig. 112) and paste or Scotch tape the ends together, but give one end a half twist before joining them together, so that *ABC* on one end is pasted along *ABC* at the other end. Now repeat Paul Bunyan's experiments by cutting the strip down the center and then cut the resulting strip a second time.

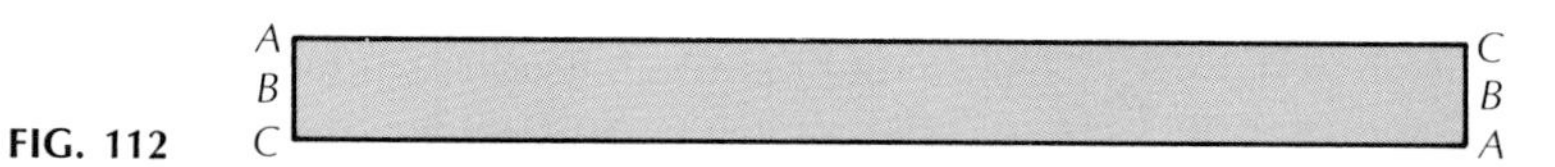

FIG. 112

The original strip which you had before cutting is called a Möbius band, named after A. F. Möbius, who explained its properties in a paper submitted to the Academy of Sciences in 1858. It has some most amazing properties, two of which we have already seen. It has only one side, since a person walking along it will return to a point just underneath his starting point without crossing the edge. If we try to paint the two "sides" of this band each a different color, we shall see that the two "sides" are really one. This is called a *one-sided surface* and is the simplest case of a whole family of such surfaces. We notice that the "two edges" are really one, and that the edge is a simple closed curve. Truly, in the words of the limerick,

A mathematician confided
That a Möbius band is one-sided
And you'll get quite a laugh

If you cut it in half,
For it stays in one piece when divided.

Can you place 12 pennies on the squares of a 36-square checkerboard so that (*a*) no square has more than 1 penny and (*b*) no line, horizontal, vertical, or diagonal, has more than 2 pennies?

Start by placing 1 penny at each end of the long diagonal, as shown in the diagram.

▲ EXERCISES

1. Instead of cutting the Möbius band in the center, cut it one-third of the distance from the edge. What happens? The smaller of the two pieces is a Möbius band. Why?
2. Make a strip as before, but give one end a full twist instead of a half twist before pasting the ends together. Is the resulting surface one-sided or two-sided? How many edges does it have? What happens if we cut it down the center? What happens if we cut it one-third of the way from the edge?

3. Make another strip, but give one end a twist and a half before pasting. How many sides to the surface? How many edges? What happens when we cut it down the center? At the one-third point?
4. Take two narrow strips of paper and paste to make two cylinders. Now paste the two cylinders together at right angles (Fig. 113). Cut both strips down the middle. What is the resulting surface?

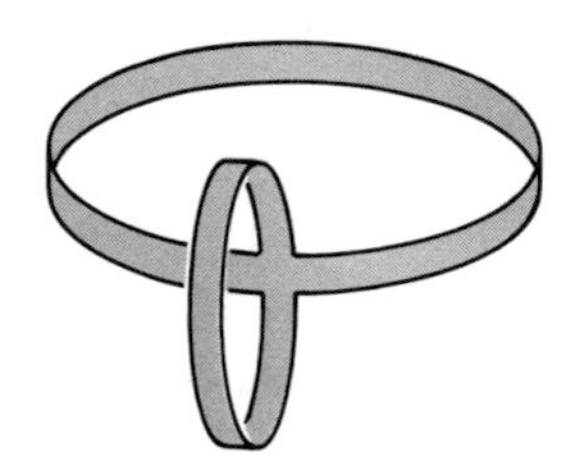

FIG. 113

5. Repeat Exercise 4 except give one cylinder a half twist to make it a Möbius strip (Fig. 114). How does the result, after cutting, differ from that of Exercise 4? How do you explain this?

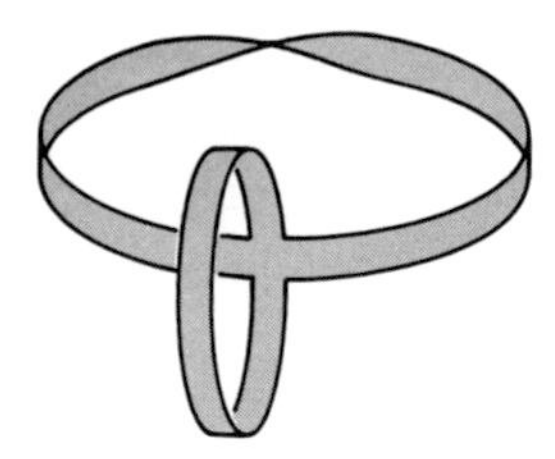

FIG. 114

Networks

Many of the elementary problems of topology depend on the properties of networks. A *network*, or *finite graph*, is a collection of points which are called *vertices*, together with a collection of *arcs* or *paths* such that:

1. Each arc has two of the vertices as end points.
2. Each vertex is an end point of at least one arc.
3. No two arcs intersect except at an end point of both.
4. The entire configuration is one connected piece.

The simplest network is a single arc and its two vertices. Another simple network is the simple closed curve, which consists of two arcs and two vertices, or we may locate n points on a simple closed curve and consider it as a network of n arcs and n vertices. Examples of more complicated networks are the lines on a checkerboard and the highway map of the United States. Here the cities are the vertices and the highways between two adjacent cities are the arcs.

Condition (3) does not prevent a situation such as that shown in Fig. 115 from occurring in a network. It requires only that the point E be designated as a vertex and that $AE, CE, DE,$ and BE be arcs of the network. It is not permissible to consider the complete paths AB and CD as arcs of the network, since they would intersect at a point which is not a vertex.

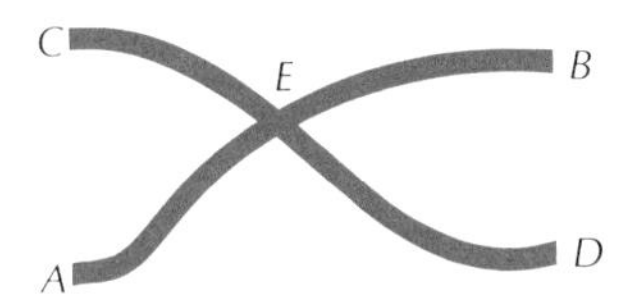

FIG. 115

In studying the properties of networks, we shall be interested, among other properties, in the question of the number of pieces or *regions* into which the network divides the plane. For example, in Fig. 115 there is only one region, for you can go from any point in the plane to any other point without crossing the network. On the other hand, there are two regions associated with any closed circuit. However, the network in Fig. 116, consisting of seven vertices and nine arcs, has four regions.

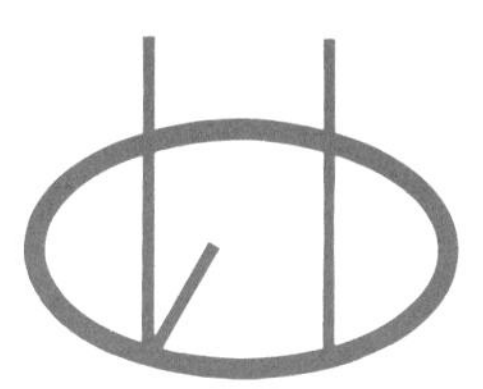
FIG. 116

Euler-Poincaré Formula

If, in any network lying in a plane, V denotes the number of vertices, A the number of arcs, and R the number of regions into which the plane is cut by the network, then

$$V-A+R=2$$

Here we have one of the deepest and most fundamental properties of networks. This relation holds, regardless of how simple or how complicated the network is.

Euler (1707–1783) observed this relation between the vertices, edges, and faces of the polyhedra. Poincaré (1854–1912) developed the extensions of the formula to apply on all types of surfaces.

We shall give a simple proof of this formula, depending only on the intuitive use of the Jordan curve theorem. First, notice that the formula is true for the simplest network, a single arc with its two end points as vertices. Here $V = 2$, $A = 1$, $R = 1$, and $2 - 1 + 1 = 2$. Now we shall consider three modifications of any given simple network to produce a more complicated one.

■ **Modification 1.** Designate a point on an arc of the network as a vertex, and divide the arc into two arcs. For example, in Fig. 117 we

have a network consisting of two vertices and three arcs. The modification indicated changes it into a network which looks the same but consists of three vertices and four arcs. Whatever may have been the values of V, A, and R in the original network, this modification makes them $V+1$, $A+1$, R, and we have

$$(V+1)-(A+1)+R=V-A+R$$

This means that the value of this relationship between the vertices, arcs, and regions is the same in the original and modified networks.

■ **Modification 2.** At a vertex of the network attach an arc which has no other point in common with the original network, as, for example, in Fig. 118. This modification always increases the value of A

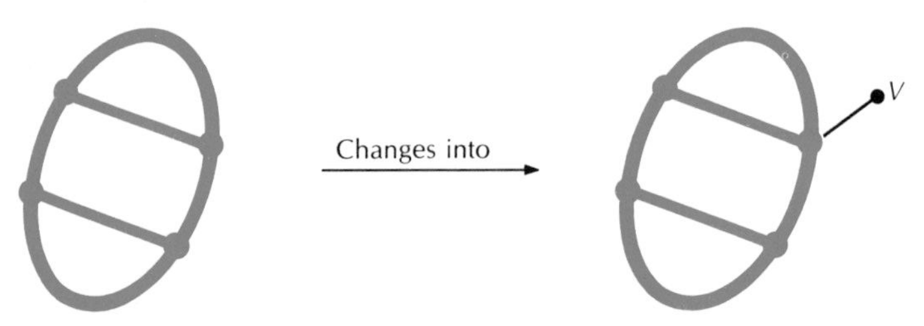

FIG. 118

by 1. It also increases the value of V by 1, namely, the vertex which is the end point of the new arc not on the original network. The quantity R remains unchanged. Since V and A are both increased by 1, the value of $V - A + R$ is the same in both the original and the modified network.

Modification 3. Join two vertices of the network by a new arc having no other points in common with the network, as, for example,

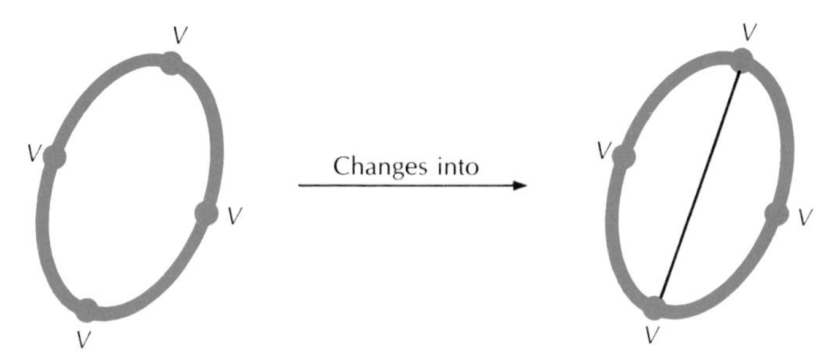

FIG. 119

in Fig. 119. This modification increases A by 1, but it also increases R by 1. The quantity V remains unchanged. Then the original and modified networks give the same value of $V - A + R$.

Now, starting with a single arc, we may build up any network whatever in the plane by using, as needed, the modifications 1, 2, and 3 discussed above. But we have shown that the network consisting of a single arc and its two end points as vertices satisfies the equation $V - A + R = 2$, and no one of the three modifications changes the value of the left-hand side. Thus this relationship is still true for whatever final network is produced.

EXERCISES

1. Compute V, A, and R for the lines of the checkerboard of 9 squares. Verify $V - A + R = 2$.
2. Repeat Exercise 1 for the usual checkerboard of 64 squares.
3. Considering the corners as vertices, the edges as arcs, and the faces as regions, verify $V - A + R = 2$ for the cube.
4. Repeat Exercise 3 for the tetrahedron shown in Fig. 120.

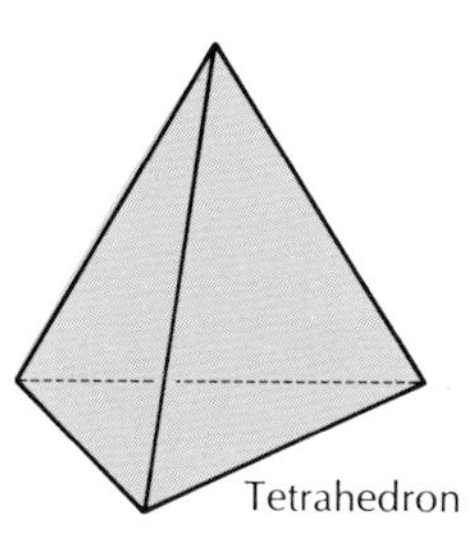

FIG. 120

5. Repeat Exercise 3 for the regular octahedron. An octahedron (Fig. 121) consists of two pyramids with square bases joined along the bases. It consists of eight faces, each of which is an equilateral triangle.

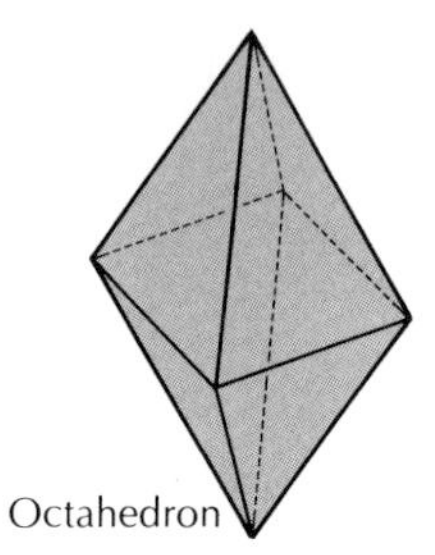

FIG. 121 Octahedron

6. If a network in the plane has 16 vertices and 23 arcs, into how many regions does it cut the plane?
7. Show that in any network in the plane the number of vertices V is never more than $A+1$.
8. Show that any network separating the plane into exactly two parts has the same number of arcs as it does vertices.

FIG. 122 Tree

9. A *tree* is a network which does not contain any simple closed curve. In Fig. 122 we see that such a network may resemble a tree with branches, hence its name. Prove that in any tree the number of vertices is always one more than the number of arcs.
10. In Fig. 116, count V and R. Compute A by the Euler-Poincaré formula. Check by counting A.
11. Repeat Exercise 10 for Fig. 121.
12. Repeat Exercise 10 for Fig. 144 on page 363.

Problems just for fun

Farmer True Patriot has a triangular plot of land and paints the corner posts one red, one white, and one blue. Suppose the inside of the plot is broken into smaller triangular pieces in any way whatever, but leaving the three sides of the original large plot unbroken. Now Mr. Patriot's sons paint the corner posts inside as they please, maybe some yellow, some pink, etc. Can you prove that there is always a subplot whose three corner posts are three different colors?

6 *The Seven Bridges of Königsberg*

At Königsberg in Germany there were once seven bridges arranged as shown in Fig. 123. The problem is to plan a Sunday afternoon stroll

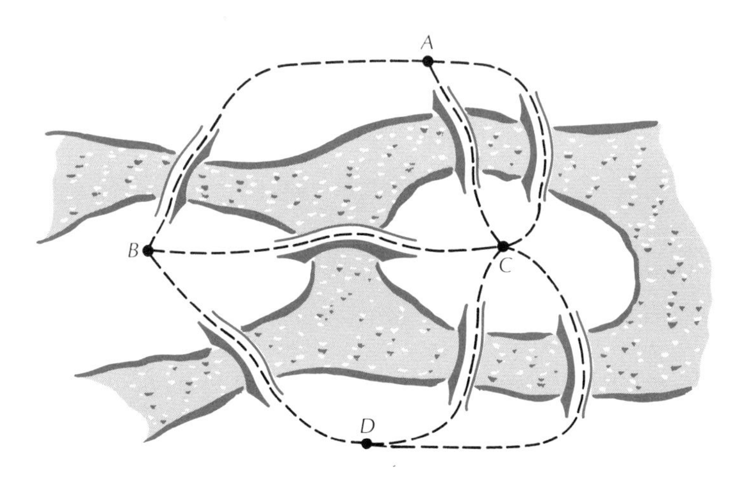

FIG. 123

so as to cross each bridge exactly once and return to the starting point. Can you plan the Sunday walk?

Traversing Networks

The Königsberg problem is one example of traversing a network. We say that a network can be walked through or *traversed* if a path exists which starts at a point, covers each arc of the network exactly once, and terminates at the beginning point. This path is permitted to cross itself if needed.

In Fig. 124 we have a network which can be traversed, and one such path is indicated by the numbers, starting with the vertex

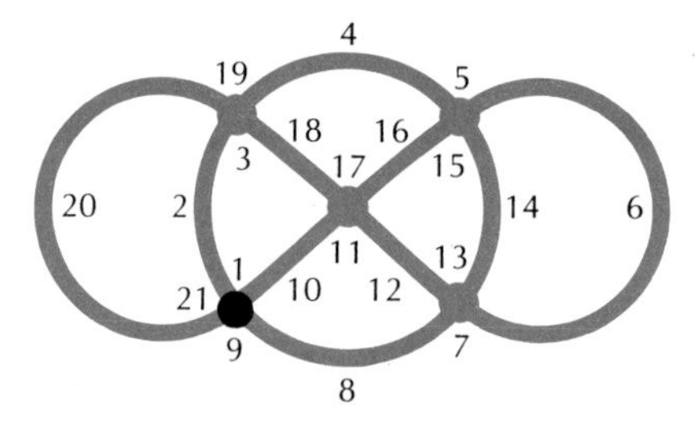

FIG. 124

marked 1 and walking as indicated by the numbers 2, 3, 4, etc. On the other hand, in Fig. 125 we have a network which cannot be traversed,

FIG. 125

for no matter how we use the three arcs from A to B, our path will terminate at the opposite point from where it started.

Returning to the Königsberg problem, we can convert the figure into a network by labeling the points marked A, B, C, D in Fig. 123 and using the dashed paths as arcs. Then we have a network of four vertices and seven arcs which crosses each bridge exactly once. If this network can be traversed, we have the plan for our Sunday walk. Conversely, the plan for a Sunday walk would give the path for traversing the network. We shall settle this Königsberg bridge problem by means of the principle of the next section.

General Traversing Principle

Statement. A network may be traversed if and only if the order of each vertex is even.

By the *order* of a vertex, we mean the number of different arcs which have that vertex as an end point. In Fig. 124 we saw a network which could be traversed. It has five vertices, and all are of order 4 in agreement with the general traversing principle. In Fig. 125, however, we have a network which cannot be traversed, and we observe that it has two vertices both of odd order in accordance with the general principle.

We shall now prove the general principle in two parts.

1. If a network can be traversed, every vertex is of even order.

This is easily seen, for each time the path approaches a particular vertex along an arc, there must also be an arc along which the path leaves the vertex. As each arc is used just once, the total number of arcs meeting at the vertex is even.

2. If the order of every vertex is even, the network can be traversed.

Select any vertex B as the beginning (Fig. 126). Select any arc meeting at B, say, BC, and proceed along it to the next vertex C. Here select any arc except the one which has been used and proceed on it to the next vertex. At this vertex choose any arc that has not been used as yet and proceed on it to the next vertex. Continue in this manner. As long as the vertex approached is not B, the total number of arcs at

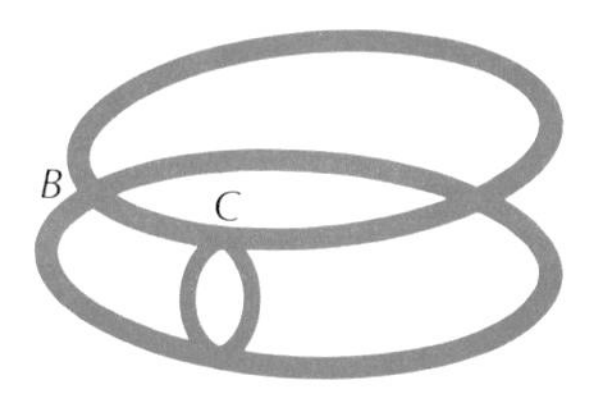

FIG. 126

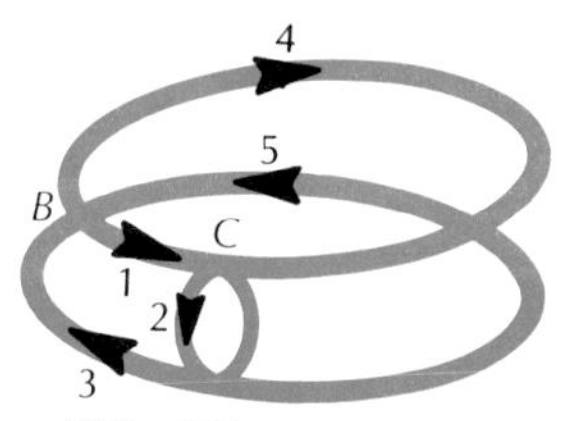

FIG. 127

the vertex already used will be odd. Since the order of this vertex is even, we will always have at least one unused arc to lead the path away from this vertex. Thus the path continues through the network until we return to the starting point B. We must return to B at some stage, for the path cannot continue indefinitely (there are only a finite number of arcs). We now start the path from B again along an unused arc and continue the previous process. At some stage on a return to B we find no unused arcs available at B. This stops the process.

We have now a path starting and ending at B (it also may pass through B at several intermediate points) and using no arc of the network more than once. If the path uses every arc of the network, it is the desired traversing path.

On the other hand, if there are arcs of the network which do not belong to this path (see Fig. 127), there is an arc which has one of its end points on the path. This follows from the fact that a network is connected. Let C denote such an end point. Since the order of C is even and the path has used an even number of arcs meeting at C, there are at least two arcs at C not used in the path. Choose one of these and proceed along it to the next vertex. Just as before, the total number of arcs used at this vertex will be odd, and we shall have at least one arc available to choose as an exit from the vertex. We can always continue this process unless the vertex approached is C. Furthermore, we must approach C at some stage of this continuing process, since the total number of arcs is finite and the process can continue unless C is approached.

Now we have a path beginning and ending at B, together with a side excursion beginning and ending at C. Let us combine them into a single path by starting at B, proceeding along the original path to C, making the side excursion and returning to C, then continuing on the original path to its ending at B.

This new path begins and ends at B and uses no arc of the network more than once. If every arc is used, it is the desired traversing path. If not, we build a new side excursion and add it to the path. By continuing to increase the path, adding side excursions, we arrive eventually at a path using all the arcs. This completes the proof of the general traversing principle.

9 *Walks with Different Beginning and End*

Traversing a network requires the path to begin and end at the same point. What is the situation if we do not require the same beginning and ending? If the number of vertices of *odd* order is *exactly two,* a path exists beginning at one odd vertex and ending at the other and using each arc of the network exactly once. For example, such a path exists in the network of Fig. 125.

On the other hand, if the number of vertices of odd order is neither zero nor two, no path whatever exists which uses each arc exactly once. In the Königsberg bridge network there are four vertices of odd order. Thus no Sunday stroll can be planned at all, even if we allow the walk to end at a different point from its start.

▲ EXERCISES

1. Take four points and join every pair of them by paths that do not cross each other. This is like a square plus its diagonals except that the diagonals do not intersect. Can you traverse this network?
2. Take five points and join every pair of them by paths that do not intersect. This cannot be done on a piece of paper, but can be done easily if we use three dimensions in drawing our paths. Can you traverse this network?
3. Can you traverse the network formed similarly with six points?
4. Suppose every pair of n points are joined. For what values of n can the network be traversed?
5. If one of the Königsberg bridges is closed, can you plan a Sunday walk crossing each open bridge exactly once? Can the stroll end at the same point at which it began?
6. Where can a new bridge be built so that a Sunday stroll will be possible (not requiring the same point as beginning and end)?
7. Which bridges must be closed if a Sunday walk beginning and ending at the same point is to be possible?
8. Where must new bridges be built if a Sunday walk with the same beginning and end point is to be possible?
9. Plan a walk through the five-room house shown in Fig. 128 so as to pass through each door exactly once. Is such a walk possible? Why?

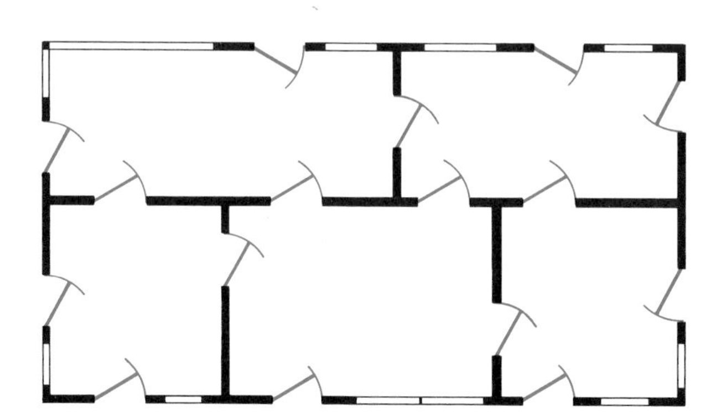

FIG. 128

10. What is the smallest number of doors in Fig. 128 that must be closed to make the walk possible if it must end at the starting point? If we permit it to end at a different point?
11. Plan a walk through the five-room house shown in Fig. 129 so as to pass through each door exactly once. In what rooms must the walk begin and end? Where could we cut two more doors so as to make a walk beginning and ending at the same place possible?

Problems just for fun

After a party at the home of Hal and Marie Dimwit, it was decided that five of the ten people there would clean the dishes, while the other five relaxed. In order to choose the five who would do the work, everyone formed a circle. Then Hal was to choose a number, point to a person, and start counting. When the count reached the chosen number, that person would leave to clean dishes and the count would begin again from that point until a second person was picked, and so on until the five kitchen help were chosen.

The men schemed to have the women chosen for the work, but poor Hal Dimwit forgot the scheme and started with the wrong person and the wrong number. He started his count as shown in the diagram, and all the men were chosen to do the work.

Can you find (*a*) the number Hal used and (*b*) the number he should have used and where he should have started counting?

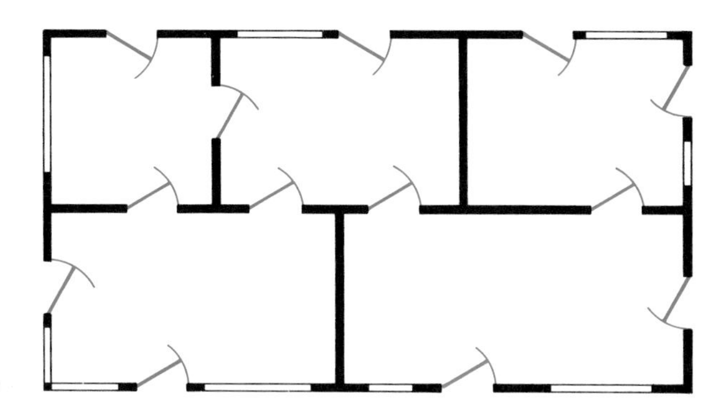

FIG. 129

Euler-Poincaré Formula for Surfaces

We proved the Euler-Poincaré formula for networks lying in the plane. However, it is easy to see that it also applies to networks drawn on the surface of a sphere. Select one region into which the network divides the surface of the sphere and cut a small hole in this region. Now, imagining the sphere to be made of very elastic rubber, we may flatten out the sphere with a hole in it until it becomes a circular flat piece with the edge of the hole as its outer edge. The network on the sphere becomes a network in the plane with the same number of vertices, arcs, and regions. The region on the sphere containing the hole becomes the outside region of the network in the plane. Then

$$V-A+R=2$$

This formula also holds for the edges of any polyhedron, for if we think of the polyhedron surface as rubber, we may deform it into a sphere, and the edges of the polyhedron become a network on the sphere.

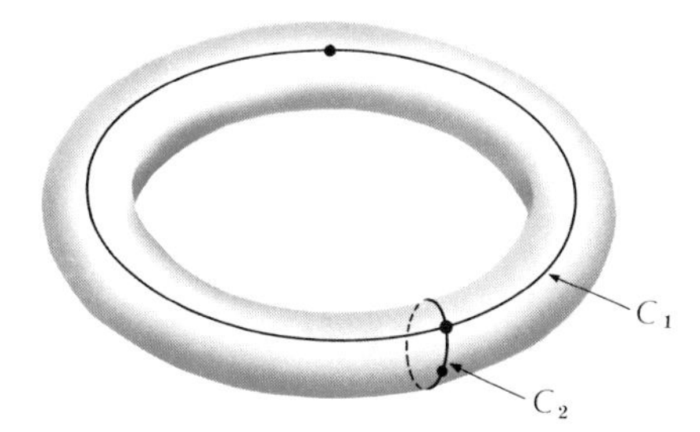

FIG. 130

Figure 130 shows the surface known as a *torus*. It resembles an automobile inner tube. On it draw a network consisting of two circles,

C_1 and C_2, one going around the torus in each direction as shown in the figure. In this network we have $V=3$, $A=4$, $R=1$, and substituting in the Euler-Poincaré formula (page 343), we get

$$V-A+R=3-4+1=0$$

instead of 2 as formerly.

But we notice a fundamental difference between the sphere and the torus. In the sphere, every cut which is a simple closed curve or closed circuit divides the spherical surface into two parts (Fig. 131).

FIG. 131

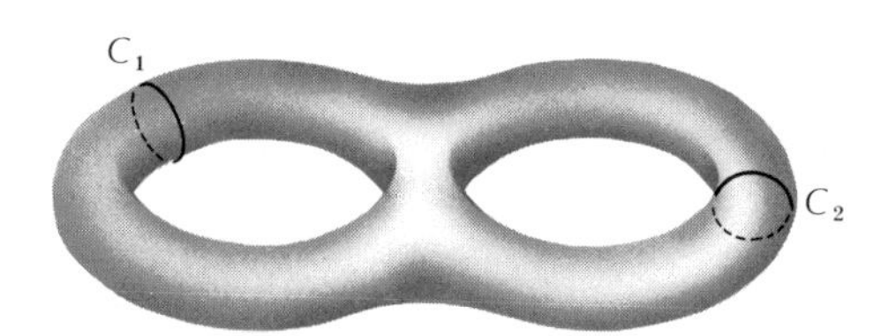

FIG. 132

On the inner tube, some circuit cuts divide the surface but some do not. Each of the circuits C_1 and C_2 shown in Fig. 130 fails to divide the surface into two parts. But any two nonintersecting circuits divide the torus surface. On the other hand, the more complicated surface shown in Fig. 132 contains two circuit cuts that do not intersect, and still the surface is in one piece.

This property of the number of circuits necessary to divide the surface is called the *genus of the surface.* The surface is said to have *genus* p if there are p nonintersecting circuits on the surface which

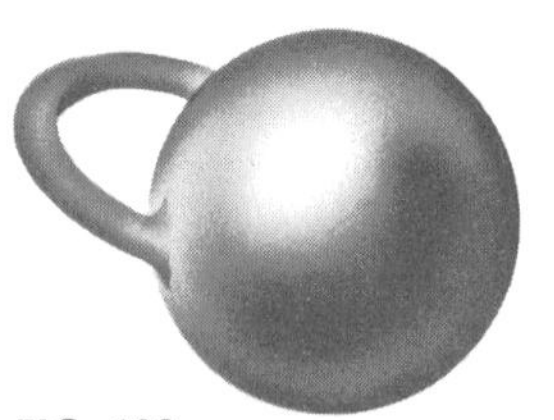

FIG. 133

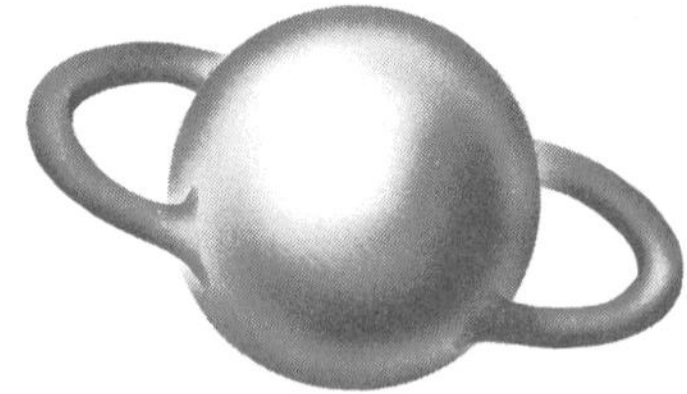

FIG. 134

leave the surface in one piece but each $p+1$ such cuts divide the surface. Thus the sphere is of genus 0, the torus of genus 1, and the surface in Fig. 132 of genus 2. A torus may be deformed into the sphere with one handle, as shown in Fig. 133. Also, the surface of genus 2 of Fig. 132 may be deformed into a sphere with two handles, as shown in Fig. 134. Thus we may consider each such two-sided surface as a sphere with p handles and the number of handles is the same as the genus.

For such a two-sided surface, the Euler-Poincaré formula becomes

$$V - A + R = 2 - 2p$$

where p is the genus. This formula applies only to networks that cut the surface into regions that can be deformed into flat disks. The proof of this generalized formula is not too difficult, but it is long, and we shall not give it here.

▲ EXERCISES

1. Compute V, A, and R for the network on the torus consisting of four circuits, two going around the torus each way. Verify that $V - A + R = 0$.
2. Compute V, A, and R for the network on a torus consisting of five circuits going around the torus one way and one circuit going the other way. Verify that $V - A + R = 0$.
3. Compute V, A, and R for the network on a torus consisting of six circuits, three going each way. Verify that $V - A + R = 0$.

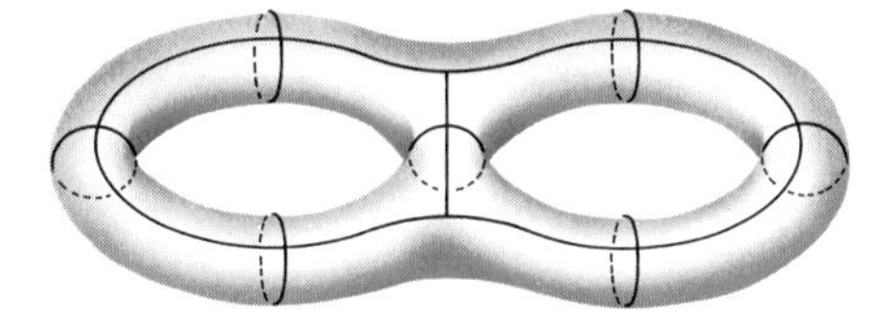

FIG. 135

4. Compute V, A, and R for the network shown in Fig. 135, and verify the generalized Euler-Poincaré formula in this case.
5. In Fig. 135, if we add one more circuit, what does this do to the values of V, A, and R found in Exercise 4? Check the Euler-Poincaré formula for these new values.
6. If the network on the torus consists of only one circuit going around the torus, we have $V = 2$, $A = 2$, $R = 1$, and $V - A + R = 2 - 2 + 1 = 1$. Why?

One-sided Surfaces

In the preceding section we discussed the two-sided surfaces without edges. Earlier, in the Möbius band, we saw a surface with only one side, but it had an edge. There are surfaces without edges, which have only one side, but it requires some imagination to visualize them. The simplest one can be visualized by imagining a Möbius band and a circular piece of rubber sewed together along their edges. This is called the *projective plane*. If we try to accomplish this sewing physically.

Problems just for fun

In the olden days pilgrims on their way to visit holy shrines spent their nights at wayside inns at short distances apart, for travel was slow in those days. They liked to while away their long winter evenings by proposing puzzles to one another.

A devout parson told of his parish, through which ran a small stream which joined the sea some hundred miles to the south. He showed a map of a part of his parish and propounded his puzzler.

"Here, my worthy pilgrims, is a strange riddle. Behold how at the branching of the stream is an island. Upon this island doth stand my own poor parsonage, and ye may all see the whereabouts of the village church. Mark ye, also, that there be eight bridges and no more over the stream in my parish. On my way to church it is my wont to visit sundry of my flock, and in the doing thereof I do pass over every one of the eight bridges once and no more. Can any of ye find the path, after this manner from the house to the church, without going out of the parish? Nay, nay, my friends, I do never cross the stream in any boat, neither by swimming nor wading, nor do I go underground like a mole, nor fly in the air as doth the eagle; but only pass over by the bridges."

Can you find the parson's path? At first it seems impossible, but the conditions stated provide a loophole.

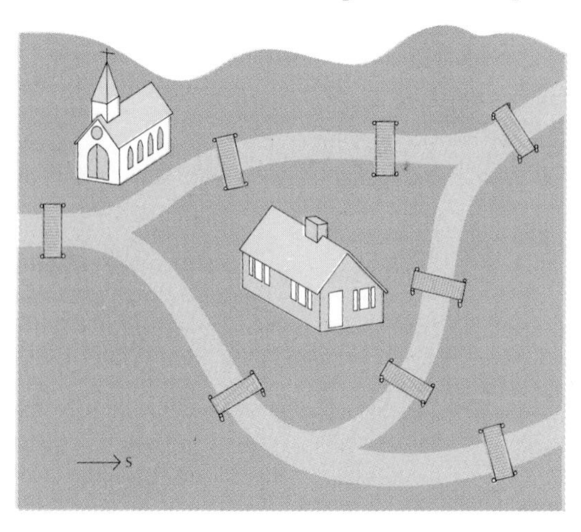

we shall find ourselves in trouble. The physical sewing can be carried out actually in four-dimensional space, but that is another story.

The next simplest one-sided surface can be visualized by imagining two Möbius bands sewn together along their edges. It is called a *Klein bottle* (Fig. 136). This sewing may also be done physically in four-

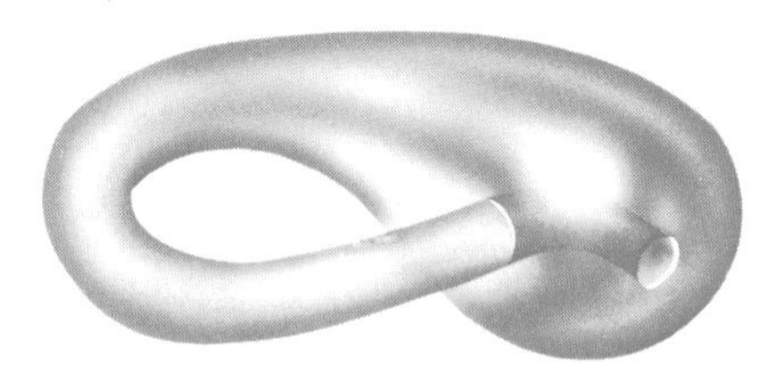

FIG. 136

dimensional space, but in three-dimensional space we can do it only in our mind.

Thus there exists a whole series of two-sided surfaces and correspondingly a whole series of one-sided surfaces, of which we have examined a few of the simpler ones.

12 *Regular Polyhedra*

A polyhedron is called *regular* if all its faces are congruent regular polygons, the same number of faces meet at each vertex, and all the dihedral angles are the same. Five regular polyhedra are known: tetrahedron (Fig. 120), cube, octahedron (Fig. 121), dodecahedron (Fig. 137), and icosahedron (Fig. 138).

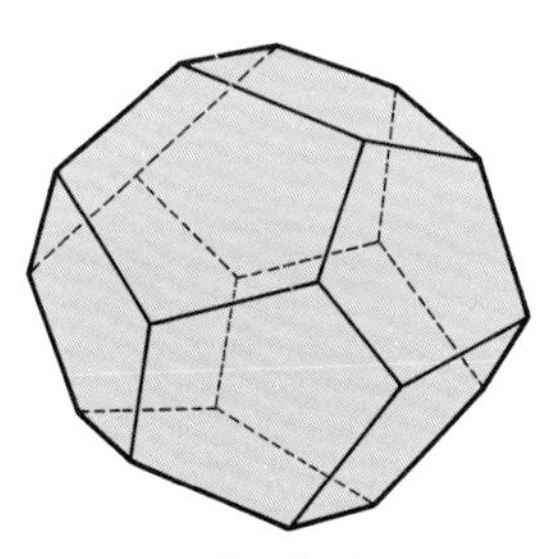

Dodecahedron
FIG. 137

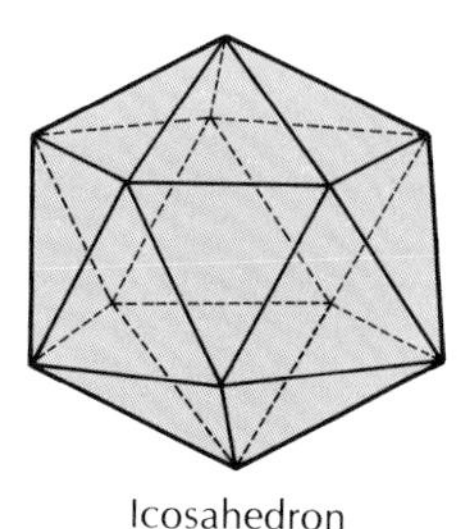

Icosahedron
FIG. 138

The sizes of face angles and the dihedral angles are not topological properties. Even ignoring these properties, we can prove by topology that there are no more regular polyhedra to be found. We shall prove that there are only five networks in which each region has the same number of sides and every vertex has the same order.

If each region has n sides, then nR counts all arcs. But each arc bounds two regions, so it has been counted twice, once from each side. Thus in a regular polyhedron

$$nR = 2A$$

Also, if every vertex is of order k, then kV counts all the arcs. But each one has been counted twice, once from each end, so in a regular polyhedron

$$kV = 2A$$

Now we substitute the values of R and V from the last two relations into the Euler-Poincaré formula

$$V - A + R = 2$$

We get

$$\frac{2A}{k} - A + \frac{2A}{n} = 2$$

or, dividing by $2A$, we obtain

$$\frac{1}{k} + \frac{1}{n} = \frac{1}{A} + \frac{1}{2}$$

Now $n \geqq 3$, since the faces are at least triangles. Also, $k \geqq 3$, since there are at least three faces at each vertex. But one of n or k must be 3, for if both k and n are 4 or larger, the left-hand side of the last equation above would be $\leqq \frac{1}{2}$, and no value of A could make the equation true.

If $n = 3$, the above equation becomes

$$\frac{1}{k} - \frac{1}{6} = \frac{1}{A}$$

and k may be 3, 4, or 5. It is not possible to have $k \geqq 6$, for this would make A negative or infinite.

Similarly, if $k = 3$, the equation becomes

$$\frac{1}{n} - \frac{1}{6} = \frac{1}{A}$$

and $n = 3$, 4, or 5 are the only possible values.

All together we have just five solutions, as $n = 3$ and $k = 3$ occurs in each case:

$n=3$	$k=3$	$A=6$	$R=4$	tetrahedron
$n=3$	$k=4$	$A=12$	$R=8$	octahedron
$n=3$	$k=5$	$A=30$	$R=20$	icosahedron
$n=4$	$k=3$	$A=12$	$R=6$	cube

$n=5 \qquad k=3 \qquad A=30 \qquad R=12$ dodecahedron

Thus we have proved by topology that there do not exist more than five regular polyhedra.

13

Coloring Maps

Father Mapper handed his son two maps to color (Figs. 139 and 140). The son studied the maps a moment and said, "I can color one of them, but I do not have enough colors to color the other." This was found to be correct.

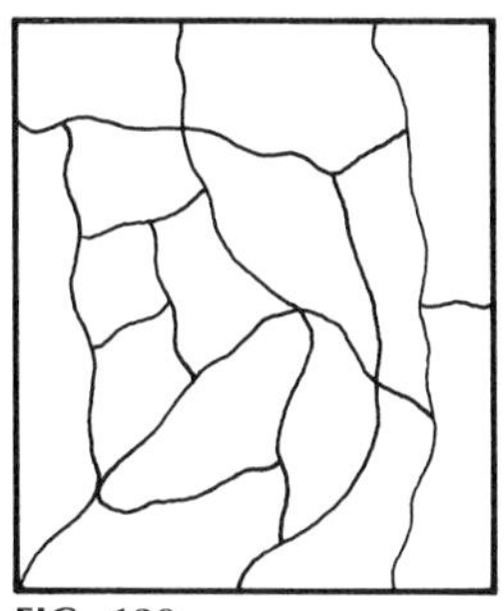

FIG. 139

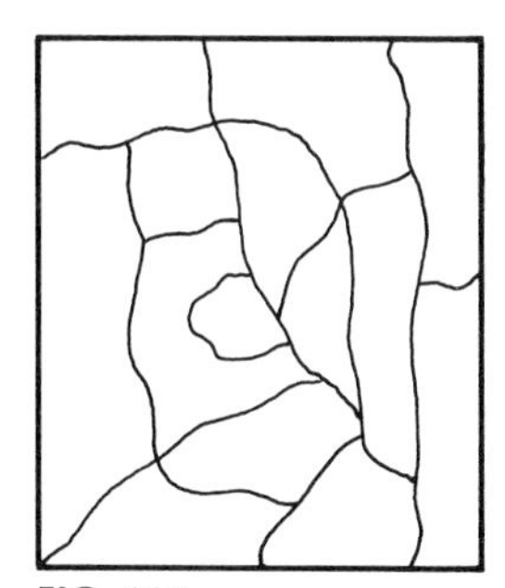

FIG. 140

How many colors did the son have, and which map did he find impossible to color? Of course, we always color a map so that two regions meeting along an arc are colored two different colors.

The Four-color Problem

The puzzle above leads us to the more general question: How many colors must a printer keep in stock in order to be able to print every possible map that may be brought to him? Or, we may say, how many colors are required to color all possible networks so that whenever two regions meet along an arc, they are colored differently. Regions which meet only at a vertex may be colored the same, since this creates no difficulty. Figure 141 shows a map which requires four colors, for every region meets all the remaining three. But no one has ever discovered a network or a map requiring more than four colors. In 1890 Heawood proved that every network in the plane can be colored in five colors. But no one to this day has proved that it can be done in four colors or has found a network requiring five. However, if a map needing five colors exists, it must be fairly complicated, for in 1946 DeBacker proved that any such map must have at least 35 regions.

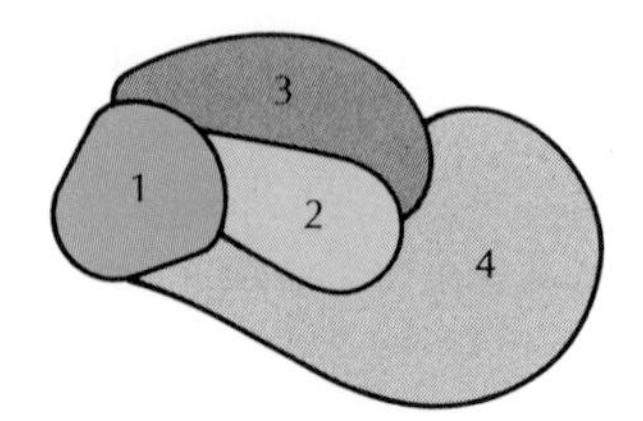

FIG. 141

The four-color problem is unsolved for maps on the simplest of all surfaces, the plane or the sphere. It might be guessed then, that there is little hope of solving the color problem on the more complicated surfaces. On the contrary, the problem is completely solved for all the other well-known surfaces such as the torus, the surface of genus 2, the Möbius strip, the Klein bottle, etc. The number of colors needed varies with the surfaces. But in each case the problem is completely solved, for we can prove that every map can be colored in q colors and can exhibit a map requiring exactly q colors. For the two-sided surfaces of genus p, we know the number of colors q in all cases below genus 8 except the sphere (Table 81).

TABLE 81

p	0	1	2	3	4	5	6	7	8
q	4(?)	7	8	9	10	11	12	12	13(?)

The torus is the surface of genus 1, and we see in Table 81 that every possible map on the torus can be colored in seven colors. In Fig. 142 we exhibit a map of seven regions on a torus in which we may check that each region touches all the other six. The figure appears to be a rectangle but it becomes a cylinder when we sew the two sides marked *ABCA* together. And this cylinder becomes a torus when we sew the two circles together along *AEFA*. Thus we cannot color this

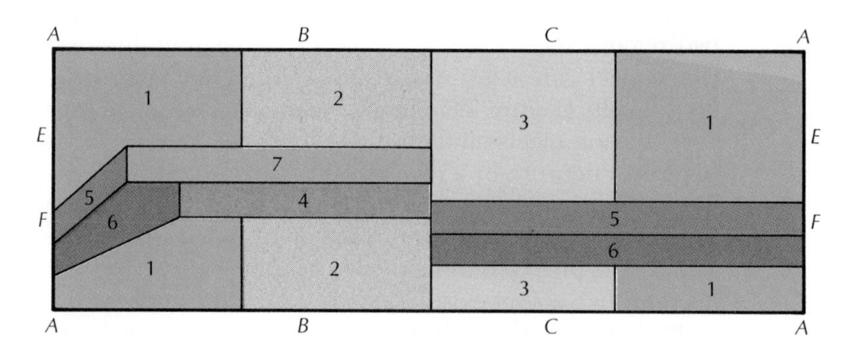

FIG. 142

map in fewer than seven colors. Since seven colors are sometimes needed and every possible map may be colored in seven colors, the color problem is completely solved for the torus.

Also, for the surface of genus 2 shown in Fig. 132 (page 354), the color problem is completely solved and the number is eight. But for the simplest of all surfaces, the sphere or plane, the problem has never been solved. Perhaps some reader of this book will become famous by solving the color problem for the plane or sphere.

A Two-color Problem

While our printer must keep at least four colors in stock in order to be able to color *any* map, there are some maps requiring fewer colors. In Fig. 139 we find a map requiring only three colors. In Fig. 143 we

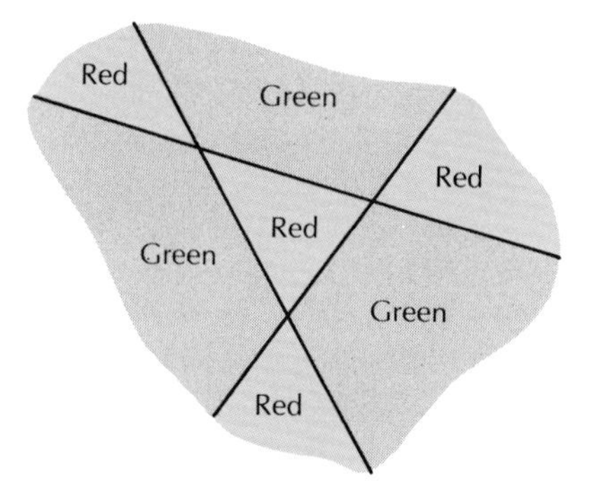

FIG. 143

see a map that can be colored in two colors. In this map we see that the boundaries of the countries are all straight lines. We shall show that every such map can be colored in two colors.

Every map which is formed by a finite number of straight lines can be colored in two colors.

Every straight line divides the plane into two parts and we shall arbitrarily assign the numbers 1 and 2 to these two sides. Do this for each of the lines of our map. Now consider any country of our map. It lies entirely on one side of each of the lines so that all points of this country have the same number, 1 or 2, assigned from the line. However, it may be on the 1 side of some lines and the 2 side of others. Add up the numbers for this country for all lines. If this total is even, we will color the country red; if odd, we will color it green.

Now consider two countries which have a side in common. This side is part of a line and the two countries are on opposite sides of this line. Hence the number assigned from this line is 1 for one

Problems just for fun

The job of a heating-tunnel inspector is to walk through and examine each day all the heating tunnels joining the 12 buildings shown in the diagram. To complete his job he finds he must walk through some tunnels twice, but naturally he wishes to avoid unnecessary walking. Each tunnel is 1 mile long, and he discovers a route which completes his job by walking 19 miles.

Can you find where he should start and how he should walk to go through all the tunnels in 19 miles?

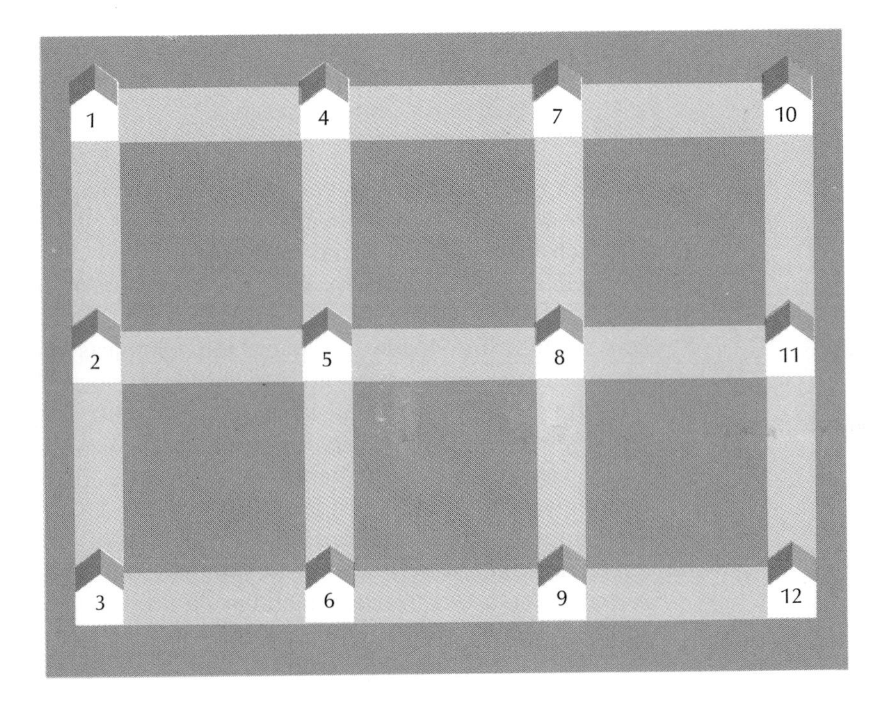

country and 2 for the other. Except for this line, the two countries are on the same side of all other lines. Then all the other numbers are the same for the two countries. Thus the sum is $K + 1$ for one country and $K + 2$ for the other. One is even and the other is odd and the two countries are colored differently.

▲ EXERCISES

1. Draw a map on the torus consisting of five countries, all of which touch the four others.
2. Draw a map on the torus in which there are six countries all touching one another.
3. Draw a map on the Möbius strip of five regions, all touching one another. What does this tell us about the number of colors needed for maps on a Möbius strip?
4. Draw a map on the surface of Fig. 132 (page 354) consisting of six countries, all of which touch one another.
5. Draw a map on the surface of Fig. 132 of seven countries, all of which touch one another.
6. The color number for the Möbius strip is 6; that is, we can prove that every map on the surface can be colored in this number of colors. Can you draw a map of 6 regions on the Möbius strip, each region touching the other 5 along an arc? This will show that the color number 6 cannot be reduced.
7. Show that every map consisting entirely of circles can be colored in two colors.

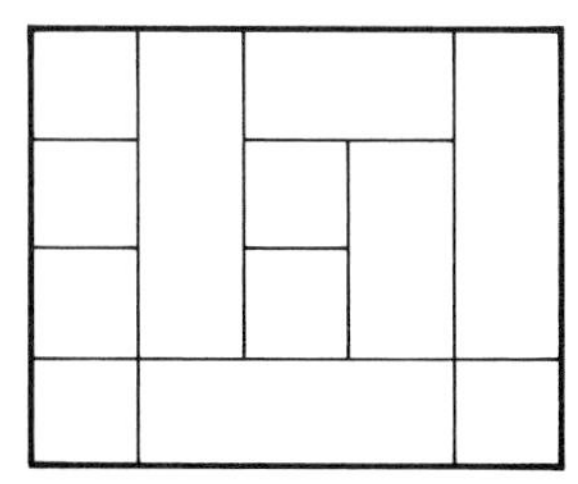

FIG. 144

8. If we do not color the "outside" country in the map of Fig. 144, show that we can color the map in three colors but not in two colors.

CHAPTER TWENTY-FOUR

Logic – The Art of Reasoning

Formal thought, consciously recognized as such, is the means of all exact knowledge; and a correct understanding of the main formal sciences, Logic and Mathematics, is the proper and only safe foundation for a scientific education.

ARTHUR LEFEVRE

1 *Reasoning*

Aristotle first remarked that the distinctive trait which distinguishes human beings from other mammals is the ability to reason. Some recent experiments by psychologists have shown that a number of animals have limited reasoning power. On the other hand, our observations of some human beings make us wonder whether they reason at all. While Aristotle's remark may not be entirely correct, we will agree that human beings do consider various bits of information and draw conclusions from them and that human beings are exhibiting their best mental talents when this reasoning is being carefully and accurately done. The study of *logic* is the study of the laws of reasoning and the principles of valid reasoning.

Mathematics is impossible without logic, for we must use correct arguments if we are to solve problems, prove theorems, etc. But logic enters into life in many other ways. In writing and speaking we must state precisely what we mean or we shall be misunderstood. Most arguments and disagreements result from inaccurate statements or the inaccurate use of words. And, correctly or not, all the people about us are continually drawing conclusions from statements and observations. A study of the principles of valid reasoning, that is, of logic, will assist us in understanding this process in which all of us take part.

However, we must not conclude that correct reasoning is always desirable or will best accomplish the desired purpose. An appeal to sentiment or emotion may fall flat if subjected to cold logical analysis. A speech that will win an election may not be logical at all. An opponent of a senator from North Carolina obtained the menu from the Washington hotel where the senator lived and used it as a campaign issue. He made speeches all over the state pointing to the "elegant vittles" the senator was eating compared to the simple fare of the voters. It had little logic, but it changed many votes. Often a politician will win more votes by arousing emotions than by a logical analysis of the issues. But such tactics will fail when the voters are intelligent and logical.

Advertisers are interested in selling their product, not in logic. Often advertising is based on illogical inferences. One example is the picture we often see of a beautiful girl or a distinguished man using the product. How does this sell the product? As readers, we are supposed to reason: It is desirable to be beautiful or distinguished. The picture shows the beautiful or distinguished one using this product. Therefore it is desirable that we use this product. Clearly, this is dubious reasoning, but evidently enough people arrive at the above conclusion so that this is good advertising.

In a great many simple cases we can recognize valid or invalid reasoning without logical principles. But in the more complicated cases, a study of logic or logical principles is of great assistance.

Statements

Before we can examine arguments for their validity, we must have a clear understanding of statements—what they mean and what they do not mean. Let us take as an example the simple statement: Not all students pass their courses. Now consider the following statements:

a. Some students fail their courses.
b. If one is a student, one must expect to fail.
c. No student passes all his courses.
d. The teacher will always fail some students.
e. The class in which all students pass does not exist.
f. Among all students will be found some who fail courses.
g. Some students pass their courses.

These statements sound rather similar, and the careless person may feel that they all state the same thing. We must think very carefully to locate the ones which state precisely the same thing. But we must first be certain of the meaning of our words. Does "fail" mean the same as "not pass"? It does if these are the only two alternatives in our grading system. But if *condition* or *incomplete* exist in our grading system, then "not pass" is not the same as "fail." So two students from different colleges may draw quite different meanings from the same statement. To discuss these statements we will assume that "fail" and "not pass" mean the same.

In examining statements, the use of circle diagrams has been found helpful. We let the points inside one circle represent the class of all students, and we let the points inside a second circle represent "persons who pass their courses" (Fig. 145). Now our original statement may be restated: Not all students belong to the class of persons who pass their courses. This tells us that some points of the *student* circle are outside the other circle, and we indicate this by placing a *T* for true in this portion. Now do some students belong to the class of persons who pass their courses? We may consider this as obvious, but if we examine our original statement, we see that it does not say so. Thus this is not answered by the statement, and we indicate this by placing the question mark in that portion. We might also question whether any part of the circle "persons who pass their courses" is outside the "student" circle. This is a matter of the definition of the word "student" and is not pertinent to the examination of the statements.

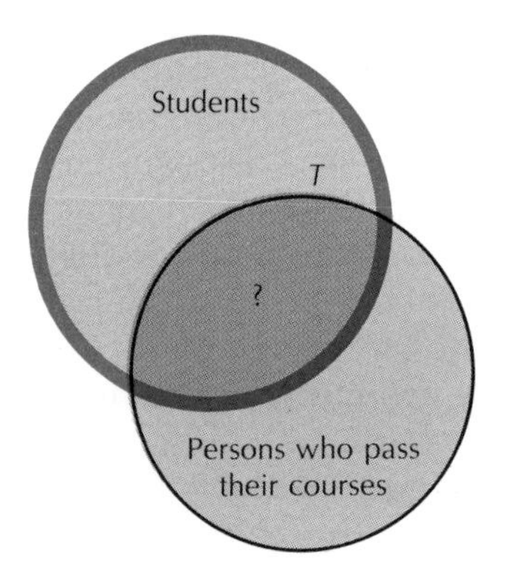

FIG. 145

Now, if we examine the various statements we see that *a* and *f* are equivalent to the original statement. Statements *b, c,* and *g* are not equivalent. Whether *d* and *e* are equivalent depends on what we mean by "all the students." If our original statement refers to all the students under one teacher, then *d* is equivalent; otherwise not. Similarly *e* is equivalent only if we refer to students of one class in the original statement.

Let us consider the four statements:

Some good people are not happy.
Some unhappy people are good.
Some unhappy people are not wicked.
Some good people are unhappy.

These four statements are all equivalent, but we may be uncertain of this as we get mixed up with the nots and somes. For instance, "some happy people are not good" and "some unhappy people are not good" sound very much like the other four but are not equivalent statements.

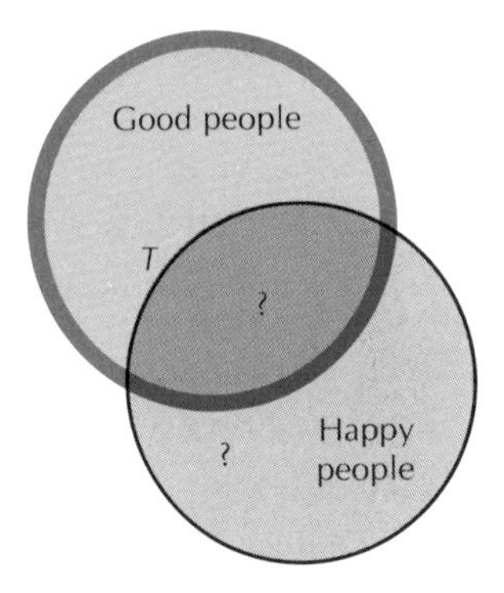

FIG. 146

The diagram will help us in this difficulty, and we draw the first statement (Fig. 146). The second statement says that some points outside the "happy" circle are inside the "good" circle, and this agrees with the diagram. Similarly, we see that the third and fourth statements are equivalent.

Consider the statement: "All men are liars." You may or may not consider this a true statement. In reasoning we are not so concerned with establishing the truth or falseness of such an assertion, but in understanding exactly what the statement says and what statements are true or false *assuming* that the original statement is true. Let us draw the circle diagram (Fig. 147).

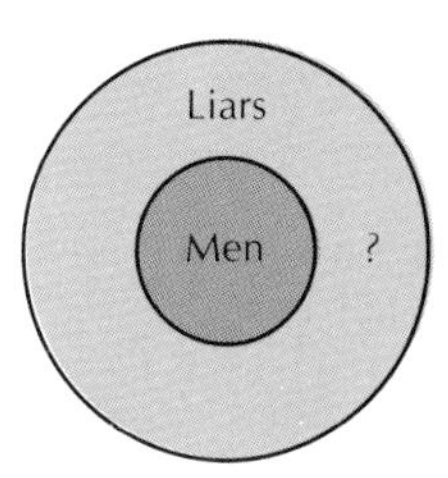

FIG. 147

From this we see that, if we *assume* the truth of the statement, the following are true:

Every man is sometimes untruthful.
A person who tells the truth every time is not a man, i.e., a woman.
Some liars are men.
No man always tells the truth.

And the following statements are false or doubtful:

Some liars are women.
Some men always lie.
All liars are men.
Some men always tell the truth.

Circle diagrams provide a useful tool in analyzing many statements. They enable us to see exactly what a statement does and does not say. We should practice using them until we are thoroughly familiar with them. Later we will study another tool, called a *truth table,* for analyzing statements and drawing conclusions.

▲ EXERCISES

1. Assuming that the first statement in each set is true, decide which of the remaining ones are equivalent and which are true, false, and doubtful. Use circle diagrams wherever possible.

   a. All Yankees are Americans.
      Some Americans are not Yankees.
      No non-American is a Yankee.
      No Yankee is not an American.
   b. No cat is a vegetarian.
      Some cats are not vegetarians.
      Either it is a cat or it is a vegetarian.
      All cats are nonvegetarians.
   c. It never rains in the desert.
      If we are in the desert, it is not raining.
      We are not in the desert, so it must rain sometime.
      Where it is raining is not desert.
   d. If we lose this game, I'll eat my hat.
      Unless the game is not lost, I'll not eat my hat.
      Either the game is won or I eat my hat.
      Unless I eat my hat, the game is won.

e. While poverty exists, the world is not perfect.
   Either the world is perfect or poverty exists.
   I know the world is not perfect, therefore poverty exists.
   Unless poverty does not exist, the world cannot be perfect.

2. What statement can you know is true if each of the following statements is false?

   a. No man lives a century.
   b. Some students are dishonest.
   c. All Cretans are liars.
   d. Every even number is divisible by 4.
   e. Some people are not happy.
   f. The stories were either dull or dirty.
   g. If it rains, we must go home.

3. Four milers on the track team all propose marriage to Susie just before the Big Ten meet. She makes the following promises:

   a. To Bob: Only if you win will I marry you.
   b. To John: If I do not marry you, you will not have won.
   c. To Sam: Either you win, or I will not marry you.
   d. To Bill: Unless you do not win, I will not marry you.

   Consider the possibility that each one may win, and state in each case what Susie must, may, and may not do, if she is to keep all her promises.

4. A football coach has four fullbacks. The following statements are true:

   a. Either he does not use Murphy or he does not win the game.
   b. Only if he does not use Brown does he win the game.
   c. Only if he does not use Smith does he not win the game.
   d. Unless he does not use Jones, he does not win the game.

   Which fullback should he use and why to be sure of winning?

A set of balls, some of which are white and some yellow, are all spotted with one or more of the colors green, blue, and red. All the balls with red spots have blue spots also. All the white balls with blue spots and all the yellow balls which do not have both red and green spots are removed. Describe the balls which remain.

3 *Hidden Implications*

There is a very old proverb which goes, "Say what you mean and mean what you say." Failure to observe this maxim is the cause of much confusion between people, some of which is accidental and some deliberate. Much malicious gossip has its origin in statements, innocent in themselves, which have double meanings or can be twisted into other interpretations. In other words, the hearer understands something different from what the speaker has said.

If we are to study logic, or the art of thinking and reasoning correctly, we must learn to phrase our statements so that they express our meaning precisely. And we must learn to be on guard for statements with hidden implications. Sometimes these statements are made deliberately, and we must be wary of the trap. There is the case of the student who had not started his term theme and who was asked by his instructor how the theme was coming. He replied, "I have not finished it yet!" The student spoke the truth, and yet the instructor probably drew an incorrect conclusion.

An anecdote from the day of sailing vessels will further illustrate our point of hidden implications. One day the captain wrote in the ship's log, "The mate was drunk today." The mate protested long and ardently but had little to reply to the captain's insistent comment, "It's the truth, isn't it?" Some days later it was the mate's turn to keep the log for the day, and he wrote in it, "The captain was sober today." To the angry captain he replied calmly, "It's the truth, isn't it?"

▲ EXERCISES

Discuss the following statements for hidden implications:

1. Have you stopped beating your wife?
2. Well, there is no reason why he should lie.
3. All pigs are created equal, but some pigs are more equal than others. (George Orwell, "The Animal Farm.")
4. More male drivers are involved in traffic accidents than women drivers.
5. If you are feeling bad, use Mrs. Springer's Tonic; you'll never feel better.
6. Those who called him a wit were half right.
7. John Hancock, the signer of the Declaration of Independence, was a graduate of Harvard College.
8. Dr. Smith, I never heard you give a better lecture.

Problems just for fun

In any group of six persons, there will be three persons, each two of whom are related, or there will be three persons, no two of whom are related. Can you prove it?

4 *Argument*

The meat of logic is the combining of several statements into a valid argument, or the drawing of a valid conclusion from several given statements. The original statements are called *premises* (or in mathematics, *hypotheses*), and the resulting statement the *conclusion*. The process of passing from the premises to the conclusion is *deduction*.

Sometimes this process is very simple and sometimes it is not. If we admit that all men are liars and all liars are evil, then we conclude easily that all men are evil. We see this also from Fig. 148, where

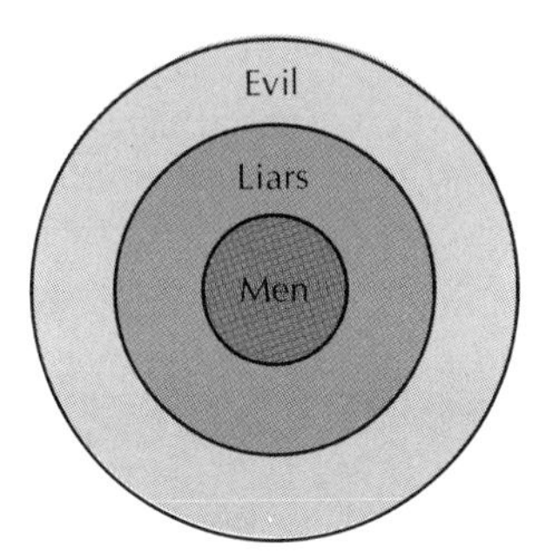

FIG. 148

the points representing men are entirely within the circle of points representing liars, and both these circles are entirely within the "evil" circle. In this simple case, the diagram seems unnecessary; but in more involved arguments, diagrams are frequently helpful.

The direct argument is the simplest type of deduction or proof. But sometimes this is not available, and we are forced to the indirect proof. This usually takes the form of a set of statements, one of which must be true. We show that all except a particular one must be false.

Thus, by this indirect way, we prove the truth of the statement. A prosecuting attorney is using this method when he argues before a jury that one of three men must have committed the murder, and two of the men have proved their presence elsewhere at the time of the murder.

The indirect argument is valid, but we must be certain that all possible alternatives have been listed and examined. If we are sitting on the jury mentioned, we must be very certain that the prosecuting attorney has proved his point that one of three men must have committed the murder. Once this is established, his conclusion that the third man is guilty is valid. But if there is a fourth man who might have committed the murder, the argument is not sound. In everyday life, indirect arguments are subject to doubt, because not all alternatives are considered. But in mathematics and many other logical inquiries, indirect argument is often necessary.

A special case of indirect argument is known under the name *reductio ad absurdum.* For example, suppose we had two statements, "A is true" and "B is true," and we wished to prove "C is true." We might proceed: Let us assume that C is false and consider together the three statements "A is true," "B is true," "C is false." If we could show that this produces a contradiction, then we should know that "C is false" led to a contradiction and thus that C must be true. This is the method of *reductio ad absurdum,* which finds frequent use in mathematical proof.

A simple example is the proof that no largest integer exists. Let us assume that this is false and that a largest integer exists. Let us call it N and consider the quantity $N + 1$. It is an integer, for any integer plus 1 is an integer; and obviously it is greater than N. Then our assumption that a largest integer exists has led to a contradiction. Thus we have proved that no largest integer exists by a *reductio ad absurdum* argument.

A simple argument consists of three parts, the premises, the conclusion, and the deduction. The premises may be true or false. The conclusion may be true or false. The deduction may be valid or invalid. Sometimes we are inclined to believe that the premises must be true because we know that the conclusion is true. But this does not follow. In fact, from a false premise, with perfectly valid reasoning, we may obtain either a true or a false conclusion. We write the following table of possibilities, where T means true, F false, V valid, and IV invalid:

Premises	T	F	F	T	T	F	F
Deductions	V	V	V	IV	IV	IV	IV
Conclusions	T	T	F	T	F	T	F

We notice that this table has every possible combination except premise T, deduction V, and conclusion F. This means that we cannot obtain a false conclusion from true premises and valid reasoning. We can give examples of all seven combinations in the table. At the end of the Junior Prom when the students are tired from dancing, we might have the following argument where one of the premises is false, the reasoning valid, and the conclusion true: The students have been studying, and whenever they study they become tired; hence they are tired.

We may conclude just two possibilities from our table:

1. If the premises are true and the reasoning is valid, our conclusion is true.
2. If the conclusion is false, either a premise is false, or the reasoning is not valid, or both.

The fact that a false premise with valid reasoning may give a true conclusion bothers students. Our instinct tells us that a false premise ought to give a false conclusion. Not so! The statement "If A then B" is true unless A is true and B false. Hence if A is false, the statement is true regardless of what B is. If A is false, then both "If A then B" and "If A then not B" are true statements. We sometimes state the situation by saying, "From a false premise we can prove anything." Consider the statement: "If the moon is made of cheese, then the earth is flat." The spaceship Surveyor showed that the premise is false. Hence the statement is a true one.

▲ EXERCISES

Discuss the arguments used in the first eleven exercises.

1. Whenever it is going to rain, my rheumatism bothers me. My rheumatism bothers me. Hence it is going to rain.
2. Whenever it is going to rain, my rheumatism bothers me. My rheumatism does not bother me. Hence it is not going to rain.
3. He is a citizen of Chicago. Every citizen of Chicago is a citizen of Illinois. Every citizen of Illinois is a citizen of the United States. Hence he is a citizen of the United States.
4. Patchpants is the bravest of tailors. All tailors are men. Hence Patchpants is the bravest of men.
5. "If anyone has failed to study hard, he'll flunk this quiz." John studied hard. He will pass.
6. If Roosevelt is elected, grass will grow in the streets. (Campaign claim, 1932 election.) Grass did not grow in the streets. Hence Roosevelt was not elected.

7. I know he doesn't love me, for men never tell the truth to a girl, and he told me he loves me.
8. Bigamy is a crime. The Mormon religion permitted bigamy. Hence the Mormon religion permitted crime.
9. If a Republican is elected president, prices will rise. A Democrat is elected. Hence prices will not rise.
10. No state can exist unless its citizens exist. Hence no state can survive for two centuries.
11. A good man does not inflict pain. But a dentist inflicts pain. Hence a dentist is not a good man.
12. Find an advertisement in a newspaper or magazine giving a nonvalid argument.

Mary, Jane, and Violet belong to the same history class. If Mary is present, Jane and Violet are either both present or both absent. If Violet is present, Mary is present. Name all the possibilities under which Jane is absent.

Truth Tables

In an earlier section we studied arguments by means of circle diagrams. We wish to consider another method of analyzing arguments called *truth tables.* It is convenient to use letters to denote statements, such as

p: All men are mortal.
q: Some students will pass this course.

We denote by $\sim p$ the negative of the statement p.

$\sim p$: All men are not mortal.

or

$\sim p$: Some men are immortal.
$\sim q$: No student will pass this course.

The symbol $p \vee q$ means "p or q" or "p is true or q is true." Note that this leaves open the question of whether or not both are true. To remind us of this it may be better to express $p \vee q$ in words as "p is true or q is true or both." The symbol $p \wedge q$ means "p and q" or "p is true and q is true." The symbol $p \rightarrow q$ means "p implies q" or "*If* p is true, *then* q is true." If $p \rightarrow q$ and $q \rightarrow p$, we write $p \leftrightarrow q$ and read it "p is equivalent to q." This means p and q are both true or both false.

As a simple example of truth tables let us consider $p \wedge q$ (Table 82).

TABLE 82

p	q	$p \wedge q$
T	*T*	*T*
T	*F*	*F*
F	*T*	*F*
F	*F*	*F*

In the two left columns we have written all possibilities for p and q. In the third column we see that $p \wedge q$ has the truth value *T* only when p and q are both *T*.

Example 1. Prove $(p \rightarrow q) \leftrightarrow (q \vee \sim p)$.

TABLE 83

p	q	$\sim p$	$p \rightarrow q$	$q \vee \sim p$
T	*T*	*F*	*T*	*T*
T	*F*	*F*	*F*	*F*
F	*T*	*T*	*T*	*T*
F	*F*	*T*	*T*	*T*

In the first two columns of Table 83 we write all possibilities for p and q. For convenience we write the values of $\sim p$ in column 3. Then we fill out columns 4 and 5 with the proper values for $p \rightarrow q$ and $q \vee \sim p$. We observe in columns 4 and 5 that $p \rightarrow q$ and $q \vee \sim p$ are always both *T* or both *F*. This proves our result.

Example 2. Given $(p \vee q) \wedge \sim p$, what can we conclude about q?

TABLE 84

p	q	$\sim p$	$p \vee q$
T	T	F	T
T	F	F	T
F	T	T	T
F	F	T	F

We are given that both $\sim p$ and $p \vee q$ are true. This means that columns 3 and 4 of Table 84 must read *TT*, or we may strike out all rows except 3. In this row q is T. Under the given statement then, q is true.

Example 3. Given $p \rightarrow r$ and $q \rightarrow \sim r$ and q is true, prove p is false.

TABLE 85

p	q	r	$p \rightarrow r$	$q \rightarrow \sim r$
T	T	T	T	F
T	T	F	F	T
T	F	T	T	T
T	F	F	F	T
F	T	T	T	F
F	T	F	T	T
F	F	T	T	T
F	F	F	T	T

Since q is true we may scratch rows 3, 4, 7, and 8. From $p \rightarrow r$ we may scratch row 2. From $q \rightarrow \sim r$ we may scratch rows 1 and 5. This leaves row 6 and we see that p is false.

Example 4. Let us prove Example 3 by using a *reductio ad absurdum* argument. To do this, we assume that p is true and show that this results in a contradiction. In the above truth table, p is true scratches rows 5, 6, 7, and 8. q is true strikes out rows 3 and 4 in addition. $p \rightarrow r$ scratches row 2. This leaves row 1 which says $q \rightarrow \sim r$ is false. But we are given that this is true.

▲ EXERCISES

Prove using truth tables:

1. $[(q \to r) \wedge \sim r] \to \sim q$
2. $[(p \to q) \wedge (q \to r)] \to (p \to r)$
3. $p \wedge \sim p$ is false
4. $\sim(\sim p) \leftrightarrow p$
5. $(p \to q) \leftrightarrow (\sim q \to \sim p)$
6. $$\begin{array}{l} q \to r \\ \sim p \to \sim r \\ \hline \sim p \to \sim q \end{array}$$
7. $$\begin{array}{l} r \to \sim p \\ q \to p \\ r \\ \hline \sim q \end{array}$$
8. $$\begin{array}{l} p \vee q \\ p \to r \\ \sim p \to \sim q \\ \hline q \to r \end{array}$$
9. Prove Exercise 7 by *reductio ad absurdum.*
10. Prove Exercise 6 by circle diagrams.
11. Prove Exercise 2 by circle diagrams.
12. Prove Exercise 8 by *reductio ad absurdum.*

Problems just for fun

We have three boxes—one containing two red balls, one two blue balls, and the third one red and one blue ball. The boxes are labeled *BB*, *RR*, and *RB* but someone has mixed the labels and all three are incorrect. We wish to determine the contents of each box by drawing one ball from a box without looking inside. How many such samplings will be required to solve the problem?

6 *Paradoxes*

The dictionary defines paradox as "a statement essentially absurd and false." The paradox has had a long and interesting history, for man has always been intrigued with the dilemma of a seemingly self-contradictory statement.

As a simple example, suppose I say, "I am now lying." Let us consider this statement true. Then I am lying when I say it and thus I must not be lying. Hence my statement is false. On the other hand, let us consider the original statement false. Then I am stating a truth when I say, "I am now lying," and the statement is true. Here we have the typical paradox. Our statement gets us into difficulties whether we consider it true or false.

Problems just for fun

Three wise men were tested for their reasoning skill by being blindfolded and having a finger rubbed on their foreheads, after being told that one or more would have smudges on their foreheads. Actually, all three were given smudges. They were told to tap once if they saw one or more spots on removing the blindfold and to tap twice if they decided that there was a spot on their own forehead. When the blindfolds were removed, all three tapped once. After a pause, one man tapped twice. What was his reasoning?

▲ EXERCISES

Analyze the following paradoxes and dilemmas:

1. The barber says, "I shave all men who do not shave themselves." Who shaves the barber?
2. Every rule has an exception.
3. Everything has a cause other than itself. God is the first cause.
4. Let us consider the class of all classes each of which does not contain itself as a member. Is this class a member of itself or not?
5. Protagorus gave his pupil Euathlus legal training, guaranteeing his train-

ing by the agreement that he should pay for it if and only if he won his first court case. When Euathlus failed to practice law at all, Protagorus brought suit against him. Protagoras argued before the court, "If I win this case, Euathlus must pay me by order of the court. If I lose it, he must pay me by the terms of our agreement. But I must win or lose this case; in either case he must pay me." But Euathlus showed he had profited by his training and argued in reply, "If I lose this case, the terms of our agreement say I do not have to pay, for I will have lost my first case. On the other hand, if I win, the court will decree that I do not have to pay. In either case I do not have to pay."

6. Little Johnny asked his father if God could do everything. Father says, "Of course." "Then can He make a rock so big that He can't roll it?" asked Johnny.

Appendix

TABLE I. POWERS AND ROOTS

No.	Square	Square root	Cube	Cube root	No.	Square	Square root	Cube	Cube root
1	1	1.000	1	1.000	**51**	2,601	7.141	132,651	3.708
2	4	1.414	8	1.260	**52**	2,704	7.211	140,608	3.733
3	9	1.732	27	1.442	**53**	2,809	7.280	148,877	3.756
4	16	2.000	64	1.587	**54**	2,916	7.348	157,464	3.780
5	25	2.236	125	1.710	**55**	3,025	7.416	166,375	3.803
6	36	2.449	216	1.817	**56**	3,136	7.483	175,616	3.826
7	49	2.646	343	1.913	**57**	3,249	7.550	185,193	3.849
8	64	2.828	512	2.000	**58**	3,364	7.616	195,112	3.871
9	81	3.000	729	2.080	**59**	3,481	7.681	205,379	3.893
10	100	3.162	1,000	2.154	**60**	3,600	7.746	216,000	3.915
11	121	3.317	1,331	2.224	**61**	3,721	7.810	226,981	3.936
12	144	3.464	1,728	2.289	**62**	3,844	7.874	238,328	3.958
13	169	3.606	2,197	2.351	**63**	3,969	7.937	250,047	3.979
14	196	3.742	2,744	2.410	**64**	4,096	8.000	262,144	4.000
15	225	3.873	3,375	2.466	**65**	4,225	8.062	274,625	4.021
16	256	4.000	4,096	2.520	**66**	4,356	8.124	287,496	4.041
17	289	4.123	4,913	2.571	**67**	4,489	8.185	300,763	4.062
18	324	4.243	5,832	2.621	**68**	4,624	8.246	314,432	4.082
19	361	4.359	6,859	2.668	**69**	4,761	8.307	328,509	4.102
20	400	4.472	8,000	2.714	**70**	4,900	8.367	343,000	4.121
21	441	4.583	9,261	2.759	**71**	5,041	8.426	357,911	4.141
22	484	4.690	10,648	2.802	**72**	5,184	8.485	373,248	4.160
23	529	4.796	12,167	2.844	**73**	5,329	8.544	389,017	4.179
24	576	4.899	13,824	2.884	**74**	5,476	8.602	405,224	4.198
25	625	5.000	15,625	2.924	**75**	5,625	8.660	421,875	4.217
26	676	5.099	17,576	2.962	**76**	5,776	8.718	438,976	4.236
27	729	5.196	19,683	3.000	**77**	5,929	8.775	456,533	4.254
28	784	5.292	21,952	3.037	**78**	6,084	8.832	474,552	4.273
29	841	5.385	24,389	3.072	**79**	6,241	8.888	493,039	4.291
30	900	5.477	27,000	3.107	**80**	6,400	8.944	512,000	4.309
31	961	5.568	29,791	3.141	**81**	6,561	9.000	531,441	4.327
32	1,024	5.657	32,768	3.175	**82**	6,724	9.055	551,368	4.344
33	1,089	5.745	35,937	3.208	**83**	6,889	9.110	571,787	4.362
34	1,156	5.831	39,304	3.240	**84**	7,056	9.165	592,704	4.380
35	1,225	5.916	42,875	3.271	**85**	7,225	9.220	614,125	4.397
36	1,296	6.000	46,656	3.302	**86**	7,396	9.274	636,056	4.414
37	1,369	6.083	50,653	3.332	**87**	7,569	9.327	658,503	4.431
38	1,444	6.164	54,872	3.362	**88**	7,744	9.381	681,472	4.448
39	1,521	6.245	59,319	3.391	**89**	7,921	9.434	704,969	4.465
40	1,600	6.325	64,000	3.420	**90**	8,100	9.487	729,000	4.481
41	1,681	6.403	68,921	3.448	**91**	8,281	9.539	753,571	4.498
42	1,764	6.481	74,088	3.476	**92**	8,464	9.592	778,688	4.514
43	1,849	6.557	79,507	3.503	**93**	8,649	9.644	804,357	4.531
44	1,936	6.633	85,184	3.530	**94**	8,836	9.695	830,584	4.547
45	2,025	6.708	91,125	3.557	**95**	9,025	9.747	857,375	4.563
46	2,116	6.782	97,336	3.583	**96**	9,216	9.798	884,736	4.579
47	2,209	6.856	103,823	3.609	**97**	9,409	9.849	912,673	4.595
48	2,304	6.928	110,592	3.634	**98**	9,604	9.899	941,192	4.610
49	2,401	7.000	117,649	3.659	**99**	9,801	9.950	970,299	4.626
50	2,500	7.071	125,000	3.684	**100**	10,000	10.000	1,000,000	4.642

TABLE II. FOUR-PLACE TRIGONOMETRIC FUNCTIONS

Degrees	*Sine*	*Tangent*	*Cotangent*	*Cosine*	
0°00′	.0000	.0000	......	1.0000	90°00′
10	.0029	.0029	343.77	1.0000	50
20	.0058	.0058	171.89	1.0000	40
30	.0087	.0087	114.59	1.0000	30
40	.0116	.0116	85.940	.9999	20
50	.0145	.0145	68.750	.9999	10
1°00′	.0175	.0175	57.290	.9998	89°00′
10	.0204	.0204	49.104	.9998	50
20	.0233	.0233	42.964	.9997	40
30	.0262	.0262	38.188	.9997	30
40	.0291	.0291	34.368	.9996	20
50	.0320	.0320	31.242	.9995	10
2°00′	.0349	.0349	28.636	.9994	88°00′
10	.0378	.0378	26.432	.9993	50
20	.0407	.0407	24.542	.9992	40
30	.0436	.0437	22.904	.9990	30
40	.0465	.0466	21.470	.9989	20
50	.0494	.0495	20.206	.9988	10
3°00′	.0523	.0524	19.081	.9986	87°00′
10	.0552	.0553	18.075	.9985	50
20	.0581	.0582	17.169	.9983	40
30	.0610	.0612	16.350	.9981	30
40	.0640	.0641	15.605	.9980	20
50	.0669	.0670	14.924	.9978	10
4°00′	.0698	.0699	14.301	.9976	86°00′
10	.0727	.0729	13.727	.9974	50
20	.0756	.0758	13.197	.9971	40
30	.0785	.0787	12.706	.9969	30
40	.0814	.0816	12.251	.9967	20
50	.0843	.0846	11.826	.9964	10
5°00′	.0872	.0875	11.430	.9962	85°00′
10	.0901	.0904	11.059	.9959	50
20	.0929	.0934	10.712	.9957	40
30	.0958	.0963	10.385	.9954	30
40	.0987	.0992	10.078	.9951	20
50	.1016	.1022	9.7882	.9948	10
6°00′	.1045	.1051	9.5144	.9945	84°00′
10	.1074	.1080	9.2553	.9942	50
20	.1103	.1110	9.0098	.9939	40
30	.1132	.1139	8.7769	.9936	30
40	.1161	.1169	8.5555	.9932	20
50	.1190	.1198	8.3450	.9929	10
7°00′	.1219	.1228	8.1443	.9925	83°00′
10	.1248	.1257	7.9530	.9922	50
20	.1276	.1287	7.7704	.9918	40
30	.1305	.1317	7.5958	.9914	30
40	.1334	.1346	7.4287	.9911	20
50	.1363	.1376	7.2687	.9907	10
8°00′	.1392	.1405	7.1154	.9903	82°00′
	Cosine	*Cotangent*	*Tangent*	*Sine*	*Degrees*

TABLE II. FOUR-PLACE TRIGONOMETRIC FUNCTIONS *(Cont'd)*

Degrees	*Sine*	*Tangent*	*Cotangent*	*Cosine*	
8°00′	.1392	.1405	7.1154	.9903	82°00′
10	.1421	.1435	6.9682	.9899	50
20	.1449	.1465	6.8269	.9894	40
30	.1478	.1495	6.6912	.9890	30
40	.1507	.1524	6.5606	.9886	20
50	.1536	.1554	6.4348	.9881	10
9°00′	.1564	.1584	6.3138	.9877	81°00′
10	.1593	.1614	6.1970	.9872	50
20	.1622	.1644	6.0844	.9868	40
30	.1650	.1673	5.9758	.9863	30
40	.1679	.1703	5.8708	.9858	20
50	.1708	.1733	5.7694	.9853	10
10°00′	.1736	.1763	5.6713	.9848	80°00′
10	.1765	.1793	5.5764	.9843	50
20	.1794	.1823	5.4845	.9838	40
30	.1822	.1853	5.3955	.9833	30
40	.1851	.1883	5.3093	.9827	20
50	.1880	.1914	5.2257	.9822	10
11°00′	.1908	.1944	5.1446	.9816	79°00′
10	.1937	.1974	5.0658	.9811	50
20	.1965	.2004	4.9894	.9805	40
30	.1994	.2035	4.9152	.9799	30
40	.2022	.2065	4.8430	.9793	20
50	.2051	.2095	4.7729	.9787	10
12°00′	.2079	.2126	4.7046	.9781	78°00′
10	.2108	.2156	4.6382	.9775	50
20	.2136	.2186	4.5736	.9769	40
30	.2164	.2217	4.5107	.9763	30
40	.2193	.2247	4.4494	.9757	20
50	.2221	.2278	4.3897	.9750	10
13°00′	.2250	.2309	4.3315	.9744	77°00′
10	.2278	.2339	4.2747	.9737	50
20	.2306	.2370	4.2193	.9730	40
30	.2334	.2401	4.1653	.9724	30
40	.2363	.2432	4.1126	.9717	20
50	.2391	.2462	4.0611	.9710	10
14°00′	.2419	.2493	4.0108	.9703	76°00′
10	.2447	.2524	3.9617	.9696	50
20	.2476	.2555	3.9136	.9689	40
30	.2504	.2586	3.8667	.9681	30
40	.2532	.2617	3.8208	.9674	20
50	.2560	.2648	3.7760	.9667	10
15°00′	.2588	.2679	3.7321	.9659	75°00′
10	.2616	.2711	3.6891	.9652	50
20	.2644	.2742	3.6470	.9644	40
30	.2672	.2773	3.6059	.9636	30
40	.2700	.2805	3.5656	.9628	20
50	.2728	.2836	3.5261	.9621	10
16°00′	.2756	.2867	3.4874	.9613	74°00′
	Cosine	*Cotangent*	*Tangent*	*Sine*	*Degrees*

TABLE II. FOUR-PLACE TRIGONOMETRIC FUNCTIONS *(Cont'd)*

Degrees	*Sine*	*Tangent*	*Cotangent*	*Cosine*	
16°00′	.2756	.2867	3.4874	.9613	74°00′
10	.2784	.2899	3.4495	.9605	50
20	.2812	.2931	3.4124	.9596	40
30	.2840	.2962	3.3759	.9588	30
40	.2868	.2994	3.3402	.9580	20
50	.2896	.3026	3.3052	.9572	10
17°00′	.2924	.3057	3.2709	.9563	73°00′
10	.2952	.3089	3.2371	.9555	50
20	.2979	.3121	3.2041	.9546	40
30	.3007	.3153	3.1716	.9537	30
40	.3035	.3185	3.1397	.9528	20
50	.3062	.3217	3.1084	.9520	10
18°00′	.3090	.3249	3.0777	.9511	72°00′
10	.3118	.3281	3.0475	.9502	50
20	.3145	.3314	3.0178	.9492	40
30	.3173	.3346	2.9887	.9483	30
40	.3201	.3378	2.9600	.9474	20
50	.3228	.3411	2.9319	.9465	10
19°00′	.3256	.3443	2.9042	.9455	71°00′
10	.3283	.3476	2.8770	.9446	50
20	.3311	.3508	2.8502	.9436	40
30	.3338	.3541	2.8239	.9426	30
40	.3365	.3574	2.7980	.9417	20
50	.3393	.3607	2.7725	.9407	10
20°00′	.3420	.3640	2.7475	.9397	70°00′
10	.3448	.3673	2.7228	.9387	50
20	.3475	.3706	2.6985	.9377	40
30	.3502	.3739	2.6746	.9367	30
40	.3529	.3772	2.6511	.9356	20
50	.3557	.3805	2.6279	.9346	10
21°00′	.3584	.3839	2.6051	.9336	69°00′
10	.3611	.3872	2.5826	.9325	50
20	.3638	.3906	2.5605	.9315	40
30	.3665	.3939	2.5386	.9304	30
40	.3692	.3973	2.5172	.9293	20
50	.3719	.4006	2.4960	.9283	10
22°00′	.3746	.4040	2.4751	.9272	68°00′
10	.3773	.4074	2.4545	.9261	50
20	.3800	.4108	2.4342	.9250	40
30	.3827	.4142	2.4142	.9239	30
40	.3854	.4176	2.3945	.9228	20
50	.3881	.4210	2.3750	.9216	10
23°00′	.3907	.4245	2.3559	.9205	67°00′
10	.3934	.4279	2.3369	.9194	50
20	.3961	.4314	2.3183	.9182	40
30	.3987	.4348	2.2998	.9171	30
40	.4014	.4383	2.2817	.9159	20
50	.4041	.4417	2.2637	.9147	10
24°00′	.4067	.4452	2.2460	.9135	66°00′
	Cosine	*Cotangent*	*Tangent*	*Sine*	*Degrees*

TABLE II. FOUR-PLACE TRIGONOMETRIC FUNCTIONS *(Cont'd)*

Degrees	*Sine*	*Tangent*	*Cotangent*	*Cosine*	
24°00′	.4067	.4452	2.2460	.9135	66°00′
10	.4094	.4487	2.2286	.9124	50
20	.4120	.4522	2.2113	.9112	40
30	.4147	.4557	2.1943	.9100	30
40	.4173	.4592	2.1775	.9088	20
50	.4200	.4628	2.1609	.9075	10
25°00′	.4226	.4663	2.1445	.9063	65°00′
10	.4253	.4699	2.1283	.9051	50
20	.4279	.4734	2.1123	.9038	40
30	.4305	.4770	2.0965	.9026	30
40	.4331	.4806	2.0809	.9013	20
50	.4358	.4841	2.0655	.9001	10
26°00′	.4384	.4877	2.0503	.8988	64°00′
10	.4410	.4913	2.0353	.8975	50
20	.4436	.4950	2.0204	.8962	40
30	.4462	.4986	2.0057	.8949	30
40	.4488	.5022	1.9912	.8936	20
50	.4514	.5059	1.9768	.8923	10
27°00′	.4540	.5095	1.9626	.8910	63°00′
10	.4566	.5132	1.9486	.8897	50
20	.4592	.5169	1.9347	.8884	40
30	.4617	.5206	1.9210	.8870	30
40	.4643	.5243	1.9074	.8857	20
50	.4669	.5280	1.8940	.8843	10
28°00′	.4695	.5317	1.8807	.8829	62°00′
10	.4720	.5354	1.8676	.8816	50
20	.4746	.5392	1.8546	.8802	40
30	.4772	.5430	1.8418	.8788	30
40	.4797	.5467	1.8291	.8774	20
50	.4823	.5505	1.8165	.8760	10
29°00′	.4848	.5543	1.8040	.8746	61°00′
10	.4874	.5581	1.7917	.8732	50
20	.4899	.5619	1.7796	.8718	40
30	.4924	.5658	1.7675	.8704	30
40	.4950	.5696	1.7556	.8689	20
50	.4975	.5735	1.7437	.8675	10
30°00′	.5000	.5774	1.7321	.8660	60°00′
10	.5025	.5812	1.7205	.8646	50
20	.5050	.5851	1.7090	.8631	40
30	.5075	.5890	1.6977	.8616	30
40	.5100	.5930	1.6864	.8601	20
50	.5125	.5969	1.6753	.8587	10
31°00′	.5150	.6009	1.6643	.8572	59°00′
10	.5175	.6048	1.6534	.8557	50
20	.5200	.6088	1.6426	.8542	40
30	.5225	.6128	1.6319	.8526	30
40	.5250	.6168	1.6212	.8511	20
50	.5275	.6208	1.6107	.8496	10
32°00′	.5299	.6249	1.6003	.8480	58°00′
	Cosine	*Cotangent*	*Tangent*	*Sine*	*Degrees*

TABLE II. FOUR-PLACE TRIGONOMETRIC FUNCTIONS *(Cont'd)*

Degrees	*Sine*	*Tangent*	*Cotangent*	*Cosine*	
32°00′	.5299	.6249	1.6003	.8480	58°00′
10	.5324	.6289	1.5900	.8465	50
20	.5348	.6330	1.5798	.8450	40
30	.5373	.6371	1.5697	.8434	30
40	.5398	.6412	1.5597	.8418	20
50	.5422	.6453	1.5497	.8403	10
33°00′	.5446	.6494	1.5399	.8387	57°00′
10	.5471	.6536	1.5301	.8371	50
20	.5495	.6577	1.5204	.8355	40
30	.5519	.6619	1.5108	.8339	30
40	.5544	.6661	1.5013	.8323	20
50	.5568	.6703	1.4919	.8307	10
34°00′	.5592	.6745	1.4826	.8290	56°00′
10	.5616	.6787	1.4733	.8274	50
20	.5640	.6830	1.4641	.8258	40
30	.5664	.6873	1.4550	.8241	30
40	.5688	.6916	1.4460	.8225	20
50	.5712	.6959	1.4370	.8208	10
35°00′	.5736	.7002	1.4281	.8192	55°00′
10	.5760	.7046	1.4193	.8175	50
20	.5783	.7089	1.4106	.8158	40
30	.5807	.7133	1.4019	.8141	30
40	.5831	.7177	1.3934	.8124	20
50	.5854	.7221	1.3848	.8107	10
36°00′	.5878	.7265	1.3764	.8090	54°00′
10	.5901	.7310	1.3680	.8073	50
20	.5925	.7355	1.3597	.8056	40
30	.5948	.7400	1.3514	.8039	30
40	.5972	.7445	1.3432	.8021	20
50	.5995	.7490	1.3351	.8004	10
37°00′	.6018	.7536	1.3270	.7986	53°00′
10	.6041	.7581	1.3190	.7969	50
20	.6065	.7627	1.3111	.7951	40
30	.6088	.7673	1.3032	.7934	30
40	.6111	.7720	1.2954	.7916	20
50	.6134	.7766	1.2876	.7898	10
38°00′	.6157	.7813	1.2799	.7880	52°00′
10	.6180	.7860	1.2723	.7862	50
20	.6202	.7907	1.2647	.7844	40
30	.6225	.7954	1.2572	.7826	30
40	.6248	.8002	1.2497	.7808	20
50	.6271	.8050	1.2423	.7790	10
39°00′	.6293	.8098	1.2349	.7771	51°00′
10	.6316	.8146	1.2276	.7753	50
20	.6338	.8195	1.2203	.7735	40
30	.6361	.8243	1.2131	.7716	30
40	.6383	.8292	1.2059	.7698	20
50	.6406	.8342	1.1988	.7679	10
40°00′	.6428	.8391	1.1918	.7660	50°00′
	Cosine	*Cotangent*	*Tangent*	*Sine*	*Degrees*

TABLE II. FOUR-PLACE TRIGONOMETRIC FUNCTIONS *(Cont'd)*

Degrees	*Sine*	*Tangent*	*Cotangent*	*Cosine*	
40°00′	.6428	.8391	1.1918	.7660	50°00′
10	.6450	.8441	1.1847	.7642	50
20	.6472	.8491	1.1778	.7623	40
30	.6494	.8541	1.1708	.7604	30
40	.6517	.8591	1.1640	.7585	20
50	.6539	.8642	1.1571	.7566	10
41°00′	.6561	.8693	1.1504	.7547	49°00′
10	.6583	.8744	1.1436	.7528	50
20	.6604	.8796	1.1369	.7509	40
30	.6626	.8847	1.1303	.7490	30
40	.6648	.8899	1.1237	.7470	20
50	.6670	.8952	1.1171	.7451	10
42°00′	.6691	.9004	1.1106	.7431	48°00′
10	.6713	.9057	1.1041	.7412	50
20	.6734	.9110	1.0977	.7392	40
30	.6756	.9163	1.0913	.7373	30
40	.6777	.9217	1.0850	.7353	20
50	.6799	.9271	1.0786	.7333	10
43°00′	.6820	.9325	1.0724	.7314	47°00′
10	.6841	.9380	1.0661	.7294	50
20	.6862	.9435	1.0599	.7274	40
30	.6884	.9490	1.0538	.7254	30
40	.6905	.9545	1.0477	.7234	20
50	.6926	.9601	1.0416	.7214	10
44°00′	.6947	.9657	1.0355	.7193	46°00′
10	.6967	.9713	1.0295	.7173	50
20	.6988	.9770	1.0235	.7153	40
30	.7009	.9827	1.0176	.7133	30
40	.7030	.9884	1.0117	.7112	20
50	.7050	.9942	1.0058	.7092	10
45°00′	.7071	1.0000	1.0000	.7071	45°00′
	Cosine	*Cotangent*	*Tangent*	*Sine*	*Degrees*

TABLE III. COMPOUND INTEREST: $(1+r)^n$
Amount of $1 principal at compound interest after n periods

n	$1^1/_2\%$	2%	$2^1/_2\%$	3%	4%	5%	6%
1	1.0150	1.0200	1.0250	1.0300	1.0400	1.0500	1.0600
2	1.0302	1.0404	1.0506	1.0609	1.0816	1.1025	1.1236
3	1.0457	1.0612	1.0769	1.0927	1.1249	1.1576	1.1910
4	1.0614	1.0824	1.1038	1.1255	1.1699	1.2155	1.2625
5	1.0773	1.1041	1.1314	1.1593	1.2167	1.2763	1.3382
6	1.0934	1.1262	1.1597	1.1941	1.2653	1.3401	1.4185
7	1.1098	1.1487	1.1887	1.2299	1.3159	1.4071	1.5036
8	1.1265	1.1717	1.2184	1.2668	1.3686	1.4775	1.5938
9	1.1434	1.1951	1.2489	1.3048	1.4233	1.5513	1.6895
10	1.1605	1.2190	1.2801	1.3439	1.4802	1.6289	1.7908
11	1.1779	1.2434	1.3121	1.3842	1.5395	1.7103	1.8983
12	1.1956	1.2682	1.3449	1.4258	1.6010	1.7959	2.0122
13	1.2136	1.2936	1.3785	1.4685	1.6651	1.8856	2.1329
14	1.2318	1.3195	1.4130	1.5126	1.7317	1.9799	2.2609
15	1.2502	1.3459	1.4483	1.5580	1.8009	2.0789	2.3966
16	1.2690	1.3728	1.4845	1.6047	1.8730	2.1829	2.5404
17	1.2880	1.4002	1.5216	1.6528	1.9479	2.2920	2.6928
18	1.3073	1.4282	1.5597	1.7024	2.0258	2.4066	2.8543
19	1.3270	1.4568	1.5987	1.7535	2.1068	2.5270	3.0256
20	1.3469	1.4859	1.6386	1.8061	2.1911	2.6533	3.2071
21	1.3671	1.5157	1.6796	1.8603	2.2788	2.7860	3.3996
22	1.3876	1.5460	1.7216	1.9161	2.3699	2.9253	3.6035
23	1.4084	1.5769	1.7646	1.9736	2.4647	3.0715	3.8197
24	1.4295	1.6084	1.8087	2.0328	2.5633	3.2251	4.0489
25	1.4509	1.6406	1.8539	2.0938	2.6658	3.3864	4.2919
26	1.4727	1.6734	1.9003	2.1566	2.7725	3.5557	4.5494
27	1.4948	1.7069	1.9478	2.2213	2.8834	3.7335	4.8223
28	1.5172	1.7410	1.9965	2.2879	2.9987	3.9201	5.1117
29	1.5400	1.7758	2.0464	2.3566	3.1187	4.1161	5.4184
30	1.5631	1.8114	2.0976	2.4273	3.2434	4.3219	5.7435
31	1.5865	1.8476	2.1500	2.5001	3.3731	4.5380	6.0881
32	1.6103	1.8845	2.2038	2.5751	3.5081	4.7649	6.4534
33	1.6345	1.9222	2.2589	2.6523	3.6484	5.0032	6.8406
34	1.6590	1.9607	2.3153	2.7319	3.7943	5.2533	7.2510
35	1.6839	1.9999	2.3732	2.8139	3.9461	5.5160	7.6861
36	1.7091	2.0399	2.4325	2.8983	4.1039	5.7918	8.1473
37	1.7348	2.0807	2.4933	2.9852	4.2681	6.0814	8.6361
38	1.7608	2.1223	2.5557	3.0748	4.4388	6.3855	9.1543
39	1.7872	2.1647	2.6196	3.1670	4.6164	6.7048	9.7035
40	1.8140	2.2080	2.6851	3.2620	4.8010	7.0400	10.2857
41	1.8412	2.2522	2.7522	3.3599	4.9931	7.3920	10.9029
42	1.8688	2.2972	2.8210	3.4607	5.1928	7.7616	11.5570
43	1.8969	2.3432	2.8915	3.5645	5.4005	8.1497	12.2505
44	1.9253	2.3901	2.9638	3.6715	5.6165	8.5572	12.9855
45	1.9542	2.4379	3.0379	3.7816	5.8412	8.9850	13.7646
46	1.9835	2.4866	3.1139	3.8950	6.0748	9.4343	14.5905
47	2.0133	2.5363	3.1917	4.0119	6.3178	9.9060	15.4659
48	2.0435	2.5871	3.2715	4.1323	6.5705	10.4013	16.3939
49	2.0741	2.6388	3.3533	4.2562	6.8333	10.9213	17.3775
50	2.1052	2.6916	3.4371	4.3839	7.1067	11.4674	18.4202

TABLE IV. COMPOUND DISCOUNT: $1/(1+r)^n$
Present value of $1 due at the end of n periods

n	$1\frac{1}{2}$%	2%	$2\frac{1}{2}$%	3%	4%	5%	6%
1	.985 22	.980 39	.97561	.97087	.96154	.95238	.94340
2	.970 66	.961 17	.95181	.94260	.92456	.90703	.89000
3	.956 32	.942 32	.92860	.91514	.88900	.86384	.83962
4	.942 18	.923 85	.90595	.88849	.85480	.82270	.79209
5	.928 26	.905 73	.88385	.86261	.82193	.78353	.74726
6	.914 54	.887 97	.86230	.83748	.79031	.74622	.70496
7	.901 03	.870 56	.84127	.81309	.75992	.71068	.66506
8	.887 71	.853 49	.82075	.78941	.73069	.67684	.62741
9	.874 59	.836 76	.80073	.76642	.70259	.64461	.59190
10	.861 67	.820 35	.78120	.74409	.67556	.61391	.55839
11	.848 93	.804 26	.76214	.72242	.64958	.58468	.52679
12	.836 39	.788 49	.74356	.70138	.62460	.55684	.49697
13	.824 03	.773 03	.72542	.68095	.60057	.53032	.46884
14	.811 85	.757 88	.70773	.66112	.57748	.50507	.44230
15	.799 85	.743 01	.69047	.64186	.55526	.48102	.41727
16	.788 03	.728 45	.67362	.62317	.53391	.45811	.39365
17	.776 39	.714 16	.65720	.60502	.51337	.43630	.37136
18	.764 91	.700 16	.64117	.58739	.49363	.41552	.35034
19	.753 61	.686 43	.62553	.57029	.47464	.39573	.33051
20	.742 47	.672 97	.61027	.55368	.45639	.37689	.31180
21	.731 50	.659 78	.59539	.53755	.43883	.35894	.29416
22	.720 69	.646 84	.58086	.52189	.42196	.34185	.27751
23	.710 04	.634 16	.56670	.50669	.40573	.32557	.26180
24	.699 54	.621 72	.55288	.49193	.39012	.31007	.24698
25	.689 21	.609 53	.53939	.47761	.37512	.29530	.23300
26	.679 02	.597 58	.52623	.46369	.36069	.28124	.21981
27	.668 99	.585 86	.51340	.45019	.34682	.26785	.20737
28	.659 10	.574 37	.50088	.43708	.33348	.25509	.19563
29	.649 36	.563 11	.48866	.42435	.32065	.24295	.18456
30	.639 76	.552 07	.47674	.41199	.30832	.23138	.17411
31	.630 31	.541 25	.46511	.39999	.29646	.22036	.16425
32	.620 99	.530 63	.45377	.38834	.28506	.20987	.15496
33	.611 82	.520 23	.44270	.37703	.27409	.19987	.14619
34	.602 77	.510 03	.43191	.36604	.26355	.19035	.13791
35	.593 87	.500 03	.42137	.35538	.25342	.18129	.13011
36	.585 09	.490 22	.41109	.34503	.24367	.17266	.12274
37	.576 44	.480 61	.40107	.33498	.23430	.16444	.11579
38	.567 92	.471 19	.39128	.32523	.22529	.15661	.10924
39	.559 53	.461 95	.38174	.31575	.21662	.14915	.10306
40	.551 26	.452 89	.37243	.30656	.20829	.14205	.09722
41	.543 12	.444 01	.36335	.29763	.20028	.13528	.09172
42	.535 09	.435 30	.35448	.28896	.19257	.12884	.08653
43	.527 18	.426 77	.34584	.28054	.18517	.12270	.08163
44	.519 39	.418 40	.33740	.27237	.17805	.11686	.07701
45	.511 71	.410 20	.32917	.26444	.17120	.11130	.07265
46	.504 15	.402 15	.32115	.25674	.16461	.10600	.06854
47	.496 70	.394 27	.31331	.24926	.15828	.10095	.06466
48	.489 36	.386 54	.30567	.24200	.15219	.09614	.06100
49	.482 13	.378 96	.29822	.23495	.14634	.09156	.05755
50	.475 00	.371 53	.29094	.22811	.14071	.08720	.05429

TABLE V. AMOUNT OF AN ANNUITY

Amount of an annuity of $1 per period after n periods

n	$1\frac{1}{2}\%$	2%	$2\frac{1}{2}\%$	3%	4%	5%	6%
1	1.0000	1.0000	1.0000	1.0000	1.0000	1.0000	1.0000
2	2.0150	2.0200	2.0250	2.0300	2.0400	2.0500	2.0600
3	3.0452	3.0604	3.0756	3.0909	3.1216	3.1525	3.1836
4	4.0909	4.1216	4.1525	4.1836	4.2465	4.3101	4.3746
5	5.1523	5.2040	5.2563	5.3091	5.4163	5.5256	5.6371
6	6.2296	6.3081	6.3877	6.4684	6.6330	6.8019	6.9753
7	7.3230	7.4343	7.5474	7.6625	7.8983	8.1420	8.3938
8	8.4328	8.5830	8.7361	8.8923	9.2142	9.5491	9.8975
9	9.5593	9.7546	9.9545	10.1591	10.5828	11.0266	11.4913
10	10.7027	10.9497	11.2034	11.4639	12.0061	12.5779	13.1808
11	11.8633	12.1687	12.4835	12.8078	13.4864	14.2068	14.9716
12	13.0412	13.4121	13.7956	14.1920	15.0258	15.9171	16.8699
13	14.2368	14.6803	15.1404	15.6178	16.6268	17.7130	18.8821
14	15.4504	15.9739	16.5190	17.0863	18.2919	19.5986	21.0151
15	16.6821	17.2934	17.9319	18.5989	20.0236	21.5786	23.2760
16	17.9324	18.6393	19.3802	20.1569	21.8245	23.6575	25.6725
17	19.2014	20.0121	20.8647	21.7616	23.6975	25.8404	28.2129
18	20.4894	21.4123	22.3863	23.4144	25.6454	28.1324	30.9057
19	21.7967	22.8406	23.9460	25.1169	27.6712	30.5390	33.7600
20	23.1237	24.2974	25.5447	26.8704	29.7781	33.0660	36.7856
21	24.4705	25.7833	27.1833	28.6765	31.9692	35.7193	39.9927
22	25.8376	27.2990	28.8629	30.5368	34.2480	38.5052	43.3923
23	27.2251	28.8450	30.5844	32.4529	36.6179	41.4305	46.9958
24	28.6335	30.4219	32.3490	34.4265	39.0826	44.5020	50.8156
25	30.0630	32.0303	34.1578	36.4593	41.6459	47.7271	54.8645
26	31.5140	33.6709	36.0117	38.5530	44.3117	51.1135	59.1564
27	32.9867	35.3443	37.9120	40.7096	47.0842	54.6691	63.7058
28	34.4815	37.0512	39.8598	42.9309	49.9676	58.4026	68.5281
29	35.9987	38.7922	41.8563	45.2189	52.9663	62.3227	73.6398
30	37.5387	40.5681	43.9027	47.5754	56.0849	66.4388	79.0582
31	39.1018	42.3794	46.0003	50.0027	59.3283	70.7608	84.8017
32	40.6883	44.2270	48.1503	52.5028	62.7015	75.2988	90.8898
33	42.2986	46.1116	50.3540	55.0778	66.2095	80.0638	97.3432
34	43.9331	48.0338	52.6129	57.7302	69.8579	85.0670	104.1838
35	45.5921	49.9945	54.9282	60.4621	73.6522	90.3203	111.4348
36	47.2760	51.9944	57.3014	63.2759	77.5983	95.8363	119.1209
37	48.9851	54.0343	59.7339	66.1742	81.7022	101.6281	127.2681
38	50.7199	56.1149	62.2273	69.1594	85.9703	107.7095	135.9042
39	52.4807	58.2372	64.7830	72.2342	90.4091	114.0950	145.0585
40	54.2679	60.4020	67.4026	75.4013	95.0255	120.7998	154.7620
41	56.0819	62.6100	70.0876	78.6633	99.8265	127.8398	165.0477
42	57.9231	64.8622	72.8398	82.0232	104.8196	135.2318	175.9505
43	59.7920	67.1595	75.6608	85.4839	110.0124	142.9933	187.5076
44	61.6889	69.5027	78.5523	89.0484	115.4129	151.1430	199.7580
45	63.6142	71.8927	81.5161	92.7199	121.0294	159.7002	212.7435
46	65.5684	74.3306	84.5540	96.5015	126.8706	168.6852	226.5081
47	67.5519	76.8172	87.6679	100.3965	132.9454	178.1194	241.0986
48	69.5652	79.3535	90.8596	104.4084	139.2632	188.0254	256.5645
49	71.6087	81.9406	94.1311	108.5406	145.8337	198.4267	272.9584
50	73.6828	84.5794	97.4843	112.7969	152.6671	209.3480	290.3359

TABLE VI. PRESENT VALUE OF AN ANNUITY

Present value of $1 per period for n periods

n	$1\frac{1}{2}\%$	2%	$2\frac{1}{2}\%$	3%	4%	5%	6%
1	.9852	.9804	.9756	.9709	.9615	.9524	.9434
2	1.9559	1.9416	1.9274	1.9135	1.8861	1.8594	1.8334
3	2.9122	2.8839	2.8560	2.8286	2.7751	2.7232	2.6730
4	3.8544	3.8077	3.7620	3.7171	3.6299	3.5460	3.4651
5	4.7826	4.7135	4.6458	4.5797	4.4518	4.3295	4.2124
6	5.6972	5.6014	5.5081	5.4172	5.2421	5.0757	4.9173
7	6.5982	6.4720	6.3494	6.2303	6.0021	5.7864	5.5824
8	7.4859	7.3255	7.1701	7.0197	6.7327	6.4632	6.2098
9	8.3605	8.1622	7.9709	7.7861	7.4353	7.1078	6.8017
10	9.2222	8.9826	8.7521	8.5302	8.1109	7.7217	7.3601
11	10.0711	9.7868	9.5142	9.2526	8.7605	8.3064	7.8869
12	10.9075	10.5753	10.2578	9.9540	9.3851	8.8633	8.3838
13	11.7315	11.3484	10.9832	10.6350	9.9856	9.3936	8.8527
14	12.5434	12.1062	11.6909	11.2961	10.5631	9.8986	9.2950
15	13.3432	12.8493	12.3814	11.9379	11.1184	10.3797	9.7122
16	14.1313	13.5777	13.0550	12.5611	11.6523	10.8378	10.1059
17	14.9076	14.2919	13.7122	13.1661	12.1657	11.2741	10.4773
18	15.6726	14.9920	14.3534	13.7535	12.6593	11.6896	10.8276
19	16.4262	15.6785	14.9789	14.3238	13.1339	12.0853	11.1581
20	17.1686	16.3514	15.5892	14.8775	13.5903	12.4622	11.4699
21	17.9001	17.0112	16.1845	15.4150	14.0292	12.8212	11.7641
22	18.6208	17.6580	16.7654	15.9369	14.4511	13.1630	12.0416
23	19.3309	18.2922	17.3321	16.4436	14.8568	13.4886	12.3034
24	20.0304	18.9139	17.8850	16.9355	15.2470	13.7986	12.5504
25	20.7196	19.5235	18.4244	17.4131	15.6221	14.0939	12.7834
26	21.3986	20.1210	18.9506	17.8768	15.9828	14.3752	13.0032
27	22.0676	20.7069	19.4640	18.3270	16.3296	14.6430	13.2105
28	22.7267	21.2813	19.9649	18.7641	16.6631	14.8981	13.4062
29	23.3761	21.8444	20.4535	19.1885	16.9837	15.1411	13.5907
30	24.0158	22.3965	20.9303	19.6004	17.2920	15.3725	13.7648
31	24.6461	22.9377	21.3954	20.0004	17.5885	15.5928	13.9291
32	25.2671	23.4683	21.8492	20.3888	17.8736	15.8027	14.0840
33	25.8790	23.9886	22.2919	20.7658	18.1476	16.0025	14.2302
34	26.4817	24.4986	22.7238	21.1318	18.4112	16.1929	14.3681
35	27.0756	24.9986	23.1452	21.4872	18.6646	16.3742	14.4982
36	27.6607	25.4888	23.5563	21.8323	18.9083	16.5469	14.6210
37	28.2371	25.9695	23.9573	22.1672	19.1426	16.7113	14.7368
38	28.8051	26.4406	24.3486	22.4925	19.3679	16.8679	14.8460
39	29.3646	26.9026	24.7303	22.8082	19.5845	17.0170	14.9491
40	29.9158	27.3555	25.1028	23.1148	19.7928	17.1591	15.0463
41	30.4590	27.7995	25.4661	23.4124	19.9931	17.2944	15.1380
42	30.9941	28.2348	25.8206	23.7014	20.1856	17.4232	15.2245
43	31.5212	28.6616	26.1664	23.9819	20.3708	17.5459	15.3062
44	32.0406	29.0800	26.5038	24.2543	20.5488	17.6628	15.3832
45	32.5523	29.4902	26.8330	24.5187	20.7200	17.7741	15.4558
46	33.0565	29.8923	27.1542	24.7754	20.8847	17.8801	15.5244
47	33.5532	30.2866	27.4675	25.0247	21.0429	17.9810	15.5890
48	34.0426	30.6731	27.7732	25.2667	21.1951	18.0772	15.6500
49	34.5247	31.0521	28.0714	25.5017	21.3415	18.1687	15.7076
50	34.9997	31.4236	28.3623	25.7298	21.4822	18.2559	15.7619

TABLE VII. COMMON LOGARITHMS, $N = 10^x$

N	0	1	2	3	4	5	6	7	8	9
1.0	0000	0043	0086	0128	0170	0212	0253	0294	0334	0374
1.1	0414	0453	0492	0531	0569	0607	0645	0682	0719	0755
1.2	0792	0828	0864	0899	0934	0969	1004	1038	1072	1106
1.3	1139	1173	1206	1239	1271	1303	1335	1367	1399	1430
1.4	1461	1492	1523	1553	1584	1614	1644	1673	1703	1732
1.5	1761	1790	1818	1847	1875	1903	1931	1959	1987	2014
1.6	2041	2068	2095	2122	2148	2175	2201	2227	2253	2279
1.7	2304	2330	2355	2380	2405	2430	2455	2480	2504	2529
1.8	2553	2577	2601	2625	2648	2672	2695	2718	2742	2765
1.9	2788	2810	2833	2856	2878	2900	2923	2945	2967	2989
2.0	3010	3032	3054	3075	3096	3118	3139	3160	3181	3201
2.1	3222	3243	3263	3284	3304	3324	3345	3365	3385	3404
2.2	3424	3444	3464	3483	3502	3522	3541	3560	3579	3598
2.3	3617	3636	3655	3674	3692	3711	3729	3747	3766	3784
2.4	3802	3820	3838	3856	3874	3892	3909	3927	3945	3962
2.5	3979	3997	4014	4031	4048	4065	4082	4099	4116	4133
2.6	4150	4166	4183	4200	4216	4232	4249	4265	4281	4298
2.7	4314	4330	4346	4362	4378	4393	4409	4425	4440	4456
2.8	4472	4487	4502	4518	4533	4548	4564	4579	4594	4609
2.9	4624	4639	4654	4669	4683	4698	4713	4728	4742	4757
3.0	4771	4786	4800	4814	4829	4843	4857	4871	4886	4900
3.1	4914	4928	4942	4955	4969	4983	4997	5011	5024	5038
3.2	5051	5065	5079	5092	5105	5119	5132	5145	5159	5172
3.3	5185	5198	5211	5224	5237	5250	5263	5276	5289	5302
3.4	5315	5328	5340	5353	5366	5378	5391	5403	5416	5428
3.5	5441	5453	5465	5478	5490	5502	5514	5527	5539	5551
3.6	5563	5575	5587	5599	5611	5623	5635	5647	5658	5670
3.7	5682	5694	5705	5717	5729	5740	5752	5763	5775	5786
3.8	5798	5809	5821	5832	5843	5855	5866	5877	5888	5899
3.9	5911	5922	5933	5944	5955	5966	5977	5988	5999	6010
4.0	6021	6031	6042	6053	6064	6075	6085	6096	6107	6117
4.1	6128	6138	6149	6160	6170	6180	6191	6201	6212	6222
4.2	6232	6243	6253	6263	6274	6284	6294	6304	6314	6325
4.3	6335	6345	6355	6365	6375	6385	6395	6405	6415	6425
4.4	6435	6444	6454	6464	6474	6484	6493	6503	6513	6522
4.5	6532	6542	6551	6561	6571	6580	6590	6599	6609	6618
4.6	6628	6637	6646	6656	6665	6675	6684	6693	6702	6712
4.7	6721	6730	6739	6749	6758	6767	6776	6785	6794	6803
4.8	6812	6821	6830	6839	6848	6857	6866	6875	6884	6893
4.9	6902	6911	6920	6928	6937	6946	6955	6964	6972	6981
5.0	6990	6998	7007	7016	7024	7033	7042	7050	7059	7067
5.1	7076	7084	7093	7101	7110	7118	7126	7135	7143	7152
5.2	7160	7168	7177	7185	7193	7202	7210	7218	7226	7235
5.3	7243	7251	7259	7267	7275	7284	7292	7300	7308	7316
5.4	7324	7332	7340	7348	7356	7364	7372	7380	7388	7396

TABLE VII. COMMON LOGARITHMS, $N = 10^x$ *(Cont'd)*

N	0	1	2	3	4	5	6	7	8	9
5.5	7404	7412	7419	7427	7435	7443	7451	7459	7466	7474
5.6	7482	7490	7497	7505	7513	7520	7528	7536	7543	7551
5.7	7559	7566	7574	7582	7589	7597	7604	7612	7619	7627
5.8	7634	7642	7649	7657	7664	7672	7679	7686	7694	7701
5.9	7709	7716	7723	7731	7738	7745	7752	7760	7767	7774
6.0	7782	7789	7796	7803	7810	7818	7825	7832	7839	7846
6.1	7853	7860	7868	7875	7882	7889	7896	7903	7910	7917
6.2	7924	7931	7938	7945	7952	7959	7966	7973	7980	7987
6.3	7993	8000	8007	8014	8021	8028	8035	8041	8048	8055
6.4	8062	8069	8075	8082	8089	8096	8102	8109	8116	8122
6.5	8129	8136	8142	8149	8156	8162	8169	8176	8182	8189
6.6	8195	8202	8209	8215	8222	8228	8235	8241	8248	8254
6.7	8261	8267	8274	8280	8287	8293	8299	8306	8312	8319
6.8	8325	8331	8338	8344	8351	8357	8363	8370	8376	8382
6.9	8388	8395	8401	8407	8414	8420	8426	8432	8439	8445
7.0	8451	8457	8463	8470	8476	8482	8488	8494	8500	8506
7.1	8513	8519	8525	8531	8537	8543	8549	8555	8561	8567
7.2	8573	8579	8585	8591	8597	8603	8609	8615	8621	8627
7.3	8633	8639	8645	8651	8657	8663	8669	8675	8681	8686
7.4	8692	8698	8704	8710	8716	8722	8727	8733	8739	8745
7.5	8751	8756	8762	8768	8774	8779	8785	8791	8797	8802
7.6	8808	8814	8820	8825	8831	8837	8842	8848	8854	8859
7.7	8865	8871	8876	8882	8887	8893	8899	8904	8910	8915
7.8	8921	8927	8932	8938	8943	8949	8954	8960	8965	8971
7.9	8976	8982	8987	8993	8998	9004	9009	9015	9020	9025
8.0	9031	9036	9042	9047	9053	9058	9063	9069	9074	9079
8.1	9085	9090	9096	9101	9106	9112	9117	9122	9128	9133
8.2	9138	9143	9149	9154	9159	9165	9170	9175	9180	9186
8.3	9191	9196	9201	9206	9212	9217	9222	9227	9232	9238
8.4	9243	9248	9253	9258	9263	9269	9274	9279	9284	9289
8.5	9294	9299	9304	9309	9315	9320	9325	9330	9335	9340
8.6	9345	9350	9355	9360	9365	9370	9375	9380	9385	9390
8.7	9395	9400	9405	9410	9415	9420	9425	9430	9435	9440
8.8	9445	9450	9455	9460	9465	9469	9474	9479	9484	9489
8.9	9494	9499	9504	9509	9513	9518	9523	9528	9533	9538
9.0	9542	9547	9552	9557	9562	9566	9571	9576	9581	9586
9.1	9590	9595	9600	9605	9609	9614	9619	9624	9628	9633
9.2	9638	9643	9647	9652	9657	9661	9666	9671	9675	9680
9.3	9685	9689	9694	9699	9703	9708	9713	9717	9722	9727
9.4	9731	9736	9741	9745	9750	9754	9759	9763	9768	9773
9.5	9777	9782	9786	9791	9795	9800	9805	9809	9814	9818
9.6	9823	9827	9832	9836	9841	9845	9850	9854	9859	9863
9.7	9868	9872	9877	9881	9886	9890	9894	9899	9903	9908
9.8	9912	9917	9921	9926	9930	9934	9939	9943	9948	9952
9.9	9956	9961	9965	9969	9974	9978	9983	9987	9991	9996

TABLE VIII. VALUES OF THE EXPONENTIAL FUNCTIONS e^x AND e^{-x}

x	e^x	e^{-x}
.05	1.051	.951
.10	1.105	.905
.15	1.162	.861
.20	1.221	.819
.25	1.284	.779
.30	1.350	.741
.35	1.419	.705
.40	1.492	.670
.45	1.568	.638
.50	1.649	.606
.6	1.822	.549
.7	2.014	.497
.8	2.226	.449
.9	2.460	.407
1.0	2.718	.368
1.1	3.004	.333
1.2	3.320	.301
1.3	3.669	.272
1.4	4.055	.247
1.5	4.482	.223
1.6	4.953	.202
1.7	5.474	.183
1.8	6.050	.165
1.9	6.686	.150
2.0	7.389	.135
2.1	8.166	.122
2.2	9.025	.111
2.3	9.974	.100
2.4	11.023	.091
2.5	12.182	.082
3.0	20.086	.050
3.5	33.115	.030
4.0	54.598	.018
4.5	90.017	.011
5.0	148.413	.0067
5.5	244.692	.0041
6.0	403.429	.0025
6.5	665.14	.0015
7.0	1096.6	.0009
7.5	1808.0	.0006
8.0	2981.0	.0003
9.0	8103.1	.0001
10.0	22026.0	.00005

Answers

Chapter 1

Section 4, page 3

1. Nearly 137 years
3. 2,759,400,000
5. Approximately 169 times a hair

Section 10, page 9

1. 99, 88, 77, 66, 55, 44, 33, 22, 11
3. 12 bars at 2 for 5 cents, 6 bars at 3 for 10 cents
5. He puts 3 dollars on each end of the ruler. If one end shows lighter, this group of 3 coins contains the counterfeit. If the two groups balance, the counterfeit lies among the 3 coins not on the ruler. In each case we locate the counterfeit among 3 coins. Now put one of these on each end of the ruler. If one end is lighter, this one is the counterfeit. If both ends balance, the unused coin of the three is the counterfeit.
7. Wrong

Chapter 2

Section 2, page 13

1. a. 3 c. $\frac{1}{2}$ e. $\frac{2}{3}$
3. a. $-1 < x < 5$ c. $x < -4$ e. $x < -4$ g. $-4 < x < 4$ i. $x > 2, x < -4$
4. a. $|x| < 3$ c. $|x+1| < 4$
5. a. $|x-2| > 3$

Section 6, page 17

1. 24 feet
3. 4.69 pounds
5. 37.2 pounds copper, 25.2 pounds zinc
7. 560, 400
9. 360 acres
11. 9,940 Coeds, 11,360 Eds

Section 7, page 19

1. a. $\frac{5}{8}$, 0.625 c. $\frac{5}{4}$, 1.25 e. $\frac{1}{8}$, 0.125
2. a. 37.5% c. $66\frac{2}{3}\%$ e. $33\frac{1}{3}\%$
3. 6 days
5. 9,960
7. Yellow vase cost \$1, red vase cost \$1.50, so the loss is 10 cents
9. 20%
11. $37\frac{1}{2}\%$

Section 11, page 25

1. $A = kd^2$
3. $I = k/d^2$
5. 6,545 cubic feet
7. $56\frac{1}{4}$ feet
9. 8.6 grams
11. 92 feet, 132 feet

Section 12, page 28

1. $A = klw$
3. $\frac{4}{9}$ pound
5. 250.2 seconds

Review Exercises, page 29

1. a. $|x-5| < 3$ c. $|2x-5| < 3$ e. $|4x-5| > 2$
3. $x < b, x > c$
4. a. $x < -6$ c. $2 < x < 3$ e. $x < -6, x > 6$
5. 84 men, 24 women
7. 340, 510
8. a. $c = kr$ c. $F = km_1m_2/d^2$
9. 56 feet per second

Chapter 3

Section 6, page 37

3. 102.3°C, $P = 781$ millimeters
5. 48 feet per second, 224 feet per second, $6\frac{1}{4}$ seconds
7. 100 feet, 2.2 seconds

Chapter 4

Section 4, page 44

2. a. 7.07 c. 5 e. 5 g. 4.12
3. 8

Section 8, page 49

1. c. All points on the line $y = x + 1$ and to the right of this line. e. All points on and to the left of the line $2y - 3x = 6$. g. All points on and to the right of the line $y = 4x + 2$.
3. a. All points in the fourth quadrant on the line $y = x$ and between this line and the negative y axis. c. All points to the left of $x = 1$ bounded by the two lines $x + y = 1$ and $y - x + 1 = 0$. The points on the line $x + y = 1$ also satisfy these inequalities.

4. a. Points in the third quadrant on the line $y=x$ and to the right of this line.

c.

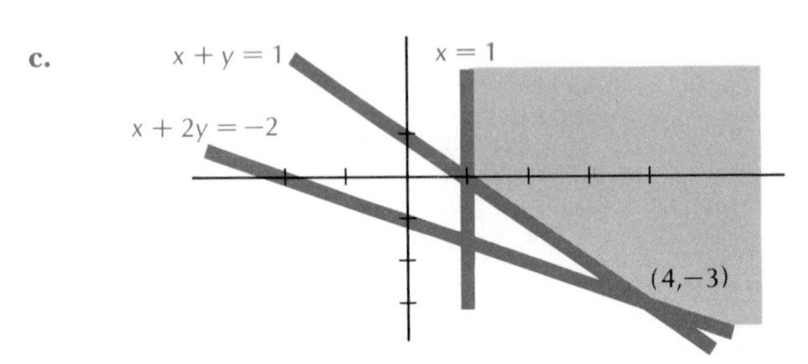

Section 9, page 53

1. $W=\frac{4}{5}T+128$
3. $P=(2{,}500/39)h$
5. $W=8g+10$, 46 pounds
7. $V=-5t+30$, 6 seconds

Section 10, page 58

1. $y=-3x$
3. $y=\frac{6}{7}x+\frac{11}{7}$
5. $x=5$
7. $y=\dfrac{y_2-y_1}{x_2-x_1}x+\dfrac{x_2y_1-x_1y_2}{x_2-x_1}$

Chapter 5

Section 3, page 62

1. 7 Cokes
3. \$3.13
5. $S=\$(50{,}000+1{,}000n)$
7. \$950
9. $206\frac{1}{4}$ pounds
11. 14 years, 11 years

Page 65

1. \$3,250
3. 8 at 20 cents, 52 at 12 cents
5. 16, 18, 20
7. 3 classical, 7 modern
9. 4,928 cubic feet
11. 63 at \$5, 37 at \$4

Section 5, page 68

1. 80 cents, 40 cents
3. 95 feet by 60 feet
5. 13 magazines, 77 newspapers
7. 50 pounds of each
9. 96
11. 60 cents
13. 15, 17
15. 13 cents per pound for sugar, 10 cents per pound for flour.

Page 70

1. 250 cubic centimeters
3. 100 pounds of each
5. \$4,545.45
7. 30 boys

9. Mother is 42, son is 12
11. 30 gallons, 5% acid, 10 gallons, 25% acid

Page 73

1. 2 hours
3. $12\frac{6}{7}$ miles, $1\frac{2}{7}$ hours
5. 3:20 P.M.
7. $4\frac{1}{2}$ miles per hour, $5\frac{1}{2}$ miles per hour
9. 4 miles per hour
11. 270 miles, 45 miles per hour going, 40 miles per hour returning

Page 75

1. $6\frac{2}{3}$ hours
3. $2\frac{5}{8}$ hours
5. 25 days, $16\frac{2}{3}$ days
7. Just over 6 hours, 58 minutes
9. $15 for the dress, $4.50 for the hat
11. $3\frac{13}{17}$ quarts

Review Exercises, page 76

1. 8 nickels, 9 dimes, 16 quarters
3. 216
5. 75 ounces of 45% alloy, 25 ounces of 25% alloy
7. 5:15 P.M., $472\frac{1}{2}$ miles
9. 60
11. $1 \leqq A \leqq 10$, $1 \leqq B \leqq 5$
12. c. 4
13. a. $A > B$, $A+B \leqq 4$ c. 4 of A and 1 of B
15.

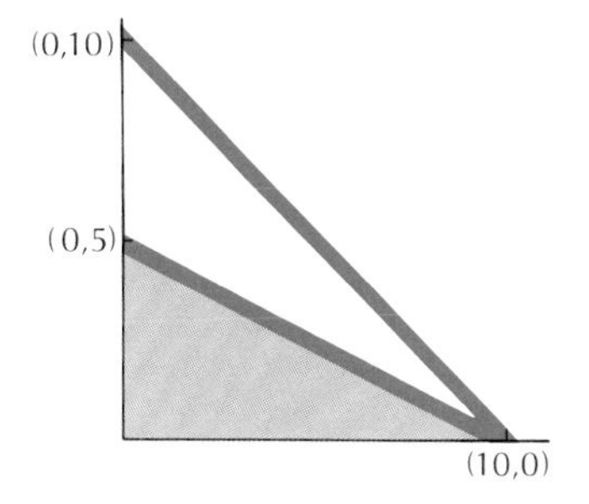

Chapter 6

Section 3, page 82

1. $s=16.1t^2$, non-negative values of t only
3. $t=0.00015625f^2$, 0.01891 second, 0.0049 second

5. $S = 12.57d^2$, positive values of d only

Section 5, page 88

1. 21, 22
3. 24 feet by 10 feet
5. 8 feet by 7 feet
7. 4 feet
9. 3
11. 30
13. 12 inches by 10 inches by 5 inches
15. 12 hours, 6.8 hours

Section 6, page 91

1. a. $x^2 - 7x + 10 = 0$ c. $15x^2 + x - 6 = 0$
3. $\frac{1}{5}$
5. 6
7. $\frac{9}{4}, \frac{4}{3}, \frac{4}{3}$

Section 8, page 94

1. $h = -16t^2 + 64t$; 64 feet; approximately 0.3 second or 3.7 seconds
3. $h = -\frac{1}{80}D^2 + 70$; $58\frac{3}{4}$ feet
5. $y = x^2 - 4x$
7. $\frac{1}{16}$ foot

Chapter 6

Section 10, page 100

1. a.

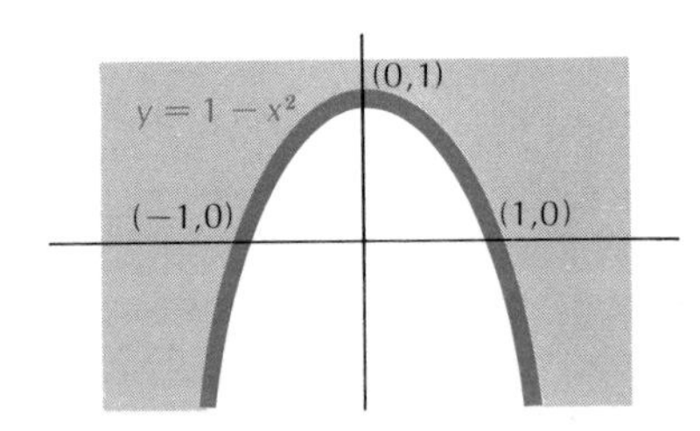

c.

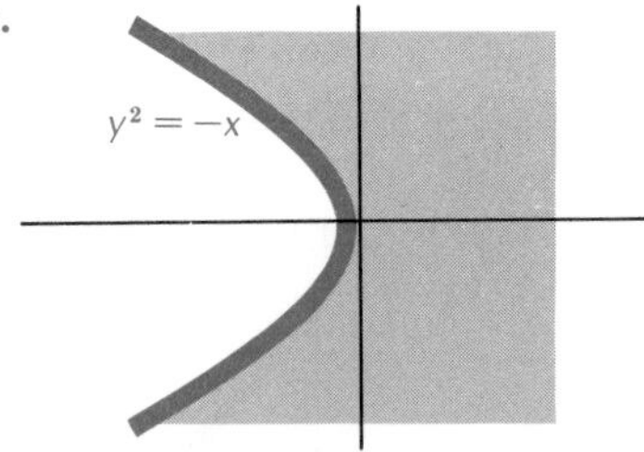

e. The points on a unit circle with center at (0,0)

g. The points on and outside a unit circle with center at (0,0)

2. a. Points on the two branches of the curve $xy = 1$

c.

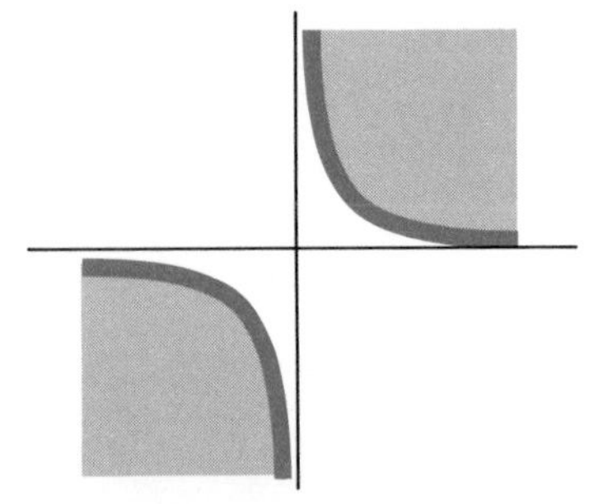

3. a.

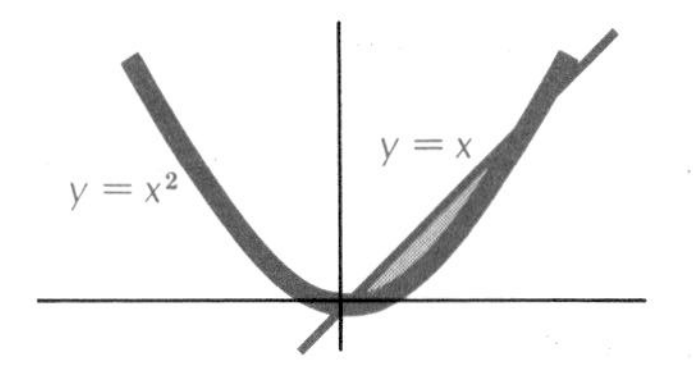

c.

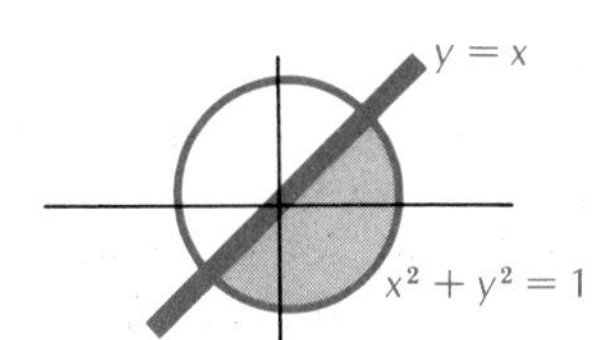

e.

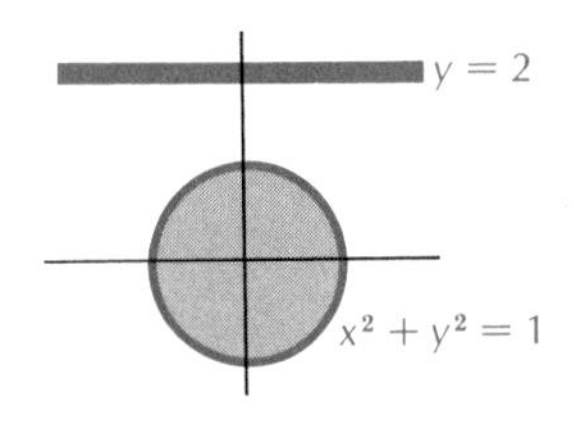

g. There are no points to satisfy the inequalities

i.

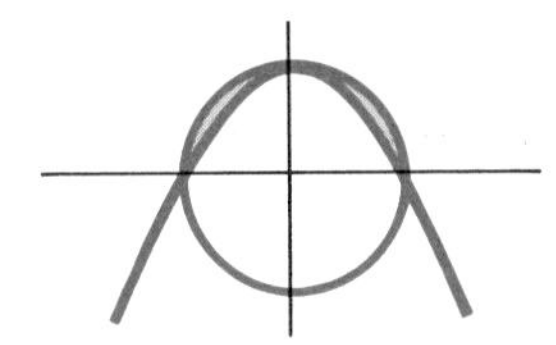

k.

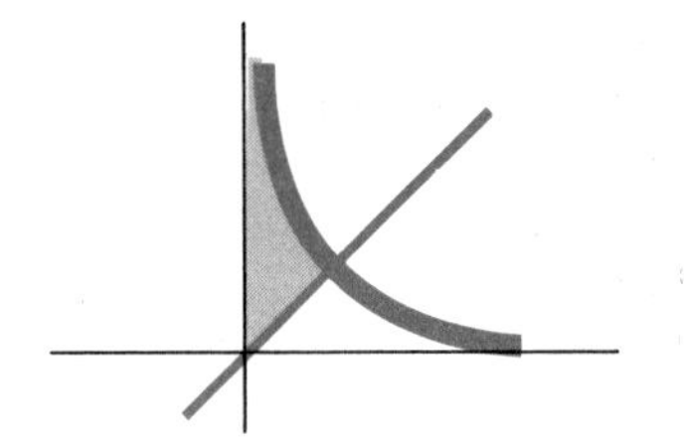

Chapter 7

Section 4, page 104

1. About 5 feet 10 inches
3. 20 feet
5. About 150 feet
7. About 225 feet
9. About 31°

Section 5, page 107

1. 60 feet
3. Approximately 51°; 12 feet 6 inches
5. About 240,000 miles
7. 5 feet 4 inches

Chapter 9

Section 2, page 119

5. 358 feet
7. 17.5 feet, 16.5 feet

Section 8, page 127

1. 40°40′
3. 2°20′

5. 11.5 inches
7. 58.9 square inches
9. 2.6 miles
11. Frontages on First Avenue are 254, 231, 185, and 115 feet; depths are 127, 242, 335, and 393 feet.
13. 33°30′ with wall and 56°30′ with ground
15. 3.1 feet to a total distance from wall of 9.7 feet
17. 90 miles
19. 3°30′; 2,000 feet from the nearer church

Review Exercises, page 132

5. 66°10′ north of east
7. 28.4 feet

Chapter 10

Section 3, page 138

1. 136 feet
3. 36°50′, 53°10′, 2,400 square feet
5. 50°40′, 25°40′
7. 8°10′ north of west
9. 10.4 miles
11. 37°20′ west of south
13. 26.4 feet
15. 9°50′ north of east, 3.4 hours

Chapter 11

Section 2, page 144

2. a. 30°, 150°, −210°, −330° c. 60°, 300°, −60°, −300° e. 45°, 225°, −135°, −315° g. 0°, 360°, −360° i. 90°, −270°
3. a. $\sin A = \frac{5}{13}$, $\tan A = \frac{5}{12}$ or $\sin A = -\frac{5}{13}$, $\tan A = -\frac{5}{12}$ c. $\cos A = \frac{3}{5}$, $\tan A = \frac{4}{3}$ or $\cos A = -\frac{3}{5}$, $\tan A = -\frac{4}{3}$ e. $\sin A = 3/\sqrt{13}$, $\cos A = -2/\sqrt{13}$ or $\sin A = -3/\sqrt{13}$, $\cos A = 2/\sqrt{13}$
4. a. 0 c. $\frac{1}{3}$ e. $-\frac{11}{10}$

Section 3, page 147

1. a. 0.5 c. 0.9397 e. −0.866 g. −0.5774 i. −0.364
2. a. 138°40′, 221°20′ c. 26°30′, 206°30′ e. 224°30′, 315°30′
3. a. 0.433 c. 1.732 e. 1.225

Section 4, page 148

1. 998 feet
3. 79°30′
5. 5°30′
7. 251 feet
9. 59° east of north, 5°30′ east of north; or the same angles west of north

Chapter 12

Section 2, page 156

5. 6
7. a. 10 amperes c. 10 amperes, 0 amperes

Chapter 13

Section 2, page 159

1. a. 27 **c.** 4 e. 16 g. 81 i. $\frac{27}{8}$
2. a. 3^3 **c.** 2^5 e. 7^2 g. 10^2 i. 2^8
3. a. a^{13} **c.** x^7 e. a^7 g. 10^2 i. 1
4. a. $\frac{10^3 \cdot 10^3}{10^2} = 10^4 = 10{,}000$ c. $\frac{3^3 \cdot 3^2}{3^4} = 3^1 = 3$ e. $\frac{2^4 \cdot 2^3 \cdot 2^2}{2^5} = 2^4 = 16$ g. $\frac{5^2 \cdot 5^4}{5^3} = 5^3 = 125$ i. $\frac{5^5 \cdot 3^5}{5^2 \cdot 3^3} = 5^3 \cdot 3^2 = 125 \cdot 9 = 1{,}125$

Section 5, page 162

1. $x^{-4} = \frac{1}{x^4}$
3. $a^0 = 1$
5. a^6
7. a^2
9. $2^4 = 16$
11. $x^0 = 1$
13. $b^{-3} = \frac{1}{b^3}$
15. $10^{-1} = \frac{1}{10} = 0.1$
17. x^2
19. a^3

Section 7, page 166

1. a. 7.3×10^{-4} c. 3.5×10^{-3} e. 6.8×10^{-9} g. 6.87×10^{11} i. 3.62×10^3
2. a. 56,300 c. 0.000000000000046 e. 0.000000832 g. 3,270,000 i. 0.00143
3. 6.696×10^8 miles
5. 2.33×10^{23} molecules per gram

Section 8, page 167

1. $10^{2.4281}$
3. $10^{3.0294}$
5. $10^{0.8932}$
7. $10^{1.2856}$
9. $10^{2.2856}$
11. $10^{-3.2856}$
13. $10^{3.8645}$
15. $10^{2.4579}$
17. $10^{1.9258}$
19. $10^{5.9258}$

Section 9, page 169

1. 673
3. 4,630
5. 0.00803
7. 0.0485
9. 6.73
11. 0.230
13. 0.000212

Section 11, page 171

1. 3,610
3. 18,750
5. 0.128

7. 0.39
9. 41.65
11. 1.14
13. 1.52
15. 1,200
17. 47.3
19. 7.17

Section 12, page 173

1. 1.51
3. 0.865
5. 0.00000491
7. 7.34
9. 249
11. 0.0000479
13. 2.48
15. 10,700
17. 538
19. 225

Review Exercises, page 174

1. a. $\frac{8}{27}$ c. 4 e. $\frac{27}{64}$
2. a. $b^{1/4}$ c. $2a^4 \cdot bc^2$ e. $a^3b^4/16$
3. a. $2.9 \cdot 10^{-5}$ c. $1.96 \cdot 10^5$ e. $3.7 \cdot 10^{-4}$ g. $3.993 \cdot 10^1$
4. a. 2920 c. 5.10 e. 0.0353 g. 1.27

Chapter 14

Section 2, page 177

1. \$436
3. \$320
5. 0.03%
7. 4,308 bricks (4,307 is the nearest whole number but this does not quite finish the job)
9. \$13.20, \$9.90
11. \$3.15 per yard
13. 225 pounds

Section 5, page 180

1. \$24, \$824
3. $\frac{6}{7}$ year
5. \$1,350
7. \$1.07 per \$100
9. 6%
11. \$6,400
13. 3.3%
15. $2\frac{1}{3}$%

Section 8, page 184

1. \$6,312.50
3. \$6,334
5. \$12,332.70
7. \$2,955
9. \$2,823.65
11. \$2,821
13. Just below 3%
15. \$3,571.93

Section 9, page 187

1. \$16,040.15
3. \$722.56
5. Present value of time payments better by \$113.10
7. \$18,462.79

Section 10, page 191

1. SCHEDULE OF PAYMENTS

Year	Principal outstanding at beginning of year	Interest at 6%	Payment	Principal repaid
1	$10,000	$600.00	$1,358.68	$ 758.68
2	9,241.32	554.48	1,358.68	804.20
3	8,437.12	506.23	1,358.68	852.45
4	7,584.67	455.08	1,358.68	903.60
5	6,681.07	400.87	1,358.68	957.81
6	5,723.26	343.40	1,358.68	1,015.28
7	4,707.98	282.48	1,358.68	1,076.20
8	3,631.78	217.91	1,358.68	1,140.77
9	2,491.01	149.46	1,358.68	1,209.22
10	1,281.79	76.91	1,358.68	1,281.77

3. $1,394.95
5. $2\frac{1}{2}$% per month
7. 2% per month
9. $29,646.14
11. $192.36 per year, $589.10

Chapter 15

Section 5, page 196

5. $x = 2$, $x = 4$, and $x =$ about -0.77
6. a. 3.32 c. 12.182 e. 0.082 g. 0.122
7. a. 0.6 c. 2.1 e. 7.0 g. −1.6

Section 6, page 199

1. 664, 80,700
3. $1,822
5. $v = 50{,}000, e^{-0.06t}$, $15,050
7. a. 29.92 inches c. 9.96 inches
9. 0.111 grams
11. About 6%
13. 1,400 years
15. $10\frac{1}{2}$ hours after they are placed in the jar

Chapter 16

Section 4, page 205

1. $\frac{1}{7}$
3. a. $\frac{1}{2}$ c. $\frac{1}{5}$
4. a. $\frac{1}{6}$
5. a. $\frac{1}{8}$ c. $\frac{3}{8}$
7. a. $\frac{1}{2}$ c. $\frac{1}{5}$
8. a. $\frac{1}{2}$
9. a. $\frac{1}{2}$
10. a. $\frac{1}{13}$ c. $\frac{1}{17}$

Section 5, page 208

1. $\frac{1}{10}$ c. $\frac{3}{10}$
2. a. $\frac{1}{2}$
3. $\frac{1}{4}$
4. a. $\frac{1}{36}$ c. $\frac{1}{12}$ e. $\frac{5}{36}$ g. $\frac{5}{36}$ i. $\frac{1}{12}$
5. a. $\frac{1}{36}$
6. a. $\frac{1}{3}$ c. $\frac{2}{3}$ e. $\frac{1}{3}$
7. a. $\frac{1}{5}$
8. a. $\frac{4}{35}$ c. $\frac{1}{35}$
9. a. $\frac{3}{10}$ c. $\frac{3}{10}$
10. a. $\frac{1}{55}$ c. $\frac{16}{55}$

Section 7, page 213

1. 8
3. 504
5. 60
7. 12
9. 30,220
10. a. 10 c. 190.

Section 9, page 217

1. a. 151,200 c. 720
2. a. 10
3. *ab; ac; ba; bc; ca; cb*
5. 259,459,200
7. a. 24
9. 120
11. a. 362
13. 27
15. 40,320

Section 10, page 219

1. $P_{3,1} \cdot P_{12,8} = 59{,}875{,}200$
3. $P_{4,4} = 24$
5. $P_{5,2} \cdot P_{7,7} = 100{,}800$
7. $2 \cdot P_{10,10} = 7{,}257{,}600$
8. a. $P_{8,3} = 336$
9. $P_{2,1} \cdot P_{4,2} = 24$

Section 12, page 222

1. a. 12,650 c. 35 e. 1
3. 10
5. 66
7. 84
9. 10
11. 117,600
13. 792
15. $\frac{52!}{39!13!} = 635{,}013{,}559{,}600$
17. 990
19. 720
21. 210

Section 13, page 226

1. $\frac{1}{17}$
3. a. $\frac{65}{253}$
4. a. $\frac{1}{6}$ c. $\frac{15}{36}$
5. a. $\frac{2}{3}$ c. 1
6. a. $\frac{1}{9}$ c. $\frac{2}{9}$
7. a. $\frac{2}{17}$ c. $\frac{13}{34}$
8. a. 99/54,145 or 1 in 547
9. $C_{49,9} \div C_{53,13}$
10. a. 90 or, if order of choice is not considered, 45 c. $\frac{2}{9}$
11. $\frac{1}{10}$
12. a. $\frac{5}{11}$ c. $\frac{1}{66}$
13. $\frac{1}{120}$

Section 14, page 231

1. $\frac{1}{8}$
3. $\frac{5}{6}$
5. $\frac{33}{95}$
7. $\frac{8}{15}$
9. $\frac{2}{9}$
11. $\frac{2}{3}$
13. a. $\frac{5}{126}$ c. $\frac{10}{21}$
14. a. $\frac{1}{3}$ c. $\frac{1}{6}$
15. a. $\frac{9}{16}$
16. a. $\frac{1}{3}$

Section 15, page 233

1. \$2
3. Worth $7\frac{1}{2}$ cents
5. Pay 1.4 cents
7. $3\frac{1}{3}$ times
9. 67 cents
11. No, probability of 8 is $\frac{5}{36}$, of 7 is $\frac{1}{6}$
13. Yes

Review Exercises, page 235

1. 120
3. a. 720 b. 120
5. 10 weeks
7. a. 24
9. $\frac{3}{10}$
11. a. $\frac{7}{12}$ c. $\frac{7}{18}$
13. \$3.82$\frac{1}{2}$
15. No, about 22 cents
16. a. 2,598,960 c. 33/66,640 or 1 in 2,019

Chapter 17

Section 1, page 239

1. 0.233 or 23.3%
3. 0.305 or 30.5%
4. a. 0.297 or 29.7%
5. a. 0.30 or 30% c. 0.09 or 9%
6. a. 2.5
7. a. 0.97 or 97% c. 0.94 or 94%
8. a. $87\frac{1}{2}$%
9. a. 4,412 c. 2,473
10. a. 1,058
11. 8,201
13. 0.006 or 1 in 170
15. a. 695 c. 172

Section 5, page 246

1. a. 36 c. 21 e. 20 g. 67
2. a. 5.5 c. 14.5 e. 11.1
3. a. 4.5 c. 3 e. 4 g. 8.5
4. a. 5.5 c. 14.5 e. 8.5
5. Median of 73 and mean of 70.8
11. 60
13. Increase 10 points

Section 6, page 251

3. Mean 360, median 410, mode none, standard deviation 140.3
7. Mean 28 cents, standard deviation 10 cents
13. Mean 67.6, median 71, mode 71
15. Much less variability between individual heights

Chapter 18

Section 5, page 262

1. TABLE 46

Class interval	Frequency	Relative frequency	Cumulative relative frequency
−0.5–0.5	−2	0.008	0.008
0.5–1.5	11	0.046	0.054
1.5–2.5	32	0.133	0.187
2.5–3.5	45	0.187	0.375
3.5–4.5	65	0.271	0.646
4.5–5.5	53	0.221	0.866
5.5–6.5	26	0.108	0.975
6.5–7.5	5	0.021	0.996
7.5–8.5	1	0.004	1.000
	240	0.999	

3.

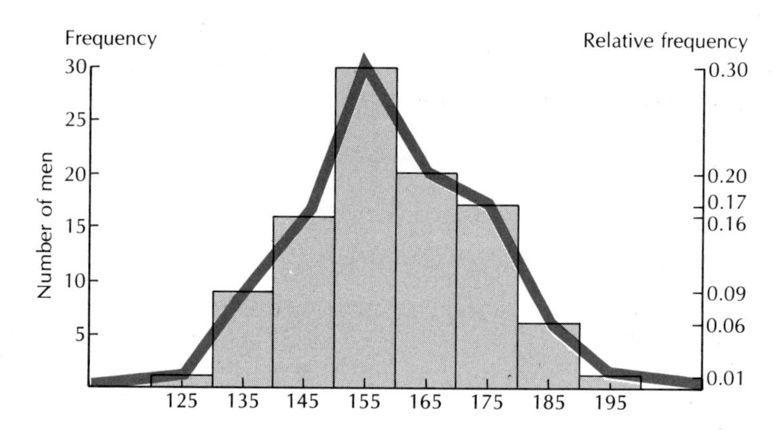

5. a. 44%

7.

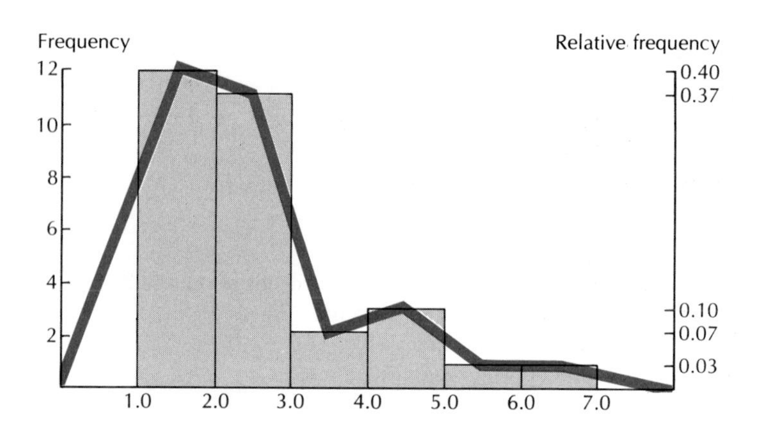

9. a. 23%

11.

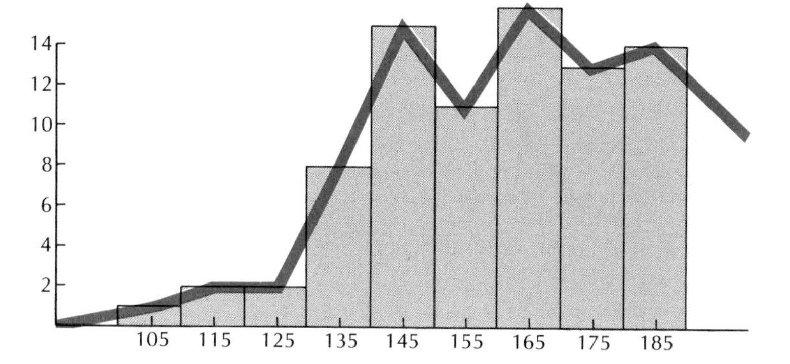

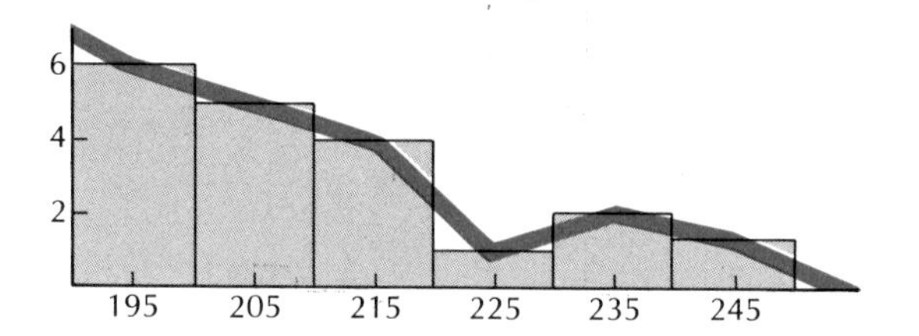

13. **TABLE 50**

Class limits	*Frequency*	*Relative frequency*	*Cumulative relative frequency*
91–100	2	0.02	0.02
101–110	15	0.15	0.17
111–120	33	0.33	0.50
121–130	26	0.26	0.76
131–140	11	0.11	0.87
141–150	5	0.05	0.92
151–160	4	0.04	0.96
161–170	1	0.01	0.97
171–180	1	0.01	0.98
181–190	1	0.01	0.99
191–200	0	0.00	0.99
201–210	1	0.01	1.00
	100	1.00	

15.

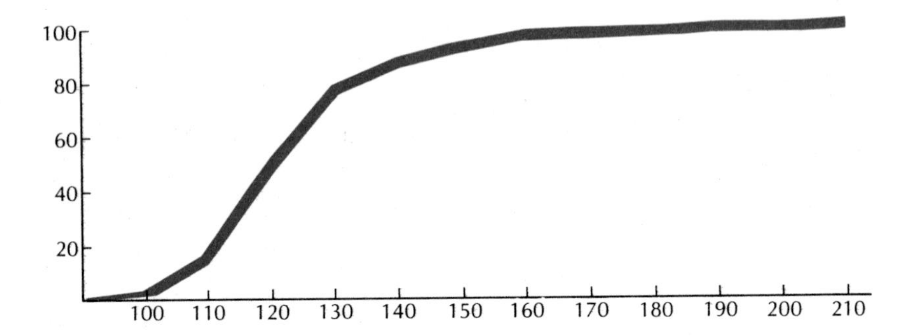

17.

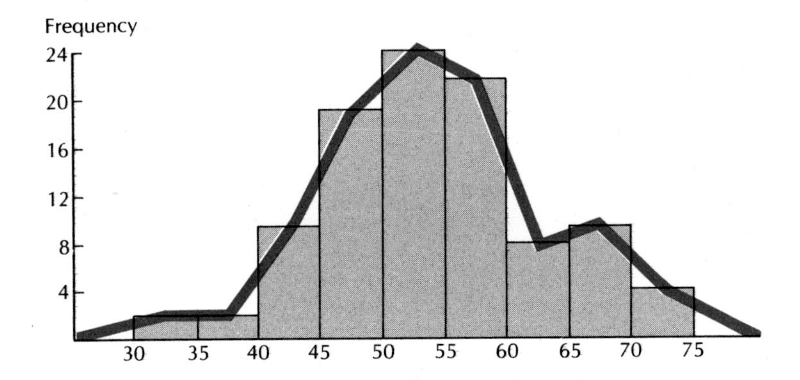

19. **TABLE 52**

Class limits	*Frequency*	*Relative frequency*	*Cumulative relative frequency*
30–35	1	0.01	0.01
36–40	9	0.09	0.10
41–45	14	0.14	0.24
46–50	21	0.21	0.45
51–55	13	0.13	0.58
56–60	21	0.21	0.79
61–65	11	0.11	0.90
66–70	8	0.08	0.98
71–75	2	0.02	1.00
	100	1.00	

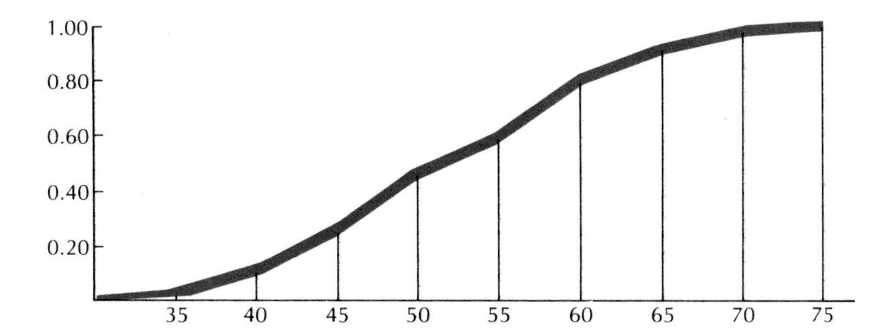

Section 6, page 267

3. Mode 155; median 153; mean 148.9
5. Modes 145 and 165; median 162; mean 168.8
7. Mode 53; median 51, mean 54.6

Section 7, page 269

1. 1.47
3. 1.3
5. 17.8
7. 9.26

Review Exercises, page 270

1. Median 14.7; mean 17; $\sigma_x = 4.3$
3. Mode 22; median 21.8, mean 30.4; $\sigma_x = 15.0$

Chapter 19

Section 5, page 283

1. 50%
2. a. 95.5%
3. 0.928
4. a. 0.308 c. 0.383
5. a. 86.6%, 433 weights c. 0.788 or 78.8% e. 150.2 pounds
7. a. 77%
9. a. 683 students c. Between 58.5 and 74.5 inches

Section 9, page 291

1. We must allow a larger range of possible values for the true mean
3. 1,012 miles
5. No, 95% confidence gives $d = 2.35$ inches
7. 588 to 612 words per minute
8. a. 44 cents to 62 cents
9. a. 80 checks c. Sample of Exercise 8 must be from another situation; probability that it came from Exercise 9 is less than 0.0005

Review Exercises, page 293

1. a. 0.106 or 10.6% c. 0.841 or 84.1%
3. a. 2%
5. No confidence statement can be made for this data

Chapter 20

Section 2, page 296

1. XLV	3. DXXXIX	5. MMXCIII
7. XL	9. CLVI	11. XXXVI

Section 4, page 301

1. 65	3. 145	5. 1,005
7. 120	9. 1,047	11. 24,714

Section 8, page 305

1. 331	3. a. 1100 c. 30	4. a. 60 c. 141
5. a. 11001 c. 11000	7. 1000111	9. 1100011

Section 10, page 309

1. a. 6 c. 3 e. 5	2. a. 6 c. 6 e. 2
3. a. 9 c. 4 e. 3	4. a. 9 c. 0 e. 11

Chapter 21

Section 4, page 320

1. 35 refrigerators and 50 stoves; profit $2,225
3. 5 of car *A*, none of *B*, 5 of *C*; profit $6,500
5. Order all 3 buses from B_1 to *A*, order 1 bus from G_2 to *A* and remainder to *B*; total time 110 minutes
7. Send all 3 buses from G_1 and all 4 buses from G_2 to *B*, send 2 buses from G_3 to *A* and 4 to *C*; total time 164 minutes
9. From 1 ship 5 tons to customer *B* and 60 tons to *C*, from 2 ship 30 tons to customer *A* and 10 tons to *B*, from 3 ship 35 tons to customer *B* (no change in total cost of $310 if we ship part of 3 to customer *A* with equal shift in 2 from customer *A* to *B*)

Chapter 22

Section 4, page 328

1. 3, 5, 17, 257
3. 2, 3, 5, 7, 11, 13, 17, 19, 23, 29, 31, 37, 41, 43, 53, 59, 61, 67, 71, 73
5. 101, 103, 107, 109; 821, 823, 827, 829; there is one more decade, student should search for it

7. $40=37+3=29+11=23+17$
   $44=41+3=37+7=31+13$
   $48=43+5=41+7=37+11=31+17=29+19$
8. $13=49-36$; $23=144-121$

Section 5, page 330

1. 6, 28

Section 6, page 331

1. $x=2\cdot3\cdot2=12$; $y=9-4=5$; $z=9+4=13$

Section 8, page 333

1. Several numbers require 5 squares

Section 8, page 335

1. $23=8+8+1+1+1+1+1+1+1$
   $42=27+8+1+1+1+1+1+1+1$
   All the others require fewer cubes

Chapter 23

Section 3, page 341

5. A four-cornered flat surface like the frame of a picture

Section 5, page 345

1. $V=16$, $A=24$, $R=10$
3. $V=8$, $A=12$, $R=6$
5. $V=6$, $A=12$, $R=8$
11. $V=4$, $R=6$, $4-A+6=2$, $A=8$

Section 9, page 351

1. No, 4 vertices order 3
3. No, network contains 6 vertices, all order 5
5. No, but closing any bridge will permit stroll beginning and ending at different points
7. Any two bridges so long as they involve all 4 points, for example, close *AB* and *CD*
9. Not possible, for associated networks contain 3 vertices of order 5 and one of order 9
11. Walk must start in one of the corner rooms with 3 doors and finish in the other corner room with 3 doors. Cut one more door in each of the corner rooms, both to the outside or both to room between the corner rooms.

Section 10, page 355

1. $V=4$, $A=8$, $R=4$
3. $V=9$, $A=18$, $R=9$
5. $V=18$, $A=27$, $R=7$

Section 15, page 363

1. The top pair (*a*) is an example of such a map covering the entire torus; the bottom pair (*b*) is an example of such a map covering only part of the torus

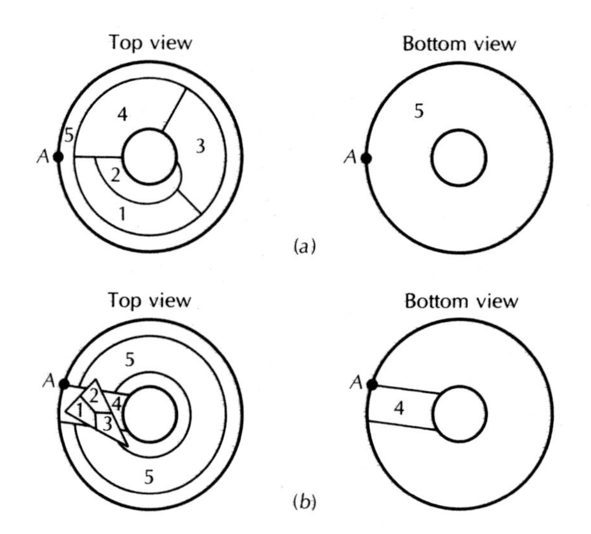

3. The Möbius band is opened out, such as

5.

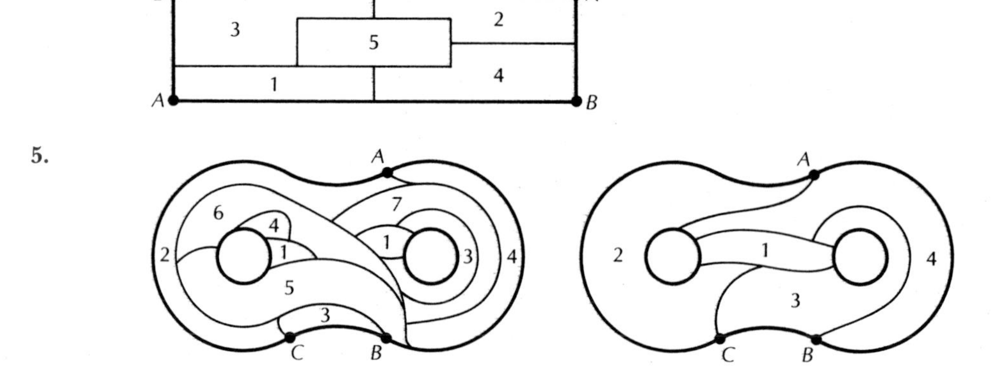

7. For each circle, let the outside be numbered 2 and inside 1. For each region add up numbers 1 and 2. If the total is odd, color red. If the total is even, color blue. For any

two regions with a border in common, one will be outside and the other inside the circle which forms the common border. On all other circles they will be the same. Hence one is even and the other odd, and one is red and the other blue.

Chapter 24

Section 2, page 368

1. **a.** Doubtful, true (equivalent), true (equivalent) **c.** True (equivalent), doubtful, true **e.** False, doubtful, true (equivalent)
2. **a.** Some man lives a century **c.** Some cretans are not liars **e.** All people are happy **g.** It may rain and we need not go home
3. **a.** If Bob wins, she *may* marry him, but she *may* marry John or Bill **c.** If Sam wins, she *may* marry him, but she *may* marry John
4. **a.** Murphy may win **c.** Smith wins

Section 3, page 370

1. Implications that you have beaten your wife in the past, whether yes or no answer
3. Implication "more equal" is better than "equal"
5. Statement is true if Mrs. Springer's tonic makes you feel worse
7. Implies John Hancock became a signer because he was a Harvard graduate

Section 4, page 373

1. Invalid deduction, win if premise is true
3. Deduction valid; hence conclusion is valid if premise "he is a citizen of Chicago" is true
5. Deduction invalid
7. Deduction valid; conclusion true if premise "men never tell the truth to girls" is true
9. Deduction invalid
11. Deduction valid; premise "a good man does not inflict pain" is false

Section 5, page 377

1.

q	r	$q \to r$
T	T	T
T	F	F
F	T	T
F	F	T

3.

p	$\sim p$
T	F
F	T

5.

p	q	$p \to q$	$\sim q$	$\sim p$	$\sim q \to \sim p$
T	T	T	F	F	T
T	F	F	T	F	F
F	T	T	F	T	T
F	F	T	T	T	T

7.

p	q	r	$q \to p$	$r \to \sim p$
T	T	T	T	F
T	T	F		
T	F	T	T	F
T	F	F		
F	T	T	F	
F	T	F		
F	F	T	T	T
F	F	F		

9.

p	q	r	$q \to p$
T	T	T	T
T	T	F	T
T	F	T	T
T	F	F	T
F	T	T	F
F	T	F	F
F	F	T	T
F	F	F	T

11.

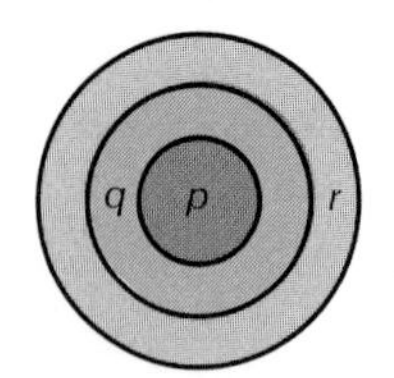

Section 6, page 378

1. If the barber shaves himself, statement is false; if barber does not shave himself, statement says he does
3. Second statement says God is first cause; first statement says something else must have caused God
5. Complete dilemma

Index

Index